FUNCTIONAL POLYMER BLENDS

Synthesis, Properties, and Performance

FUNCTIONAL POLYMER BLENDS

Synthesis, Properties, and Performance

Edited by
VIKAS MITTAL

CRC Press
Taylor & Francis Group
Boca Raton London New York

CRC Press is an imprint of the
Taylor & Francis Group, an **informa** business

CRC Press
Taylor & Francis Group
6000 Broken Sound Parkway NW, Suite 300
Boca Raton, FL 33487-2742

First issued in paperback 2017

© 2012 by Taylor & Francis Group, LLC
CRC Press is an imprint of Taylor & Francis Group, an Informa business

No claim to original U.S. Government works

Version Date: 20120224

ISBN 13: 978-1-4398-5669-7 (hbk)
ISBN 13: 978-1-138-07434-7 (pbk)

Library of Congress Cataloging-in-Publication Data

Functional polymer blends : synthesis, properties, and performances / editor, Vikas
   Mittal.
     p. cm.
   Includes bibliographical references and index.
   ISBN 978-1-4398-5669-7 (hardcover : alk. paper)
   1. Polymeric composites. 2. Polymer engineering. 3. Polymerization. I. Mittal, Vikas.

TP1087.F86 2012
668.9'2--dc23   2012000325

Visit the Taylor & Francis Web site at
http://www.taylorandfrancis.com

and the CRC Press Web site at
http://www.crcpress.com

Contents

Preface

Polymer blends constitute a very important class of materials where the components are either physically or chemically mixed with each other to achieve a certain set of functional properties. The physical means of compatibilization of the phases in the blends are addition of compatibilizers (viz. polyethylene-co-maleic anhydride), whereas chemical means include chemical reactions with the phases for chemical linking. Polymer blends are widely used materials in the modern polymer industry because of their wider range of properties as compared to the individual polymers and their ease in tailoring of properties. As a result, polymer blends find applications in numerous fields such as adhesion, colloidal stability, design of composite and biocompatible materials, and so forth. The science and technology of polymer blends has gained tremendous developments in the recent years. A large number of polymer blend systems have been reported along with further developments on their means of synthesis, microstructure, and properties characterization. These functional polymer blend systems have helped to further enhance the application spectrum of polymers in general. The current book proposes to provide these advances in synthesis and characterization methodologies for the generation of different polymer blends systems.

Chapter 1 provides an introduction as well as overview of the polymer blend systems. A number of examples of functional polymer blends have been demonstrated. Chapter 2 provides information on the miscibility enhancements in polymer blends through multiple hydrogen binding interactions. Chapter 3 presents the component dynamics prevailing in polymer blend systems. The dynamic environment differs for respective components, thereby giving rise to a difference of the segmental relaxation time of these components in both magnitude and temperature dependence. Concepts of shape memory polymer blends are presented in Chapter 4.

A shape memory polymer (SMP) is a smart material that can memorize its original shape after being deformed into a temporary shape when it is heated or receives any other external stimuli such as light, electric field, magnetic field, chemical, moisture, and pH change. Ethylene methyl acrylate (EMA) copolymer toughened polymethyl methacrylate (PMMA) blends have been detailed in Chapter 5. Along with synthesis, extensive characterization of the blends for thermal, mechanical, optical, and morphological properties has been reported. Chapter 6 provides theoretical insights through molecular dynamics simulation studies for binary blend miscibility. Calculation of some important parameters, such as Flory Huggins parameter, solubility parameter, and glass transition temperature to discuss the compatibility of polymers has been presented. Chapter 7 reports on conformation and topology of cyclic linear polymer blends (CLBs). The bond fluctuation model and results obtained from this model on the statics and dynamics of CLBs have been reported. Results on the conformational free energy of cyclic polymers (CPs) are also summarized. Strain hardening in polymer blends with fibril morphology is the focus of Chapter 8. Strain hardening is induced by nonlinear molecular structure such as chain branching. In polymer blends, strain hardening can be induced by controlling the morphology such that the fibrillation of the dispersed phase is maintained. Modification of polymer blends by irradiation techniques has been demonstrated in Chapter 9. In the case of polymer blends, such phenomena could occur not only in the bulk of each polymer but also at the interface. Therefore, another parameter could influence the phenomena induced by irradiation: the morphology of the blend. Directed assembly of polymer blends using nanopatterned chemical surfaces has been reported in Chapter 10. Two interesting approaches are discussed to create highly ordered polymer nanostructures using directed assembly of polymer blends on chemically functionalized nanopatterned surfaces.

It gives me immense pleasure to thank CRC Press for their kind acceptance to publish this book. I dedicate this book to my mother for being a constant source of inspiration. I express heartfelt thanks to my wife Preeti for her continuous help in coediting the book, as well as for her ideas to improve the manuscript.

Vikas Mittal

About the Editor

Dr. Vikas Mittal studied chemical engineering from Punjab Technical University in Punjab, India. He later obtained his Masters of Technology in polymer science and engineering from Indian Institute of Technology Delhi, India. Subsequently, he joined the Polymer Chemistry group of Professor U. W. Suter at the Department of Materials at Swiss Federal Institute of Technology, Zurich in Switzerland. There he there worked for his doctoral degree with focus on the subjects of surface chemistry and polymer nanocomposites. He also jointly worked with Professor M. Morbidelli at the Department of Chemistry and Applied Biosciences for the synthesis of functional polymer latex particles with thermally reversible behaviors.

After completion of his doctoral research, he joined the Active and Intelligent Coatings section of Sun Chemical Group Europe in London. He worked for the development of water and solvent based coatings for food packaging applications. He later joined as Polymer Engineer at BASF Polymer Research in Ludwigshafen, Germany where he worked as laboratory manager responsible for the physical analysis of organic and inorganic colloids. He is currently an assistant professor in the Chemical Engineering department at The Petroleum Institute, Abu Dhabi.

His research interests include organic-inorganic nanocomposites, novel filler surface modifications, thermal stability enhancements, polymer latexes with functionalized surfaces, etc. He has authored more than 40 scientific publications, book chapters, and patents on these subjects.

Contributors

Kyung Hyun Ahn
School of Chemical and Biological
　Engineering
Seoul National University
Seoul, Korea

Young-Wook Chang
Department of Chemical
　Engineering
Hanyang University
Ansan, Korea

Johnsy George
Defence Food Research Laboratory
Siddhartha Nagar, Mysore, India

Joung Sook Hong
Department of Chemical
　Engineering
Soongsil University
Seoul, Korea

Shiao-Wei Kuo
Department of Materials and
　Optoelectronic Science
Center for Nanoscience and
　Nanotechnology
National Sun Yat-Sen University
Kaohsiung, Taiwan

Seung Jong Lee
School of Chemical and Biological
　Engineering
Seoul National University
Seoul, Korea

Hyung Tag Lim
School of Chemical and Biological
　Engineering
Seoul National University
Seoul, Korea

Joey Mead
NSF Nanoscale Science and
　Engineering Center for High-Rate
　Nanomanufacturing
Department of Plastics Engineering
University of Massachusetts–Lowell
Lowell, Massachusetts

Vikas Mittal
Chemical Engineering Department
The Petroleum Institute
Abu Dhabi, UAE

P. Poomalai
Central Institute of Plastics
　Engineering and Technology
Debhog City Centre
Purba Medinipur, Haldia, West
　Bengal, India

Sophie Rouif
IONISOS SA
Dagneux, France

Sachin Shanbhag
Department of Scientific Computing
Florida State University
Tallahassee, Florida

Siddaramaiah
Department of Polymer Science and
    Technology
Sri Jayachamarajendra College of
    Engineering
Mysore, Karnataka, India

Rodolphe Sonnier
Centre des Matériaux
Ecole des Mines d'Alès
6, avenue de Clavières
Alès, France

Gopinath Subramanian
Scientific Computation Research
    Center
Rensselaer Polytechnic Institute
Troy, New York

Aurélie Taguet
Centre des Matériaux
Ecole des Mines d'Alès
6, avenue de Clavières
Alès, France

Osamu Urakawa
Department of Macromolecular
    Science
Graduate School of Science
Osaka University
Toyonaka, Osaka, Japan

Hiroshi Watanabe
Institute for Chemical Research
Kyoto University
Uji, Kyoto, Japan

Ming Wei
NSF Nanoscale Science and
    Engineering Center for High-Rate
    Nanomanufacturing
Department of Plastics Engineering
University of Massachusetts–Lowell
Lowell, Massachusetts

Hua Yang
College of Chemistry
Tianjin Key Laboratory of Structure
    and Performance for Functional
    Molecules
Tianjin Normal University
Tianjin, People's Republic of China

1

Functional Polymer Blends: Synthesis and Microstructures

Vikas Mittal

The Petroleum Institute
Abu Dhabi, UAE

CONTENTS

1.1 Introduction

Polymer blends are generated by the mixing of two or more polymers together in order to combine the characteristics of individual polymers. The compatibilization of the phases in the blends is also achieved by a number of physical (e.g., addition of compatibilizers viz. polyethylene-co-maleic anhydride) or chemical (e.g., chemical reactions with the phases for chemical linking) means. Polymer blends are widely used materials in the modern polymer industry owing to their wider range of properties as compared to individual polymers and ease of tailoring of properties. As a result, polymer blends find applications in numerous fields such as adhesion, colloidal stability, and design of composite and biocompatible materials [1]. The dispersed phase in certain systems can also acquire a specific morphology beneficial for applications like platy morphology, useful for barrier applications. Both solution as well as melt mixing techniques to generate the polymer blends have found their application, though melt mixing is more environmentally friendly as no solvent is required. On the other hand, melt mixing required the use of high temperature and shear for the generation of blends, which may degrade the heat-sensitive polymers, therefore requiring optimum use of mixing conditions. The studies on the polymer blends have significantly increased in number in recent years leading to generation of further functional blend systems with superior properties. Both miscible and immiscible blends have been studied along with new pathways to achieve compatibilization of the phases in immiscible blends. Some of these developments have been discussed in the following paragraphs. As an example, Figure 1.1 shows polyvinyl chloride/polymethyl methacrylate (PVC/PMMA) blends with varying amounts of components and the resulting effects on the strength and transparency of the generated blends [2]. The change in morphology as well as strength and transparency are observed by changing the composition of the blend constituents. The PMMA is observed to form finely dispersed particles of 300 to 400 nm in diameter at low concentrations in the PVC matrix. However, when PMMA is the continuous phase and PVC is the dispersed

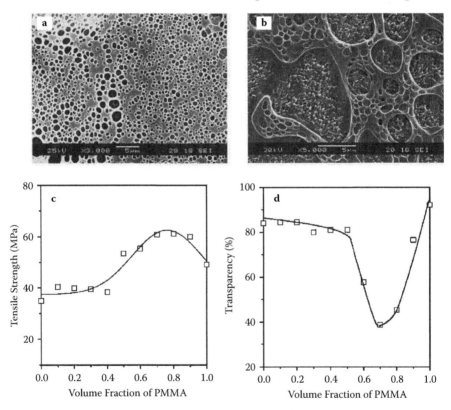

FIGURE 1.1
Scanning electron micrographs (SEMs) of (a) polyvinyl chloride/polymethyl methacrylate (PVC/PMMA) blend containing 20% PMMA, (b) PVC/PMMA blend containing 80% PMMA, (c) tensile strength of PVC/PMMA blends as a function of composition, and (d) transparency of PVC/PMMA blends as a function of blend composition. (Reproduced from Fekete, E., Foldes, E., and Pukanszky, B. 2005. Effect of molecular interactions on the miscibility and structure of polymer blends. *European Polymer Journal* 41:727–736 with permission from Elsevier.)

phase, the PVC domains are much bigger in size. The strength was observed to increase with increasing PMMA concentration until 70% to 80% after which a sharp decrease was observed. The transparency, on the other hand, decreased at different rates with increasing PMMA content until 70% after which it increased sharply indicating a strong relation between the morphology and the properties.

1.2 Functional Polymer Blends: Examples

Litmanovich et al. [3] reviewed chemical reactions in polymer blends for achieving compatibility and enhanced properties. The various compatibilizing reactions discussed were reactions of functionalized blend components to produce in situ copolymers like reactions of the end groups, reactions with polymeric compatibilizers, addition of low molecular weight components that promote a copolymer formation, and so forth. In the context of such reactions, Sundararaj and colleagues [4,5] also concluded that the main effect of using either preformed block copolymers or synthesizing them in situ by the chemical reactions is the suppression of coalescence of the fine dispersion particles formed from blend pellets. The diameter of the particles increased in the PS/PMMA blend as a function of annealing time at 180°C when no copolymer was used. On the other hand, the diameter remained unchanged when 5% copolymer of PS-b-PMMA was added to the blend as shown in Figure 1.2.

Fekete et al. [2] reported the effect of polymer–polymer interactions on the miscibility and macroscopic properties of PVC/PMMA, PVC/PS and PMMA/PS blends. The blends were prepared by mixing the components in an internal mixer at 185 to 190°C and 50 rpm for 10 min. The compounds were then compression molded into 1 mm thick plates at 190 to 200°C. The authors observed that the degree of miscibility was the highest in the PMMA/PVC blends. PVC/PS and PMMA/PS blends displayed two glass transitions in the entire composition range as shown in Figure 1.3. It was also observed that in the case of the PMMA/PS blends, the shift in the glass transition temperatures was smaller (2 to 4°C) as compared to the PVC/PS blends (7 to 10°C).

The authors also reported that specific interaction was formed between the carbonyl groups of PMMA and the hydrogen atom of the CHCl group in PVC in the PVC/PMMA blends. Carbonyl absorption band of PMMA shifted from 1735 to 1732 cm^{-1}, which was also accompanied by a change in the half width of the vibration.

Rameshwaram et al. [6] investigated the structure–property relationships and the effects of a viscosity ratio on the rheological properties of polymer blends using oscillatory and steady shear rheometry and optical microscopy.

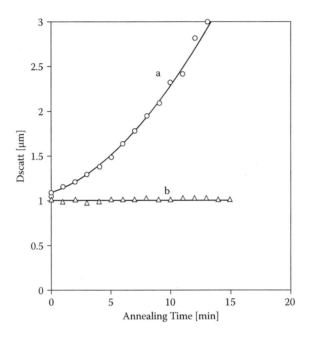

FIGURE 1.2
Particle diameter versus annealing time at 180°C for (a) 70/30 polystyrene/polymethyl methacrylate (PS/PMMA) bend without addition of a copolymer and (b) 70/30 PS/PMMA blend with 5% PS-b-PMMA copolymer. (Reproduced from Macosko, C. W., Guegan, P., Khandpur, A. K., Nakayama, A., Marechal, P., and Inoue, T. 1996. Compatibilizers for melt blending: Premade block copolymers. *Macromolecules* 29:5590–5598 with permission from American Chemical Society.)

These immiscible blends consisted of ultrahigh viscous polybutadiene (PB1), high viscous polybutadiene (PB2), and low viscous polydimethylsiloxane (PDMS). The authors studied the effect of viscosity ratio on the rheological properties of the immiscible polybutadiene (PB1)/polydimethylsiloxane (PDMS) blend with an extremely high viscosity ratio (162,000) and the immiscible polybutadiene (PB2)/PDMS blend with a high viscosity ratio (37) as a function of various compositions. Figure 1.4 showed the storage and loss moduli for the PB2/PDMS blend at the various weight fractions of PB2. An increase in both G′ and G″ was observed for the blends until the PB2 content of 0.7 was reached. The storage modulus was observed to decrease while the loss modulus slightly increased when the content of PB2 was increased from 0.7 to 0.9. The viscoelastic properties of the PB1/PDMS blends also increased systematically with increasing the weight fraction of PB1, and were observed to exhibit plateau values above a certain maximum weight fraction of PB1. Also, the viscoelastic properties of the PB1/PDMS blends were not affected by the change of blend morphology or phase inversion; on the other hand, the viscoelastic properties of the PB2/PDMS blends were significantly affected by phase inversion.

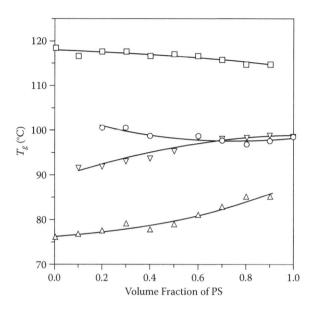

FIGURE 1.3
Glass transition temperature of the components in polymethyl methacrylate/polystyrene (PMMA/PS) (□, PMMA; ○, PS) and polyvinyl chloride (PVC)/PS (△, PVC; PS) blends. (Reproduced from Fekete, E., Foldes, E., and Pukanszky, B. 2005. Effect of molecular interactions on the miscibility and structure of polymer blends. *European Polymer Journal* 41:727–736 with permission from Elsevier.)

Wang et al. [7] reported effective strategy to compatibilize three immiscible polymers, polyolefins, styrene polymers, and engineering plastics, by using a polyolefin-based multiphase compatibilizer. A model ternary immiscible polymer blend consisting of polypropylene (PP)/polystyrene (PS)/polyamide (PA6), and a multiphase compatibilizer (PP-g-(MAH-co-St)) as prepared by maleic anhydride (MAH) and styrene (St) dual monomers melt grafting PP was used for the analysis. The authors reported that the multiphase compatibilizer, PP-g-(MAH-co-St) showed effective compatibilization in the PP/PS/PA6 blends as the particle sizes of both PS and PA6 dispersed phases greatly decreased after the addition of the compatibilizer. Interfacial adhesion in immiscible pairs was observed to increase under the influence of the compatibilizer. The morphology of PP/PS/PA6 (70/15/15) uncompatibilized blend revealed that the blend was constituted from the PP matrix as continuous phase with dispersed composite droplets of PA6 core encapsulated by PS phase; however, a different morphology was observed in the compatibilized blend. The three components interacted strongly with each other in the presence of compatibilizer, and PS did not encapsulate PA6 as before, as shown in Figure 1.5. Similarly, for the 40/30/30 blend, the morphology was observed to change from a three-phase cocontinuous morphology (uncompatibilized) to the dispersed droplets of PA6 and PS in the PP matrix (compatibilized).

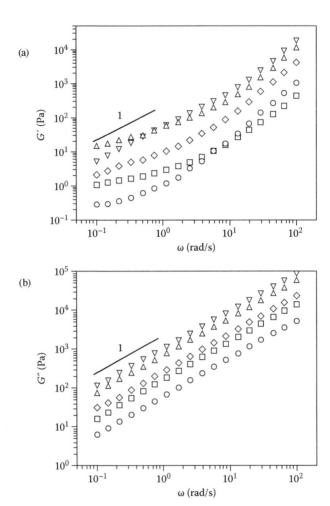

FIGURE 1.4
(a) Dynamic storage modulus (G′) of high viscous polybutadiene/low viscous polydimethyl-siloxane (PB2/PDMS) blends without preshear and (b) dynamic loss modulus (G″) of PB2/PDMS blends without preshear. (Symbols: PB2 concentration of 0.1 [circles], 0.3 [squares], 0.5 [diamonds], 0.7 [triangles], and 0.9 [reversed triangles], respectively.) (Reproduced from Rameshwaram, J. K., Yang, Y.-S., and Jeon, H. S. 2005. Structure–property relationships of nanocomposite-like polymer blends with ultrahigh viscosity ratios. *Polymer* 46:5569–5579 with permission from Elsevier.)

Higgins et al. [8] reported the application of Born–Green–Yvon (BGY) theory to predict behavior that is outside the typical range of experimental conditions used to obtain the parameters. As an example, the authors used the characteristic parameters obtained by characterization of the blend from one set of molecular weights to predict the behavior of a different molecular

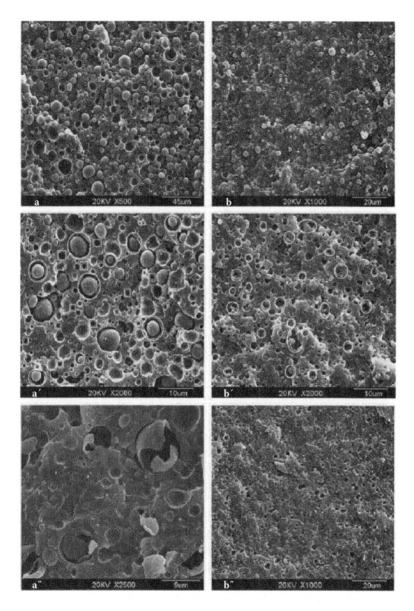

FIGURE 1.5
Morphology of the uncompatibilized (a, a′, a″) and compatibilized (b, b′, b″) ternary poly-propylene/polystyrene/polyamide (PP/PS/PA6) (70/15/15) blends: (a,b) cryofractured surface, (a′, b′) cryofractured and PS phase extracted using tetrahydrofuran (THF), (a″, b″) cryofractured and PA6 phase extracted using HCOOH. (Reproduced from Wang, D., Li, Y., Xie, X.-M., and Guo, B.-H. 2011. Compatibilization and morphology development of immiscible ternary polymer blends. *Polymer* 52:191–200 with permission from Elsevier.)

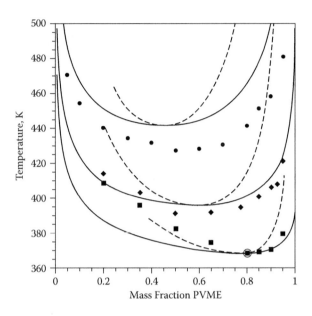

FIGURE 1.6
Coexistence results for polystyrene/poly(vinyl methyl ether) (PS/PVME) of varying molecular weight ratios. The PVME component had a molecular weight of 51,500 g/mol, whereas the PS molecular weight changed from 20,400 (solid circles) to 51,000 (diamonds) to 200,000 (squares) g/mol. (Reproduced from Higgins, J. S., Tambasco, M., and Lipson, J. E. G. 2005. Polymer blends; stretching what we can learn through the combination of experiment and theory. *Progress in Polymer Science* 30:832–843 with permission from Elsevier.)

weight combinations as shown in Figure 1.6. Experimental coexistence data for three different PS/PVME (poly(vinyl methyl ether)) blends are reported in Figure 1.6. PVME had a molecular weight of 51,500 g/mol, whereas the PS molecular weight changed from (top to bottom) 20,400 to 51,000 to 200,000 g/mol. The temperature of the circled point shown in the figure was used to determine the mixed interaction parameter. The solid lines indicate the calculated binodal curves, whereas the dashed lines represent the spinodal curves. It was clearly observed that the coexistence behavior was represented well by the theory. The prediction for the lowest molecular weight blend though was not as close to the experimental observations in comparison to the other higher molecular weight blends; however, the agreement was still quite reasonable.

Menyhard et al. [9] reported the generation of polymer blends based on the β-modification of polypropylene. The authors studied the melting and crystallization characteristics as well as the structure and polymorphic composition of the blends by polarized light microscopy (PLM) and differential scanning calorimetry (DSC). It was observed that the most important factor of the formation of the blend with β-crystalline phase when semicrystalline polymers were added to isotactic polypropylene (iPP) was the α-nucleation

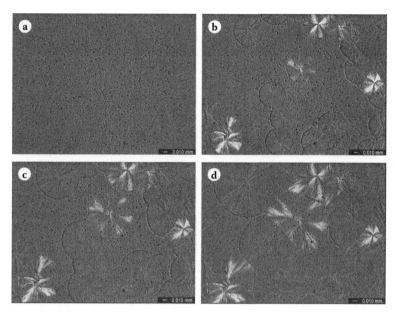

FIGURE 1.7
Polarized light microscopy micrographs of a β-nucleated isotactic polypropylene/polyamide-6 (iPP/PA-6) blend containing 5 wt% PA-6. Isothermal crystallization was carried out at $T_c = 135°C$ for (a) $t_c = 0$ min, (b) $t_c = 21$ min, (c) $t_c = 45$ min, and (d) $t_c = 60$ min. (Reproduced from Menyhard, A., Varga, J., Liber, A., and Belina, G. 2005. Polymer blends based on the β-modification of polypropylene. *European Polymer Journal* 41:669–677 with permission from Elsevier.)

effect of the second polymer. Also, in the case of polymers with an α-nucleating effect, the temperature range of their crystallization needs to be lower than that of β-iPP. The authors also concluded that the β-nucleated iPP/PVDF and iPP/PA-6 blends were extreme examples showing that the β-iPP matrix could not completely form even in the presence of a highly effective β-nucleant because of the strong α-nucleating ability and higher crystallization temperature range of PVDF and PA-6. Figure 1.7 shows the polarized light microscopy micrographs of a β-nucleated iPP/PA-6 blend containing 5 wt% PA-6. Isothermal crystallization was carried out at $T_c = 135°C$ for times 0 min, 21 min, 45 min, and 60 min. The micrographs indicated the generation of blends with heterogeneous structure. PA-6 with a higher melting point crystallized first at around 190°C. A polymorphic structure consisting of α- and β-spherulites was formed.

Gopakumar et al. [10] reported the in situ compatibilization of poly(phenylene sulfide) (PPS)/wholly aromatic thermotropic liquid crystalline polymer (TLCP) Vectra A950 blends by reactive extrusion. The authors prepared the in situ compatibilized PPS/TLCP blends in a twin-screw extruder by reactive blending of PPS and TLCP in the presence of dicarboxyl-terminated poly(phenylene sulfide) (DCTPPS). Block copolymer was formed during reactive blending, by transesterification reaction between carboxyl

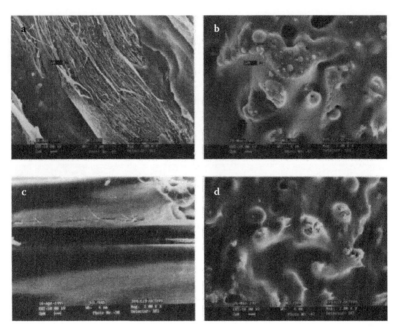

FIGURE 1.8
Scanning electron micrograph (SEM) of compatibilized poly(phenylene sulfide) PPS/Vectra A950 blends. (a) Perpendicular to injection flow direction: (a) skin region, (b) core region. Parallel to injection molded direction: (c) skin region and (d) core region. (Reproduced from Menyhard, A., Varga, J., Liber, A., and Belina, G. 2005. Polymer blends based on the β-modification of polypropylene. *European Polymer Journal* 41:669–677 with permission from Elsevier.)

groups of DCTPPS and ester linkages of TLCP. The heat of melting, crystallization temperature, and heat of crystallization of the PPS phase was observed to decrease in PPS/Vectra A950 blends on compatibilization that indicated the presence of favorable interaction between the blend components. Apart from that, the tensile and impact properties of the compatibilizer blends were observed to enhance, which also indicated better interfacial adhesion between the components. Both uncompatibilized and compatibilized PPS/Vectra A950 blends were observed to exhibit skin-core morphology with the Vectra A950 fibers present more in the skin region but less in the core region. Figure 1.8 shows the scanning electron micrograph (SEM) of compatibilized PPS/Vectra A950 blends. Improved interfacial adhesion between the blend constituents was clear as fracture of the Vectra A950 fibrils was observed instead of their pulling out of the matrix. Enhanced interfacial bonding between the PPS matrix and Vectra A950 fibrils was also observed.

Patlazhan et al. [11] studied the shear-induced fractal morphology of immiscible reactive polymer blends. The example of grafting and cross-linking multilayer systems of statistic terpolymer of ethylene, butyl acrylate, and maleic anhydride (MAH) and statistic copolymers (CPA) including

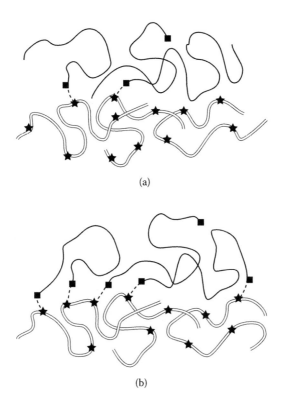

(a)

(b)

FIGURE 1.9

(a) Grafting maleic anhydride–statistic copolymers (MAH-CPA$_G$) interface and (b) cross-linking MAH-CPA$_C$ interface (b). The dotted lines represent chemical links. CPA$_G$ includes one acid group at the one end and one amine function at the other, whereas CPA$_C$ included two amine functions. (Reproduced from Patlazhan, S., Schlatter, G., Serra, C., Bouquey, M., and Muller, R. 2006. Shear-induced fractal morphology of immiscible reactive polymer blends. *Polymer* 47:6099–6106 with permission from Elsevier.)

polyamide and acid groups terminated by acid or amine groups were used for the study. The interfaces obtained in the study included grafted interface formed by anchoring CPA$_G$ chains to reactive functional groups of MAH terpolymer (Figure 1.9a) and cross-linked interface formed by the chemical bonding of CPA$_C$ copolymer functional ends to MAH reactive groups (Figure 1.9b). The reactive polymer systems were observed to display considerable hydrodynamic instabilities followed by the branched finger-like formations. The morphologies developed in the reactive polymer blends corresponded to fractal structures, and the fractal dimensions of the cross-linked and grafted systems were 1.84 and 1.75, respectively. These values were close to the fractal dimension of the Laplacian growth patterns.

Ravati and Favis [12] generated a low percolation threshold conductive device prepared through the control of multiple encapsulation and multiple percolation effects in a five-component polymer blend system.

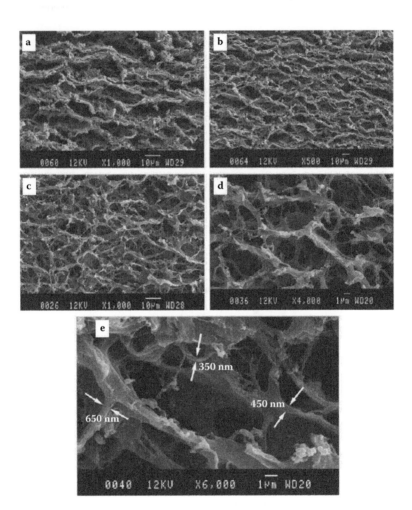

FIGURE 1.10
Scanning electron micrographs (SEMs) of (a,b) polyaniline (PANI) network in 25/25/25/25 polystyrene/polymethyl methacrylate/poly(vinylidene fluoride)/polyaniline (PS/PMMA/PVDF/PANI) blend after extraction of all phases by dimethylformamide (DMF) followed by freeze drying, and (c,d,e) PANI network in 15/20/15/25/25 PS/PS-co-PMMA/PMMA/PVDF/PANI blend after extraction of all phases by DMF followed by freeze drying. (Reproduced from Ravati, S., and Favis, B. D. 2010. Low percolation threshold conductive device derived from a five-component polymer blend. *Polymer* 51:3669–3684 with permission from Elsevier.)

Conductive polyaniline (PANI) formed the core of the five-component continuous system that also included high-density polyethylene (HDPE), polystyrene (PS), poly(methyl methacrylate) (PMMA), and poly(vinylidene fluoride) (PVDF) along with PS-co-PMMA copolymer. Figure 1.10 shows the SEM micrographs of PANI network in PS/PMMA/PVDF/PANI and PS/PS-co-PMMA/PMMA/PVDF/PANI blends after extraction of all phases by DMF followed by freeze drying.

Tsuneizumi et al. [13] studied the chemical recycling of poly(lactic acid)–based polymer blends using environmentally benign catalysts, clay catalysts, and enzymes. Poly(L-lactic acid) (PLLA)–based polymer blends (e.g., PLLA/polyethylene [PE] and PLLA/poly(butylenes succinate) [PBS]) were degraded into repolymerizable oligomer.

For the case of PLLA/PE, the first method chosen was the direct separation of PLLA and PE first by their different solubilities in toluene, followed by the chemical recycling of PLLA. Another method based on the selective degradation of PLLA in the PLLA/PE blend in a toluene solution at 100°C for 1 h was also performed. This led to the generation of lactic acid oligomer with a reduced molecular weight of 200 to 300 g/mol. The PE fraction had no change in the molecular weight and was recovered by precipitation. In the case of the PLLA/PBS blend, direct separation of PLLA and PBS by solubility in toluene was performed. Sequential degradation of the PLLA/PBS blend was also used. A lipase was used to first degrade PBS into cyclic oligomer, which was then repolymerized to produce a PBS. Subsequently, PLLA was degraded into repolymerizable oligomer. Figure 1.11 schematically presents these methods used for the recycling of blends.

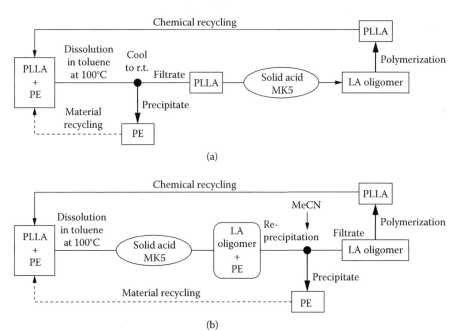

FIGURE 1.11
Recycling of (a) poly(L-lactic acid)/polyethylene (PLLA/PE) blend and (b) PLLA/poly(butylenes succinate) (PBS) blend. (Reproduced from Tsuneizumi, Y., Kuwahara, M., Okamoto, K., and Matsumura, S. 2010. Chemical recycling of poly(lactic acid)-based polymer blends using environmentally benign catalysts. *Polymer Degradation and Stability* 95:1387–1393 with permission from Elsevier.)

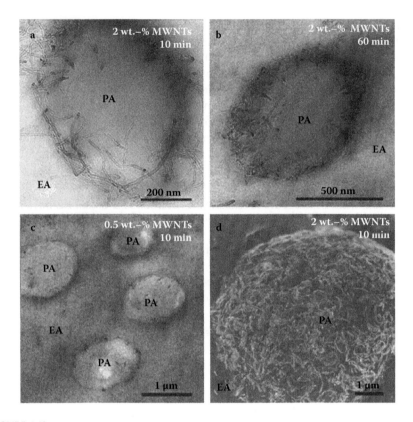

FIGURE 1.12

(a,b,c) Transmission electron micrographs (TEMs) of blends of ethyl acrylate (EA), polyamide (PA), and multiwalled carbon nanotubes (MWNTs): (a) 2 wt% MWNTs, 10 min mixing; (b) 2 wt% MWNTs, 60 min mixing; (c) 0.5 wt% MWNTs, 10 min mixing; and (d) scanning electron micrograph (SEM) of a cryofractured blend of EA, PA, and 2 wt% MWNTs, 10 min mixing. (Reproduced from Baudouin, A.-C., Auhl, D., Tao, F. F., Devaux, J., and Bailly, C. 2011. Polymer blend emulsion stabilization using carbon nanotubes interfacial confinement. *Polymer* 52:149–156 with permission from Elsevier.)

Baudouin et al. [14] demonstrated the effect of interfacial confinement of unfunctionalized multiwalled carbon nanotubes (MWNTs) on coalescence suppression in an immiscible polymer blend comprising of polyamide (PA)/ethylenemethyl acrylate random copolymer (ethyl acrylate, EA). Figure 1.12 shows the transmission electron micrographs (TEMs) of the blends generated by a twin-screw mini-compounder. No EA subinclusions were observed in the presence of MWNTs, even after 60 min of mixing. The nanotubes were observed to be mainly localized at the interface. The authors also confirmed the interfacial localization as is evident from the SEM micrograph after cryofracture (Figure 1.12d), which demonstrated a PA droplet covered with nanotubes. Such particle (nanotubes) stabilized blends were suggested to be good alternatives to the blends compatibilized by block copolymers.

Busche et al. [15] reported the properties of polystyrene (PS)/poly(dimethyl siloxane) (PDMS) blends partially compatibilized with star polymers containing a γ-cyclodextrin core and polystyrene arms. The mechanism of compatibilization was observed to be threading of the cyclodextrin core by PDMS and subsequent solubilization in the PS matrix facilitated by the star arms.

The authors pointed out that there was ample evidence for partial compatibilization of PS and PDMS by the star polymer in solution-cast films. The glass transition temperatures of PS and PDMS shifted toward each other, and the shift was greater when the star polymer was present as compared to the case where no star polymer was used. Also, in the presence of the star polymer, a significant amount of PDMS was retained in the films, which otherwise leaches out of the films in the absence of the star polymer. As indicated by Figure 1.13, the compatibilized samples exhibited a higher storage and higher amount of retained PDMS indicating restricted molecular mobility of PDMS in the presence of the star polymer.

Virgilio et al. [16] reported in situ Neumann triangle–focused ion beam–atomic force microscopy (NT-FIB-AFM) method to measure modified PS/HDPE interfacial tensions in ternary PS/PP/HDPE blends demonstrating partial wetting. The ternary blend was also modified with styrene-(ethylene-butylene), styrene-butadiene, and styrene-(ethylene-butylene)-styrene triblock copolymers. Figure 1.14 shows the morphology of the PS/PP/HDPE 10/45/45 blends after 30 min of quiescent annealing used for Neumann triangle analysis. The concentration of styrene-(ethylene-butylene) (SEB) was gradually increased from 0% to 2% based on the PS content. It was observed that a gradual relocalization of the PS droplets from the PP side of the PP/HDPE interface to the HDPE side occurred as the concentration of SEB increased. At 0.5% SEB concentration, already a significant relocalization of the PS droplets on the HDPE side of the PP/HDPE interface was observed that continued with 1% and 2% concentrations, which led to a decrease in the PS droplet size.

Sohn et al. [17] studied the surface properties of comb-like polymer blends of poly(oxyethylene)s having CH_3-terminated and CF_3-terminated alkylsulfonylmethyl side chains. Figure 1.15 shows the atomic force microscopy (AFM) images of the pure polymers as well as blends with different compositions. The surface of the homopolymers was observed to be flat, whereas the blend surfaces had comparatively rough morphologies. In the blends, the side chains were found to be well oriented on the surface, while being phase separated, forming various surface morphologies. Morphologies like holes, islands, and interconnected islands were generated. For example, the surface of the 20/80 blend showed holes that had a diameter distribution of 300 to 700 nm. When the content of CF_3-10SE was more than 40 mol% in the blends, the surface morphologies were observed to change from islands (40/60 blend) and interconnected islands (60/40 blend) located in a somewhat lower phase to protruded islands (80/20 blend) that resembled an inverse form of the 20/80 blend morphology. It was thus believed that the CH_3-10SE–rich

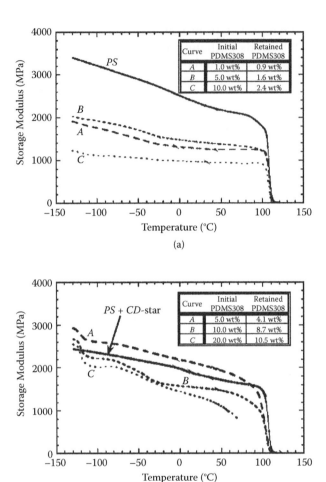

FIGURE 1.13
Storage modulus curves for polystyrene (PS)/poly(dimethyl siloxane) (PDMS) films (a) without and (b) with 1.0 wt% cyclodextrin (CD) core. The solid lines represented samples with no PDMS. The initial and retained amounts of PDMS are shown in the insets for the other samples. (Reproduced from Busche, B. J., Tonelli, A. E., and Balik, C. M. 2010. Properties of polystyrene/poly(dimethyl siloxane) blends partially compatibilized with star polymers containing a γ-cyclodextrin core and polystyrene arms. *Polymer* 51:6013–6020 with permission from Elsevier.)

domains constituted the higher parts in the topography (brighter area) and that the CF$_3$-10SE–rich domains were present in the lower parts (darker area). The authors observed that the polar CF$_3$-terminal groups in the lower regions controlled the surface properties, because the contact angles and stick–slip behaviors of the blends containing more than 40 mol% of poly(oxyethylene) with CF$_3$-terminated side chains were similar.

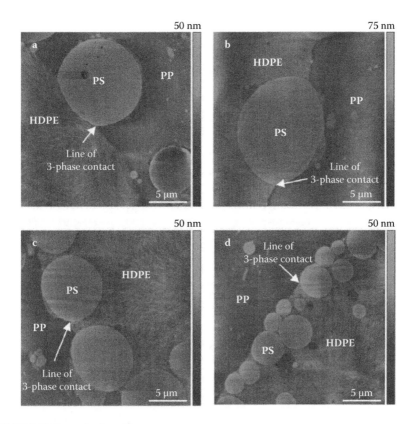

FIGURE 1.14 (See color insert.)
Focused ion beam–atomic force microscopy (FIB-AFM) micrographs of polystyrene/polypropylene/high-density polyethylene (PS/PP/HDPE) 10/45/45 blends after 30 min of quiescent annealing showing the gradual migration of the PS droplets toward the HDPE phase as a function of the increasing concentration of SEB. (a) 0.1% SEB, (b) 0.5% SEB, (c) 1% SEB, and (d) 2% SEB. (Reproduced from Virgilio, N., Desjardins, P., L'Esperance, G., and Favis, B. D. 2010. Modified interfacial tensions measured in situ in ternary polymer blends demonstrating partial wetting. *Polymer* 51:1472–1484 with permission from Elsevier.)

Park et al. [18] reported the synthesis of dye-sensitized solar cells (DSSCs) based on electrospun polymer blend nanofibers as electrolytes. Electrospun poly(vinylidenefluoride–co-hexafluoropropylene) (PVDF–HFP) and PVDF–HFP/polystyrene (PS) blend nanofibers were prepared as shown in Figure 1.16. The authors reported that the photovoltaic performance of dye-sensitive solar cell (DSSC) devices using electrospun PVDF–HFP/PS (3:1) nanofiber was much better as compared to DSSC devices using electrospun PVDF–HFP nanofiber. It was further observed that the overall power conversion efficiency of the DSSC device using PVDF–HFP nanofiber had a lower value than that of the DSSC device using electrospun PVDF–HFP/PS blend nanofibers

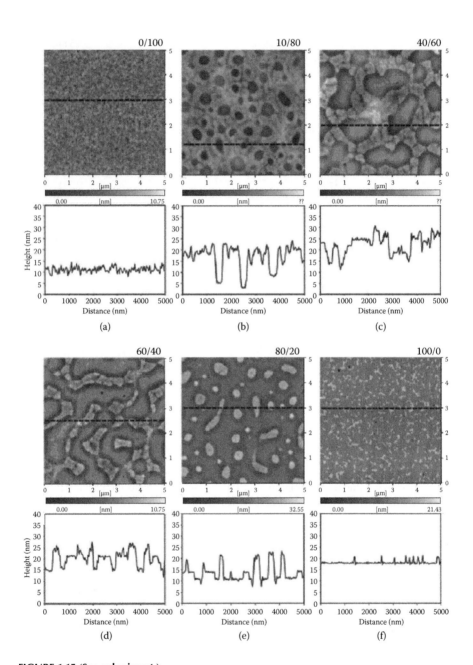

FIGURE 1.15 (See color insert.)
AFM images and line profiles of the surface topography of (a) CH₃-10SE, (b) 20/80 blend, (c) 40/60 blend, (d) 60/40 blend, (e) 80/20 blend, and (f) CF₃-10SE. (Reproduced from Sohn, E.-H., Kim, B. G., Chung, J.-S., and Lee, J.-C. 2010. Comb-like polymer blends of poly(oxyethylene)s with CH₃-terminated and CF₃-terminated alkylsulfonylmethyl side chains: Effect of terminal CF₃ moiety on the surface properties of the blends. *Journal of Colloid and Interface Science* 343:115–124 with permission from Elsevier.)

FIGURE 1.16
Scanning electron microscope (SEM) images of (a) poly(vinylidenefluoride–co-hexafluoro-propylene)/polystyrene (PVDF–HFP/PS) (1:1), (b) PVDF–HFP/PS (2:1), (c) PVDF–HFP/PS (3:1), and (d) PVDF–HFP nanofibers generated by electrospinning. (Reproduced from Park, S.-H., Won, D.-H., Choi, H.-J., Hwang, W.-P., Jang, S.-i., Kim, J.-H., Jeong, S.-H., Kim, J.-U., Lee, J.-K., and Kim, M.-R. 2011. Dye-sensitized solar cells based on electrospun polymer blends as electrolytes. *Solar Energy Materials and Solar Cells* 95:296–300 with permission from Elsevier.)

indicating that the porosity of electrospun PVDF–HFP/PS blend nanofibers was higher, ion transfer occurred well, and regular nanofiber morphology helped in the transfer of ions produced by the redox mechanism.

Huang et al. [19] reported preliminary investigations on the phase morphology development of a polymer blend in water-assisted injection molding unit. Experiments were carried out using polypropylene/polyamide-6 (PP/PA-6) blends with PP as the continuous phase. The authors evaluated the morphology generated in the blends for spatial distribution of the dispersed phase as well as for size of the dispersed phase. It was observed that the morphology developed at the position near the water inlet was induced mainly by the melt filling, whereas that near the end of the water channel was mainly ascribed to the high-pressure water-assisted filling (Figure 1.17). The authors investigated processing parameters like water pressure, melt temperature, and injection speed from their impact on the resulting morphology and concluded that higher water pressure, adequate melt temperature, and higher injection speed resulted in a more obvious deformation of the dispersed PA-6 phase.

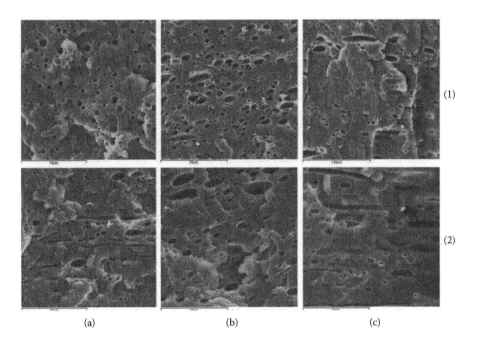

(1)

(2)

(a)   (b)   (c)

FIGURE 1.17
Scanning electron micrographs (SEMs) of (a) inner layer, (b) core region, and (c) outer layer at two positions of water-assisted injection-molded polypropylene (PP)/polyamide-6 (PA-6) blend. Water pressure, 10 MPa; melt temperature, 270°C; injection speed, 50%. (Reproduced from Huang, H.-X., and Zhou, R.-H. 2010. Preliminary investigation on morphology in water-assisted injection molded polymer blends. *Polymer Testing* 29:235–244 with permission from Elsevier.)

Chang et al. [20] reported the preparation and characterization of shape memory polymer networks based on carboxylated telechelic poly(ε-caprolactone) (XPCL)/epoxidized natural rubber (ENR) blends. It was observed that the XPCL/ENR blends could form cross-linked structure via interchain reaction between the reactive groups of each polymer during molding at high temperature. Degree of cross-linking and crystalline melting transition temperatures were observed to be dependent upon the blend compositions as well as the molecular weight of the XPCL segment in the blends. Figure 1.18 shows the strain recovery for the XPCL-1.2k, XPCL-2.0k, and XPCL-3.6k based blend networks measured as a function of temperature. Strain recovery occurred above T_m of each sample, and the final recovery rate was observed to be dependent upon the gel content of each sample.

FIGURE 1.18 (See facing page.)
The strain recovery curves for (a) XPCL-1.2k/ENR, (b) XPCL-2.0k/ENR, and (c) XPCL-3.6k/ENR blends. (Reproduced from Chang, Y.-W., Eom, J.-P., Kim, J.-G., Kim, H.-T., and Kim, D.-K. 2010. Preparation and characterization of shape memory polymer networks based on carboxylated telechelic poly(ε-caprolactone)/epoxidized natural rubber blends. *Journal of Industrial and Engineering Chemistry* 16:256–260 with permission from Elsevier.)

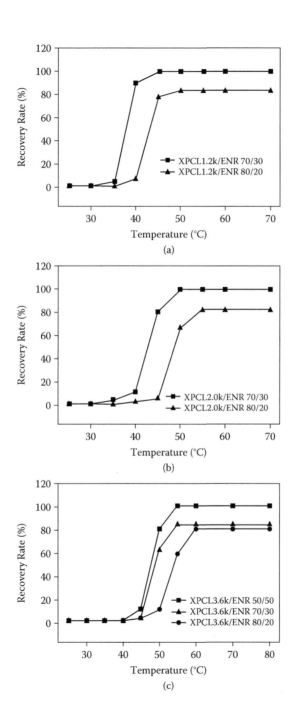

FIGURE 1.18

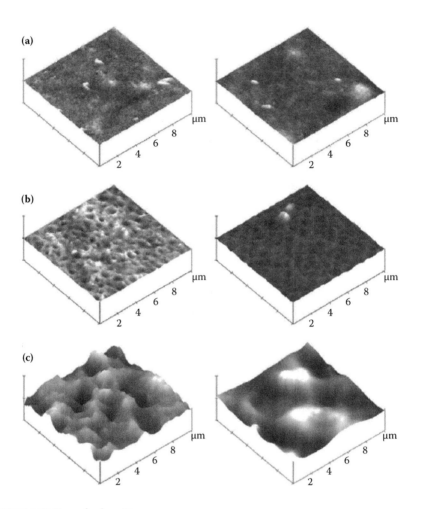

FIGURE 1.19 (See color insert.)
Atomic force microscopy height images of samples before (left) and after 30 h ultraviolet irradiation (right): (a) pectin (PEC), (b) PEC/polyvinylpyrrolidone (PVP) (80/20), and (c) PEC/PVP (60/40). (Reproduced from Kowalonek, J., and Kaczmarek, H. 2010. Studies of pectin/polyvinylpyrrolidone blends exposed to ultraviolet radiation. *European Polymer Journal* 46:345–353 with permission from Elsevier.)

When the gel content was low (e.g., 40% or less), the samples did not recover fully upon heating. On the other hand, when the gel content was high, the complete final recovery rate was obtained indicating the requirement of a network structure to achieve the property of shape memory effect.

Kowalonek et al. [21] studied pectin/polyvinylpyrrolidone (PVP) blends exposed to ultraviolet radiation and analyzed changes in the chemical structure. These biodegradable blends of natural pectin and synthetic polyvinylpyrrolidone were obtained by casting from aqueous solutions. Figure 1.19

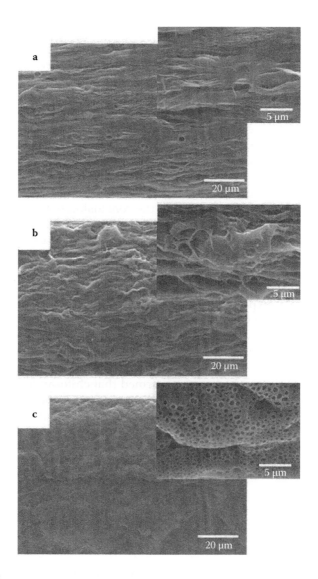

FIGURE 1.20
Scanning electron micrographs (SEMs) of the (a) 75CHT (chitosan), (b) 50CHT, and (c) 25CHT fiber surfaces after extracting the poly(ε-caprolactone) (PCL) component of the blend. (Reproduced from Malheiro, V. N., Caridade, S. G., Alves, N. M., and Mano, J. F. 2010. New poly(ε-caprolactone)/chitosan blend fibers for tissue engineering applications. *Acta Biomaterialia* 6:418–428 with permission from Elsevier.)

shows the AFM images of pure pectin polymer as well as two blend systems. Pure PVP film was smooth on the surface, whereas unexposed pectin film was characterized by the more corrugated surface. The presence of tiny particles of nanometer size on the pectin film surface was observed. In the unexposed films of blends, the surface roughness was observed to increase owing to component immiscibility. Ultraviolet (UV) irradiation led to significant modification of all surfaces, and the changes were observed to be dependent on initial surface quality. In some blends surface roughness was increased owing to the relaxation processes and rearrangement of macrochains in formed films under nonequilibrium conditions. In other cases, surface roughness decreased owing to efficient photodegradation.

Malheiro et al. [22] reported the production of poly(ε-caprolactone) (PCL)/chitosan (CHT) blend fibers for future application as tissue engineering scaffolds. The authors generated fibers of chitosan and poly(ε-caprolactone) by wet spinning from blend solutions, using a formic acid/acetone 70:30 vol% mixture as common solvent and methanol as coagulant. The spectroscopic characterization of the systems led to the conclusion that a certain degree of interaction between the phases existed, though it was not a chemical interaction. Swelling of the fibers was studied by measuring the change in diameter after immersion in a physiological solution for 24 h. Pure PCL fibers did not show significant change in the diameter, whereas CHT fibers exhibited significant swelling in the physiological solution. In the case of blend fibers, an unexpected enhancement in swelling was observed, when compared with pure CHT. This was ascribed to an increase in porosity of the fibers with increasing PCL content. Morphology of the blend fibers was investigated after eliminating the PCL component with a suitable solvent (e.g., chloroform) for 24 h. All the blends were observed to retain their dimensional stability, which also confirmed that chitosan formed the continuous phase in the blend. As seen in Figure 1.20, a smoother surface and small holes left by the extracted PCL were observed for all blends.

The authors concluded that if phase separation existed in the blends, it would only be at a very fine scale (<10 μm). The bled fibers were rigid, and the modulus was measured to be in the range of 1 to 3 GPa.

References

1. Lipatov Y. S. 2002. Polymer blends and interpenetrating polymer networks at the interface with solids. *Progress in Polymer Science* 27:1721–1801.
2. Fekete, E., Foldes, E., and Pukanszky, B. 2005. Effect of molecular interactions on the miscibility and structure of polymer blends. *European Polymer Journal* 41:727–736.
3. Litmanovich, A. D., Plate, N. A., and Kudryavtsev, Y. V. 2002. Reactions in polymer blends: Interchain effects and theoretical problems. *Progress in Polymer Science* 27:915–970.
4. Sundararaj, U., and Macosko, C. W. 1995. Drop breakup and coalescence in polymer blends: The effects of concentration and compatibilization. *Macromolecules* 28:2647–2657.

5. Macosko, C. W., Guegan, P., Khandpur, A. K., Nakayama, A., Marechal, P., and Inoue, T. 1996. Compatibilizers for melt blending: Premade block copolymers. *Macromolecules* 29:5590–5598.
6. Rameshwaram, J. K., Yang, Y.-S., and Jeon, H. S. 2005. Structure–property relationships of nanocomposite-like polymer blends with ultrahigh viscosity ratios. *Polymer* 46:5569–5579.
7. Wang, D., Li, Y., Xie, X.-M., and Guo, B.-H. 2011. Compatibilization and morphology development of immiscible ternary polymer blends. *Polymer* 52:191–200.
8. Higgins, J. S., Tambasco, M., and Lipson, J. E. G. 2005. Polymer blends; stretching what we can learn through the combination of experiment and theory. *Progress in Polymer Science* 30:832–843.
9. Menyhard, A., Varga, J., Liber, A., and Belina, G. 2005. Polymer blends based on the β-modification of polypropylene. *European Polymer Journal* 41:669–677.
10. Gopakumar, T. G., Ponrathnam, S., Lele, A., Rajan, C. R., and Fradet, A. 1999. In situ compatibilisation of poly(phenylene sulphide)/wholly aromatic thermotropic liquid crystalline polymer blends by reactive extrusion: Morphology, thermal and mechanical properties. *Polymer* 40:357–364.
11. Patlazhan, S., Schlatter, G., Serra, C., Bouquey, M., and Muller, R. 2006. Shear-induced fractal morphology of immiscible reactive polymer blends. *Polymer* 47:6099–6106.
12. Ravati, S., and Favis, B. D. 2010. Low percolation threshold conductive device derived from a five-component polymer blend. *Polymer* 51:3669–3684.
13. Tsuneizumi, Y., Kuwahara, M., Okamoto, K., and Matsumura, S. 2010. Chemical recycling of poly(lactic acid)-based polymer blends using environmentally benign catalysts. *Polymer Degradation and Stability* 95:1387–1393.
14. Baudouin, A.-C., Auhl, D., Tao, F. F., Devaux, J., and Bailly, C. 2011. Polymer blend emulsion stabilization using carbon nanotubes interfacial confinement. *Polymer* 52:149–156.
15. Busche, B. J., Tonelli, A. E., and Balik, C. M. 2010. Properties of polystyrene/poly(dimethyl siloxane) blends partially compatibilized with star polymers containing a γ-cyclodextrin core and polystyrene arms. *Polymer* 51:6013–6020.
16. Virgilio, N., Desjardins, P., L'Esperance, G., and Favis, B. D. 2010. Modified interfacial tensions measured in situ in ternary polymer blends demonstrating partial wetting. *Polymer* 51:1472–1484.
17. Sohn, E.-H., Kim, B. G., Chung, J.-S., and Lee, J.-C. 2010. Comb-like polymer blends of poly(oxyethylene)s with CH_3-terminated and CF_3-terminated alkylsulfonylmethyl side chains: Effect of terminal CF_3 moiety on the surface properties of the blends. *Journal of Colloid and Interface Science* 343:115–124.
18. Park, S.-H., Won, D.-H., Choi, H.-J., Hwang, W.-P., Jang, S.-I., Kim, J.-H., Jeong, S.-H., Kim, J.-U., Lee, J.-K., and Kim, M.-R. 2011. Dye-sensitized solar cells based on electrospun polymer blends as electrolytes. *Solar Energy Materials and Solar Cells* 95:296–300.
19. Huang, H.-X., and Zhou, R.-H. 2010. Preliminary investigation on morphology in water-assisted injection molded polymer blends. *Polymer Testing* 29:235–244.
20. Chang, Y.-W., Eom, J.-P., Kim, J.-G., Kim, H.-T., and Kim, D.-K. 2010. Preparation and characterization of shape memory polymer networks based on carboxylated telechelic poly(ε-caprolactone)/epoxidized natural rubber blends. *Journal of Industrial and Engineering Chemistry* 16:256–260.

21. Kowalonek, J., and Kaczmarek, H. 2010. Studies of pectin/polyvinylpyrrolidone blends exposed to ultraviolet radiation. *European Polymer Journal* 46:345–353.
22. Malheiro, V. N., Caridade, S. G., Alves, N. M., and Mano, J. F. 2010. New poly(ε-caprolactone)/chitosan blend fibers for tissue engineering applications. *Acta Biomaterialia* 6:418–428.

2

Miscibility Enhancement of Polymer Blends through Multiple Hydrogen Bonding Interactions

Shiao-Wei Kuo

Department of Materials and Optoelectronic Science
National Sun-Yat-Sen University
Kaohsiung, Taiwan

CONTENTS

2.1 Introduction

Polymer blending is to combine two or more components and has superior mechanical, optical, or thermal properties than these individual polymers. From the practical and economical points of view, polymer blending from existing polymers is the most effective and convenient route to create new and useful materials with greater versatility and flexibility than the development of new polymers. Basically, three different types of blends can be distinguished; completely miscible, immiscible, and partially miscible blends [1,2] as shown in Figure 2.1.

However, most randomly selected polymer pairs usually are immiscible and incompatible, resulting in products with more inferior properties than the average of the base polymers. In these compatible polymer blends, the control of interfacial tension plays an important role to govern the blend morphology and associated mechanical properties. The dispersed phase size reduction with decreasing interfacial tension is important for obtaining uniform blend properties, while retaining the physical properties of both of the homopolymers. During the last two decades, research activities on polymer compatibilization have grown at an exponential rate and have been

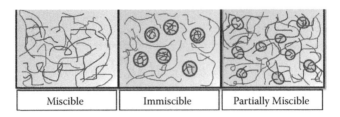

| Miscible | Immiscible | Partially Miscible |

FIGURE 2.1
Typical polymer blend classification.

the subject of some reviews [3,4]. The block and graft copolymers are well known as efficient compatibilizers to reduce the interfacial tension of polymer blends. In addition, the compatibilizer plays an important role in suppressing coalescence of the dispersed domains. However, the microphase separation inherent to most block copolymers has high viscosity, making it difficult to disperse into binary homopolymer blends. In addition, block copolymer with high molecular weight may prefer to micellize within one of the homopolymers phases rather than reside at the interface. In order to overcome many of the shortcomings of block copolymer, research activities in the area of the reactive-type compatibization have been stimulated and clearly become the mainstream in compatibilizing incompatible polymer during the past few years. The main advantage of using compatibilizer in polymer blends is the suppression of coalescence achieved through stabilizing the interface. An effective compatibilizer can improve the interfacial adhesion of a blend and consequently enhance its mechanical properties. However, effectiveness of a compatibilizer should consider both interfacial adhesion and matrix intrinsic property change of the base polymers. Most literature tends to emphasize only the interfacial properties but ignores the change of the matrix intrinsic properties induced by the compatibilizer [3,4]. In a reactively compatibilized blending system, a fraction of the compatibilizer (unreacted, partially reacted, or fully reacted), more or less, is expected to be distributed and dissolved in both base matrices that certainly will affect the intrinsic properties of the base polymers. It may increase or decrease the matrix intrinsic properties depending on the systems or the reaction mechanism involved in the compatibilizer with matrices.

However, the compatible polymer blend is still immiscible due to a high degree of polymerization, thus the entropy term becomes vanishingly small and the miscibility is increasingly dependent on the nature of the enthalpic term contribution. To enhance the formation of a one-phase miscible system in polymer blends, it is necessary to ensure that favorable specific intermolecular interaction exists between the two base components of the blend. Ideally, one polymer possesses donor sites and the other possesses acceptor sites on the chain. The most commonly observed interactions are the general acid–base type (i.e., hydrogen bonding, ion-dipole, π-π interaction, or charge transfer interaction) [5–11]. A great deal of work in recent years has been involved with intermolecular hydrogen bonding. The miscibility of polymer blends, self-assembly and supramolecular nanostructures, nanocomposites, and low surface energy materials mediated by hydrogen bonding strength have been discussed [12–19]. Although there are already several excellent reviews on hydrogen bond polymer blends [20–26], this chapter will mainly discuss the recent research approaches in polymer blends through multiple hydrogen bonding interaction.

2.2 Hydrogen Bonding Interaction

Specific interactions have been a topic of intense interest in polymer science recently, such as dipole-dipole interaction, hydrogen bonding, and ionic interaction. A hydrogen bond results from a dipole-dipole force between an electronegative atom and a hydrogen atom bonded to nitrogen, oxygen, or fluorine. The nature of hydrogen bonding and its effect on the microstructure and physical properties of various materials have received attention from scientists in recent years [27–32]. Hydrogen bonding is an intensively studied interaction in physics, chemistry, and biology, and its significance is conspicuous in various real-life examples. An understanding of H-bonding interaction calls for input from various branches of science leading to a broad interdisciplinary research. The hydrogen bond as a directed attractive interaction between electron-deficient hydrogen and a region of high electron density have been reviewed intensely [33,34]. H-bond is a noncovalent, attractive interaction between a proton donor A-H and a proton acceptor B in the same or in a different molecule. In general, the definition of a hydrogen bond is that the proton usually lies on a line joining the A, B atoms (i.e., the hydrogen bond is linear; A-H...B), and the distance between the nuclei of A and B atom is considerably less than the sum of the van der Waals radii of A and B and the diameter of the proton (i.e., the formation of the hydrogen bond leads to a contraction of the A-H...B system). The A and B atoms are usually only the most electronegative such as F, O, and N atoms. However, the experimental and theoretical results reveal that even C–H can be involved in H-bonds, and p electrons can act as proton acceptors in the stabilization of weak H-bonding interaction in many chemical systems [29,30]. Most other authors also define a hydrogen bond by its effect on the properties of a material or by its molecular characteristic. Covalent bonds have strengths of the order of 50 kcal/mole; van der Waals attractions may be in the order of 0.2 kcal/mole, while hydrogen bonds most often possess strength in the range of 1 to 40 kcal/mole. The strength of the strong H-bonding interactions ranges from 15 to 40 kcal/mol [27–30]. For the moderate (conventional) and weak H-bonds, the strengths vary from 4 to 15 to 1 to 4 kcal/mol, respectively.

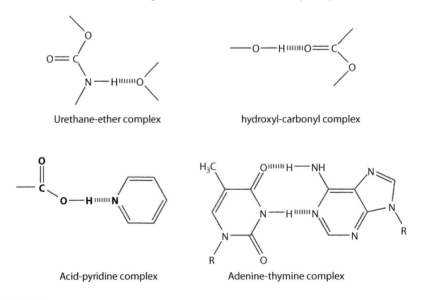

FIGURE 2.2
Some molecules exhibiting intramolecular hydrogen bonds.

Furthermore, the hydrogen bonding can exist in intermolecular or intramolecular bonds, and this range of energies is like a liquid at room temperature. The hydrogen bond donor and the hydrogen bond acceptor can belong to the same molecule or to two different molecules: the former case is known as "intramolecular hydrogen bond" (Figure 2.2).

The latter as "intermolecular hydrogen bond" is shown in Figure 2.3. Obviously, the intramolecular hydrogen bond is necessarily a bent bond, whereas the intermolecular one is generally linear or nearly linear. However, the intramolecular hydrogen bonding (same functional group) in polymer chains will have two different types: one is from interchain hydrogen bonding interaction and the other is intrachain hydrogen bonding interaction. For example, the formation of α-helix and β-sheet structures of polypeptides is stabilized through intrachain and interchain hydrogen bonds, respectively [16]. In addition, the strength of each individual hydrogen bond is also

FIGURE 2.3
Some molecules exhibiting intermolecular hydrogen bonds.

strongly dependent on what kinds of solvents you use, and it is generally believed that the addition of a polar solvent would significantly decrease the hydrogen bond strength over many orders of magnitude because solvent molecules can also participate in hydrogen bonding interaction. Therefore, the supramolecular chemistry of hydrogen-bonded polymers is mostly done in nonpolar solvents such as linear and cyclic alkanes, toluene, dichloromethane, and chloroform.

In addition, the most widely used experimental methods to characterize the hydrogen bond are (1) infrared and Raman techniques, which provide information about the stretching and deformation vibrations of A-H bonds and acceptor groups; (2) electronic absorption and fluorescence spectroscopies in the ultraviolet and visible regions, which show the effect of hydrogen bond formation on the electronic levels of the participating molecules; (3) solid or liquid state magnetic resonance spectroscopy, which can be used to study the effect of hydrogen bond formation on the chemical shift of the A-H; and (4) X-ray photoelectron spectroscopy has also been used to study the specific interaction between the metal and ligand or hydrogen bonding interaction in polymer blends [35–37]. Among these methods, by far the most sensitive and inexpensive is infrared spectroscopy.

2.3 Miscibility Enhancement through Multiple Hydrogen Bonds

In general, the hydrogen-bonded polymer blend is generally predicted by the Painter–Coleman association model due to exact prediction in most systems. Painter and Coleman suggested adding an additional term to the simple Flory–Huggins expression to account for the free energy of hydrogen bond formation upon the mixing of two polymers [38–40]:

$$\frac{\Delta G_N}{RT} = \frac{\phi_A}{N_A}\ln\varphi_B + \frac{\phi_B}{N_B}\ln\phi_B + \phi_A\phi_B\chi_{AB} + \frac{\Delta G_H}{RT} \qquad (2.1)$$

where ϕ_A and ϕ_B are the volume fractions of polymers A and B, respectively, in the blend, and N_A and N_B are the corresponding degrees of polymerization. Thus, the free energy of mixing is dominated by the balance of the last two terms, $\chi\phi_A\phi_B$, which is an unfavorable contribution derived from physical forces, and $\Delta G_H/RT$, a favorable contribution derived from hydrogen bonding or "chemical" forces. The positive contribution from the physical forces is determined using a Flory-type χ parameter that is, in turn, estimated from solubility parameters calculated from the molar attraction and molar volume constants of non-hydrogen-bonded groups. The negative contribution from chemical forces is determined from equilibrium constants and enthalpies of hydrogen bond formation, which are derived from infrared (IR) spectroscopic data to describe the self- and interassociations and the distribution of hydrogen-bonded species in the polymer blend. Although these self- and interassociation equilibrium constants cannot be obtained independently from their mixtures, fortunately, the relative magnitudes of the inter- and self-association equilibrium constants are more important and dominantly determine the contribution of the free energy of mixing rather than their individual absolute values. If the interassociation is strongly favored over the self-association, the polymer blend is expected to be miscible, such as the poly(vinyl phenol)/poly(vinyl pyrrolidone) blend system ($K_A/K_B \doteqdot 100$) [41,42]. Conversely, if the self-association is stronger than the interassociation, the blend tends to be immiscible or partially miscible as the poly(vinyl phenol)/poly(acetoxystyrene) blend system [43].

In general, to obtain a one-phase system in polymer blends, it is usually necessary to ensure that favorable specific intermolecular interaction exists between two base components of the blend. There has been much interest in miscible polymer blends for which one or both polymers are random copolymers. Many studies have shown that several copolymer–homopolymer and copolymer–copolymer blends may be miscible in a certain range of compositions and temperatures even though the respective constituent homopolymers are pairwise immiscible and no specific interaction exists between these blend systems due to the "copolymer repulsion effect" [44,45]. The overall interaction energy in these blend systems can be obtained by the binary interaction model based on Flory–Huggins lattice theory that can predict the effect of the copolymer composition on these miscible blends [46]. In addition, the formation of hydrogen bonding usually also induces the miscibility of the polymer blends as it is always a significant contribution to the free energy of mixing. In general, the miscibility of an immiscible blend can be enhanced by introducing a functional group to one polymer capable of forming an intermolecular association with another polymer [46]. In general, there are three methods to enhance the miscibility of an immiscible blend through a hydrogen bond: (1) incorporation of hydrogen-bonding monomer on main chain, (2) inert diluent segment effect, and (3) ternary polymer blend, which have been discussed in detail previously [19,22].

For example, the polystyrene is immiscible with many polymers due to lack of a functional group capable of interacting with other polymers. In previous studies of the roles of intermolecular association in miscibility enhancement, we found that the incorporation of a large number of hydrogen bond acceptors (ca. 45 mol% of polyacetoxystyrene) or donors (ca. 13 mol% of polyvinylphenol) into a polystyrene (PS) chain renders the modified polymer miscible with phenolic resin (a well-known hydrogen bonding donor) [47] or poly(ε-caprolactone) (a well-known hydrogen bonding acceptor) [48], respectively. If the monomers possess relatively weak hydrogen-bonding moieties (e.g., hydroxyl, carboxyl, pyridyl, or ether groups), then the corresponding weak intermolecular interactions require a relatively high mole percentage of the copolymer to induce miscibility, resulting in properties of the polymer blend that differ substantially from those of the unmodified polymer [49,52]. Ideally, adding low mole percentages of the recognition units into the two immiscible phases would result in a miscible phase. According to Painter–Coleman association model (PCAM) prediction, the relative magnitudes of inter- and self-association equilibrium constants are more important that dominantly determine the contribution of the free energy of mixing rather than their individual absolute values. Based on our knowledge, the multiple hydrogen bonding is an easy approach to enhance the miscibility of the polymer blend. Multiple hydrogen-bonded arrays play a fundamental role in complex biological systems (e.g., DNA complexation). DNA is a very influential structure in polymer science, where it is often presented as a defined macromolecule possessing a nearly perfect molecular structure. As a result, the preparation of synthetic polymers that mimic DNA remains a very important challenge in polymer science [53]. The self-assembly of pairs of DNA strands is mediated by intermolecular hydrogen bonding between complementary purine (A and G) and pyrimidine (T and C) bases attached to a phosphate sugar backbone: G binds selectively to C, and A binds selectively to T [54]. Taking this cue from nature, supramolecular structures can be prepared from synthetic polymers possessing nucleotide bases on their side chains. The binding force of multiple hydrogen-bonding systems can be tuned leading to association constants from several M^{-1} up to more than 10^6 M^{-1} as shown in Figure 2.4 [55].

Liu et al. investigated the coaggregation of PtBA-*b*-poly(2-cinnamyloxyethyl methacrylate) (PCEMA) and PS-*b*-PCEMA in a mixture of $CHCl_3$ and hexane. To ensure coaggregation, the PCEMA block was tagged with the hydrogen-bonding DNA base pairs T and A [56]. Lutz et al. demonstrated that the adenine functionalized copolymer self-assembles with its thymine-functionalized counterpart into supramolecular aggregates, which show a temperature-dependent "melting" behavior in nonpolar solvents [57–59]. Rotello et al. demonstrated the thermally reversible formation of micron-size gel-like spherical aggregates through noncovalent polymer cross-linking. This cross-linking occurred as a direct result of specific three-point hydrogen bonding between thymine and diacyldiamidopyridine functionalities [60–62]. The spherical aggregates are stable indefinitely at ambient temperature, dissociate at 50°C, and reform upon cooling; this heating–cooling cycle can be repeated multiply with no decomposition [60]. Long et al. synthesized a nucleobase-functionalized triblock copolymer featuring A- and T-containing blocks through nitroxide-mediated radical polymerization [64]. The blending of complementary polymers led to dramatic increases in viscosities and glass transition temperatures as a result of hydrogen-bonding interactions between the A and T units [64,65]. Sleiman and Bazzi used ring-opening metathesis polymerization to synthesize A-containing block copolymers that self-assembled into cylindrical morphologies through self-complementarily hydrogen bonding the adenine units in the molecular core [66].

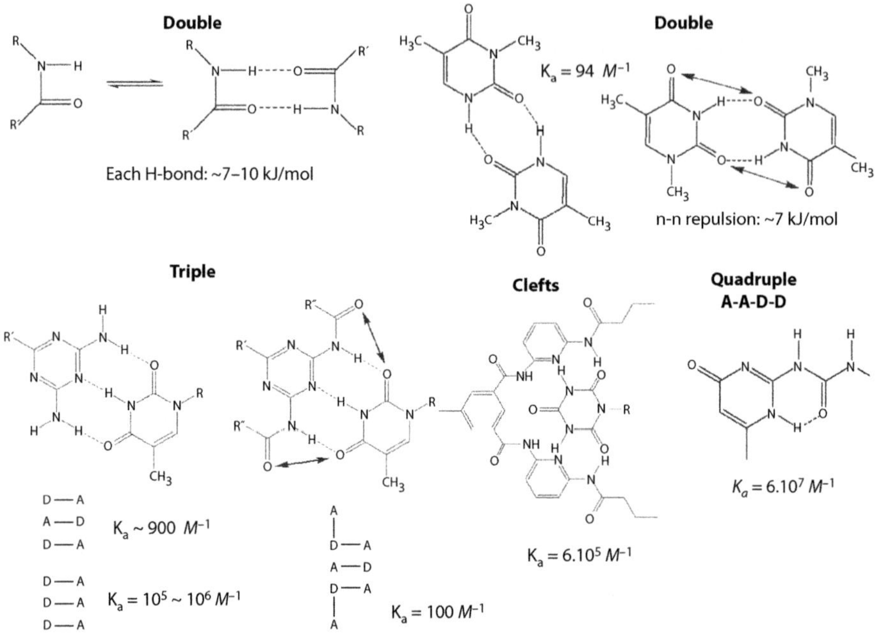

FIGURE 2.4

Hydrogen bonds and their association constants (A: hydrogen-bonding acceptor; D: hydrogen-bonding donor; n: lone pair repulsion). (Reprinted with permission from Binder, W. H. 2005. Polymeric ordering by H-bonds. mimicking nature by smart building blocks. *Monatshefte Chemie* 136:1–19. Copyright 2005, Springer.)

Park and Zimmerman reported a supramolecular polymer blend consisting of a pair of immiscible polymers, poly(butyl methacrylate) (PBMA) and PS. A urea derivative of guanosine (UG) and 2,7-diamido-1,8-naphthyridine (DAN) form an exceptionally strong quadruply hydrogen-bonding complex as shown in Figure 2.5 [67]. Size exclusion chromatography (SEC), dynamic light scattering (DLS), and viscosity analyses have been used to provide evidence for the formation of supramolecular network structures in these binary blend systems [67].

We investigated the miscibility behavior, specific interactions, and supramolecular structures of blends of the DNA-like copolymers poly(vinylbenzylthymine-*co*-butyl methacrylate) (T-PBMA) and poly(vinylbenzyladenine-*co*-styrene) (A-PS) with respect to their vinylbenzylthymine (VBT) and vinylbenzyladenine (VBA) contents through free radical copolymerizations as shown in Figure 2.6 [68].

Figure 2.7 displays differential scanning calorimetry (DSC) thermograms of A-PS/T-PBMA = 50/50 blends, where the A-PS and T-PBMA components contain various contents of VBA and VBT, respectively [68]. The binary blend of PS and PBMA exhibits two glass transition temperatures located at the same temperatures as those of their respective pure polymers, revealing

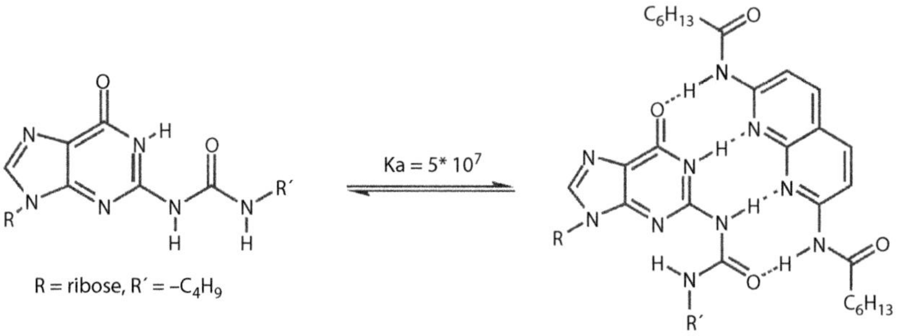

FIGURE 2.5
Formation of a strong complex between 2,7-diamido-1,8-naphthyridine (DAN) and urea derivative of guanosine (UG). (Reprinted with permission from Park, T., and Zimmerman, S. C. 2006. Formation of a miscible supramolecular polymer blend through self-assembly mediated by a quadruply hydrogen-bonded heterocomplex. *J. Am. Chem. Soc.* 128:11582–11590. Copyright 2006, American Chemical Society, USA.)

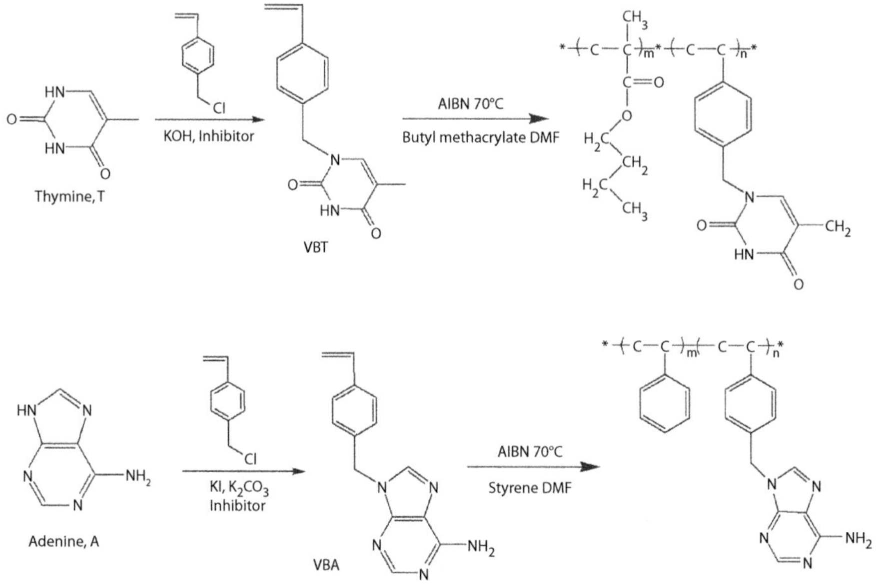

FIGURE 2.6
Syntheses of PVBT-*co*-PBMA (poly(vinylbenzylthymine-*co*-butyl methacrylate) [T-PBMA]) and PVBA-*co*-PS (poly(vinylbenzyladenine-*co*-styrene) [A-PS]) random copolymers prepared through free radical polymerization. (Reprinted with permission from Kuo, S. W., and Cheng, R. S. 2009. DNA-like interactions enhance the miscibility of supramolecular polymer blends. *Polymer* 50:177–188. Copyright 2009, Elsevier Science Ltd., UK.)

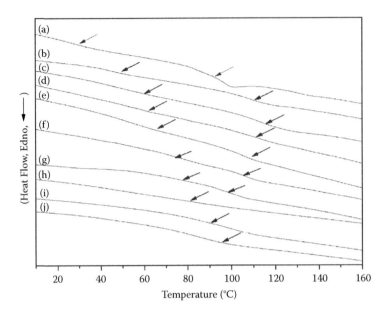

FIGURE 2.7
Differential scanning calorimetry (DSC) curves of the binary blends (a) PS/PBMA, (b) A5-PS/T07-PBMA, (c) A5-PS/T11-PBMA, (d) A5-PS/T24-PBMA, (e) A08-PS/T7-PBMA, (f) A08-PS/T11-PBMA, (g) A11-PS/T7-PBMA, (h) A08-PS/T11-PBMA, (i) A11-PS/T11-PBMA, and (j) A11-PS/T24-PBMA. (Reprinted with permission from Kuo, S. W., and Cheng, R. S. 2009. DNA-like interactions enhance the miscibility of supramolecular polymer blends. *Polymer* 50:177–188. Copyright 2009, Elsevier Science Ltd., UK.)

that they are completely immiscible. We found, however, that the value of T_g shifted upon increasing the VBA and VBT contents in the copolymers; when 8 mol% or more of VBA and 11 mol% or more of VBT were incorporated into the PS and PBMA main chains, respectively, the PS/PBMA binary blends formed miscible pairs exhibiting a single value of T_g through strong multiple hydrogen-bonding interactions between the A and T units. Meanwhile, the single values of T_g of the copolymer blends fall between those of the two parent polymers (PS and PBMA), but they are significantly higher than the values predicted by the Fox equation, again indicating the presence of strong multiple hydrogen-bonding interactions between the A and T segments in the copolymers.

In addition, Figure 2.8 shows atomic force microscopy (AFM) images for providing microscopic evidence for the homogeneous mixing without phase separation [68]. The thin film prepared by casting a 4 g/dL solution of PBMA and PS in chloroform on glass exhibited islands with lateral dimensions ca. 100 nm (Figures 2.8a and 2.8b). However, the thin film of a mixture of A-11 PS and T-24 PBMA prepared in the same way as that above was smooth, with no features evident on the nanometer scale (Figures 2.8c and 2.8d). These observations are consistent with the formation of a miscible blend driven

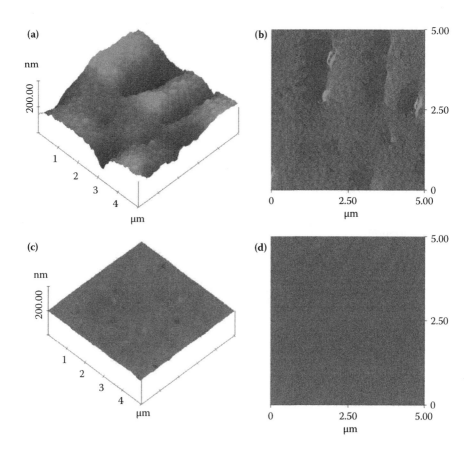

FIGURE 2.8 (See color insert.)
Tapping mode atomic force microscopy (AFM) images of polystyrene/poly(butyl methacrylate) (PS/PBMA) blend (a) height and (b) phase images, and A11-PS/T24-PBMA blend (c) height and (d) phase images. (Reprinted with permission from Kuo, S. W., and Cheng, R. S. 2009. DNA-like interactions enhance the miscibility of supramolecular polymer blends. *Polymer* 50:177–188. Copyright 2009, Elsevier Science Ltd., UK.)

by the A-T recognition. Although the structure of the polymer assembly is not known, it is likely that a supramolecular network is formed. In addition, SEC, DLS, and viscosity analyses provided evidence for the formation of supramolecular network structures in these binary blend systems.

In our system, pure PVBA and PVBT dissolve only in high-polarity solvents, such as dimethylformamide (DMF) and dimethyl sulfoxide (DMSO), which interfere with self- and interassociation hydrogen bonding and would, therefore, provide incorrect equilibrium constants [38]. Thus, we synthesized two low-molecular-weight model compounds, 9-hexadecyladenine (AC-16) and 9-hexadecylthymine (TC-16), to determine the interassociation equilibrium constants (K_a) through proton nuclear magnetic resonance (^{1}H NMR) spectroscopic titration experiments in $CDCl_3$ at room temperature, based on

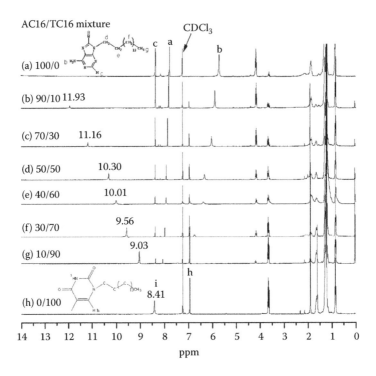

FIGURE 2.9

Proton nuclear magnetic resonance (^{1}H NMR) spectra (in *d*-chloroform CDCl$_3$, room temperature) of A-16/T-16 mixtures featuring various A/T ratios. (Reprinted with permission from Kuo, S. W., and Cheng, R. S. 2009. DNA-like interactions enhance the miscibility of supramolecular polymer blends. *Polymer* 50:177–188. Copyright 2009, Elsevier Science Ltd., UK.)

the method developed by Benesi and Hildebrand [69]. Figure 2.9 displays ^{1}H NMR spectra of AC-16/TC-16 mixtures at various ratios. The addition of AC-16 to a TC-16 solution led to a downfield shift of the signal of the thymine group that appeared initially at 8.41 ppm, indicating that strong intermolecular hydrogen bonding occurred between the T and A groups [68].

A plot of this chemical shift versus the reciprocal of the concentration (Figure 2.10) allowed us to calculate the interassociation equilibrium constant (K_a = 534 M^{-1}). We transformed the value of K_a from the model compound into K_A by dividing by the molar volume of the VBA repeat unit (0.1125 L mol^{-1} at 25°C) [24], providing a value for the interassociation equilibrium constant K_A of 4750. Likewise, we determined that the self-association equilibrium constant (K_B) of adenine was 32, after dividing the value of K_b of 3 M^{-1} for the model compound 2-ethyladenine by the molar volume of the VBA repeat unit [70].

In addition, we proposed another approach for further decreasing lower mole percentages of the recognition units to enhance the miscibility

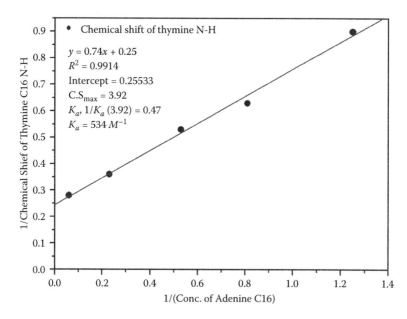

The graph shows:
- y-axis: 1/Chemical Shief of Thymine C16 N-H (ranging 0.0 to 0.9)
- x-axis: 1/(Conc. of Adenine C16) (ranging 0.0 to 1.4)
- Legend: • Chemical shift of thymine N-H
- $y = 0.74x + 0.25$
- $R^2 = 0.9914$
- Intercept = 0.25533
- $C.S_{max} = 3.92$
- $K_a, 1/K_a (3.92) = 0.47$
- $K_a = 534\ M^{-1}$

FIGURE 2.10
Determination of the interassociation equilibrium constant for the A···T interaction using the Benesi and Hildebrand method. (Reprinted with permission from Kuo, S. W., and Cheng, R. S. 2009. DNA-like interactions enhance the miscibility of supramolecular polymer blends. *Polymer* 50:177–188. Copyright 2009, Elsevier Science Ltd., UK.)

behavior by using stronger multiple hydrogen-bonding strength than A-T interaction into the PS and PBMA main chains. Komiyama et al. used free radical polymerization to prepare a series of copolymers based on 2-vinyl-4,6-diamino-1,3,5-triazine (VDAT) [71–73], which can form triply hydrogen-bonded complexes with thymine adducts in nonpolar solvents. A ^{1}H NMR spectroscopic titration experiment suggested that the interassociation equilibrium constant between diamino-1,3,5-triazine (DAT) and thymine (T) [74] is ca. 890 M^{-1} (i.e., it is stronger than the interassociation equilibrium constant between A and T, ca. 530 M^{-1}, as mentioned above). As a result, we chose 2-vinyl-4,6-diamino-1,3,5-triazine (VDAT) and 1-(4-vinylbenzyl) thymine (VBT) as monomers for independent copolymerization with styrene and butyl methacrylate monomer, noting that VDAT and T have low self-association equilibrium constants (K_{dim} = ca. 2–3 M^{-1}) [75] but form very strong complexes together (K_a = ca. 890 M^{-1}) (Figure 2.11).

Figures 2.12a and 2.12b display the miscibility window for the VDAT-PS/T-PBMA and A-PS/T-PBMA blends as predicted theoretically using the PCAM [40], respectively. The *x*-axis represents the weight percentage of VDAT and VBA in the PS copolymer, respectively; the *y*-axis represents the weight percentage of VBT in the T-PBMA copolymer. Figure 2.12a shows that the binary blends of copolymers would be completely miscible if the VDAT and VBT

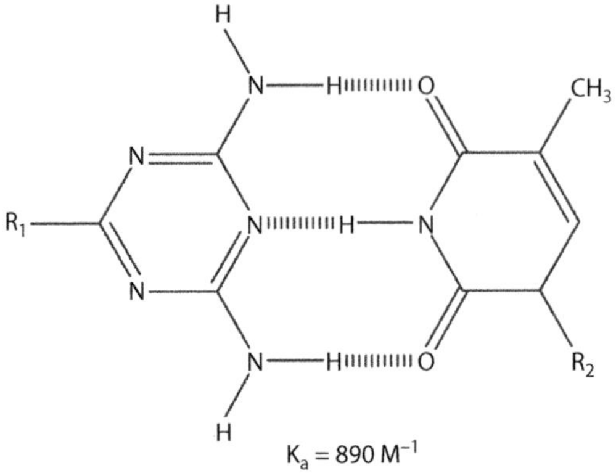

FIGURE 2.11
Formation of strong multiple hydrogen-bonding interactions from between adenine A and thymine T units (A-T). (Reprinted with permission from Kuo, S. W., and Hsu, C. H. 2010. Miscibility enhancement of supramolecular polymer blends through complementary multiple hydrogen bonding interactions. *Polymer International* 59:998–1005. Copyright 2010, Wiley-VCH, Germany.)

content were greater than 8 and 12 wt% (ca. 7 and 11 mol%, respectively). Clearly, the miscibility window of Figure 2.12a is larger than the VBA-PS/T-PBMA binary blend in Figure 2.12b [76].

In addition, the model's predicted miscibility window compares favorably with our experimental results derived from DSC analyses. The reason is that the interassociation equilibrium constant between (DAT) and thymine (T) is ca. 890 M^{-1} (i.e., it is stronger than the interassociation equilibrium constant between A and T, ca. 530 M^{-1}). In other words, the triple hydrogen-bonding strength of DAT-T is stronger than double hydrogen bonding of A-T interactions, thus stronger hydrogen-bonding strength into the PS and PBMA main chains for decreasing lower mole percentages of the recognition units to enhance the miscibility behavior.

Multiple hydrogen bonding could not only enhance the miscibility behavior of the polymer blend but also its thermal properties. For instance, poly(methyl methacrylate) (PMMA) is a colorless, commercially mass-produced, transparent polymeric material exhibiting high light transmittance, chemical resistance, and weathering corrosion resistance and good insulation [77]. These properties make PMMA a valuable substitute for glass in optical device applications (e.g., compact discs, CDs, optical glasses, and optical fibers) [78]. Because the glass transition temperature of PMMA is, however, relatively low (T_g = ca. 100°C), its applications in the optical-electronic industry are limited because it undergoes distortion when used in an inner glazing material [79,80]. To raise the value of T_g, PMMA copolymers incorporating rigid or bulky monomer structures (to overcome the miscibility problem) and monomers that can form hydrogen bonds with the carbonyl groups of PMMA have been reported widely

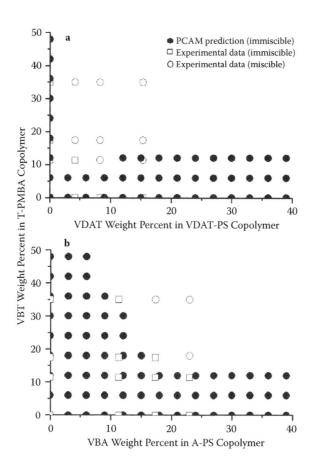

FIGURE 2.12
Theoretical miscibility windows for (a) VDAT-PS/T-PBMA and (b) A-PS/T-PBMA blends obtained from the Painter–Coleman association model PCAM; (•) spinodal curve and experimental data: (□) two-phase system; (○) one-phase system. (Reprinted with permission from Kuo, S. W., and Hsu, C. H. 2010. Miscibility enhancement of supramolecular polymer blends through complementary multiple hydrogen bonding interactions. *Polymer International* 59:998–1005. Copyright 2010, Wiley-VCH, Germany.)

[81–86]. Previously, we suggested an approach to raise the value of T_g of PMMA through copolymerization with methacrylamide (MAAM) because hydrogen-bonding interactions exist between these two monomer segments [87,88]. The values of T_g of such copolymers are generally higher than those of the corresponding polymer blends, because, as reported widely, compositional heterogeneities exist in hydrogen-bonded copolymers [89,90]. In all previous studies, the monomers have possessed relatively weak (single) hydrogen-bonding moieties (e.g., hydroxyl, carboxyl, pyridyl, or ether groups). Such intermolecular interactions must be present at relatively high mole percentages to enhance the thermal behavior of the copolymers; therefore, the structures of the copolymers differ substantially from those of their unmodified polymers [19]. Ideally, adding low mole percentages of the recognition units into the copolymers would enhance their thermal and mechanical properties.

As a result, we prepare strong multiple hydrogen-bonding interactions between the 2,4-diaminotriazine groups of the PVDAT units and the T groups of the poly(vinylbenzylthymine) (PVBT) units in blends of poly(2-vinyl-4,6-diamino-1,3,5-triazine-*co*-methyl methacrylate) (PVDAT-*co*-PMMA) and poly(vinylbenzyl thymine-*co*-methyl methacrylate) (PVBT-*co*-PMMA) as shown in Figure 2.13 [91].

Figure 2.14 displays DSC thermograms of PVDAT-*co*-PMMA/PVBT-*co*-PMMA 50/50 blends, where the D-PMMA and T-PMMA components contain various contents of VDAT and VBT, respectively. Each of these binary blends exhibited a single glass transition temperature, indicating that the blends were miscible on the range 20 to 40 nm; the value of T_g shifted upon increasing the VDAT or VBT content in these two PMMA-based copolymers. It is notable that blending only 12 mol% of VBT and 20 mol% of VDAT into the PMMA copolymer chain increased the value of T_g by 50°C relative to that of pure PMMA, providing a glass transition temperature similar to that of polycarbonate.

Fourier transform infrared (FTIR) and solid-state nuclear magnetic resonance (NMR) spectroscopic analyses both provided positive evidence for hydrogen-bonding interactions within these copolymer systems. We obtained

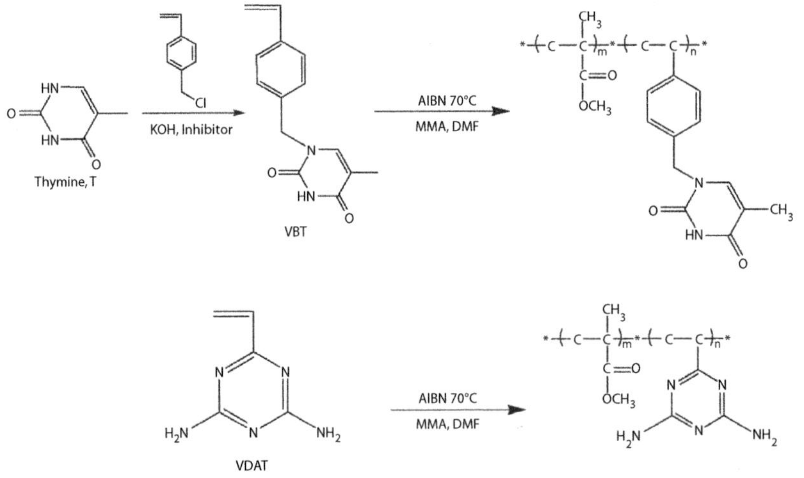

FIGURE 2.13
Syntheses of PVDAT-*co*-PMMA and PVBT-*co*-PMMA random copolymers through free radical polymerization. (Reprinted with permission from Kuo, S. W., and Tsai, S. T. 2009. Complementary multiple hydrogen-bonding interactions increase the glass transition temperatures to PMMA copolymer mixtures. *Macromolecules* 42:4701–4711. Copyright 2009, American Chemical Society, USA.)

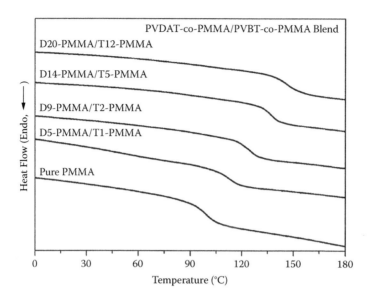

FIGURE 2.14
Differential scanning calorimetry (DSC) thermograms of D-PMMA/T-PMMA binary blends. (Reprinted with permission from Kuo, S. W., and Tsai, S. T. 2009. Complementary multiple hydrogen-bonding interactions increase the glass transition temperatures to PMMA copolymer mixtures. *Macromolecules* 42:4701–4711. Copyright 2009, American Chemical Society, USA.)

a single value of the spin-lattice relaxation times in the rotating frame for the copolymer blend that was lower than those of the pure copolymers, suggesting a decrease in the free volume of the blend. Thus, significant increases in the value of T_g of PMMA can be achieved through copolymerization of methyl methacrylate individually with complementary nucleobase monomers and then mixing of the resulting copolymers to form multiple hydrogen-bonding interactions.

We further characterized this copolymer mixture by using an Ubbelohde viscometer to measure the solution viscosity of a mixture of D20-PMMA and T12-PMMA in tetrahydrofuran (THF). The formation of supramolecular polymers in the D20-PMMA/T12-PMMA blends provided a higher solution viscosity relative to that of pure PMMA as shown in Figure 2.15; in addition, the viscosity increased upon increasing the concentrations of the copolymers. This supramolecular polymer also could be observed macroscopically—in the form of a gel—from a 1:1 mixture of D20-PMMA/T12-PMMA in THF at a concentration of 30 g/dL; at the same concentration, a solution of PMMA, which lacked any specific interpolymer hydrogen-bonding interactions, flowed freely (see the inset to Figure 2.15).

Figure 2.16 shows the possible chain behaviors of D-PMMA/T-PMMA blends through multiple hydrogen-bonding units into PMMA, which enchanced the thermal properties and dramatically increased the viscosity as a result of the formation of supramolecular polymers.

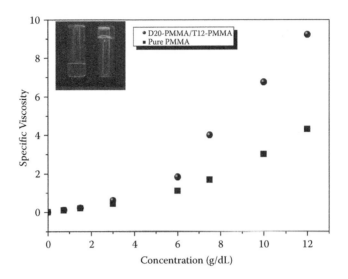

FIGURE 2.15
Plots of specific viscosity of D20-PMMA/T12-PMMA blend and pure PMMA in THF solution versus with respect to concentration. Inset: corresponding photographs of the corresponding blends at the same concentration (of 30 g/dL). (Reprinted with permission from Kuo, S. W., and Tsai, S. T. 2009. Complementary multiple hydrogen-bonding interactions increase the glass transition temperatures to PMMA copolymer mixtures. *Macromolecules* 42:4701–4711. Copyright 2009, American Chemical Society, USA.)

In addition, Weck et al. synthesized random copolymers containing cyanuric acid recognition units via ring-opening metathesis polymerization (ROMP) and studied their cross-linking behavior through complementary hydrogen bonding in a nonpolar solvent [92–94]. Scherman and Celiz performed controlled anionic ring-opening polymerization of poly(ε-caprolactone) (PCL) in toluene using self-complementary quadruply hydrogen-bonding 2-ureido-4[1H]-pyrimidinone (UPy)-functionalized initiators, which led to a marked increase in viscosity [95]. We also prepared a series of poly(methyl methacrylate) (PMMA)-based copolymers through free radical copolymerization of methyl methacrylate in the presence of 2-ureido-4[1H]-pyrimidinone methyl methacrylate (UPyMA) as shown in Figure 2.17 [96].

Figure 2.18 displays DSC curves recorded at temperatures ranging from 40 to 160°C, of the P(MMA-co-UPyMA) copolymers. Pure PMMA exhibits a single glass transition temperature at ca. 100°C; the glass transition temperatures of the P(MMA-co-UPyMA) copolymers increased upon increasing their UPyMA contents due to the incorporation of the strongly interacting UPyMA units.

The absence of two glass transition temperatures for these copolymers indicates a random copolymerization of these two monomers. Compared with P(MMA-co-MAAM) copolymers, the Tg for 8.4-MAAM was 126°C [87],

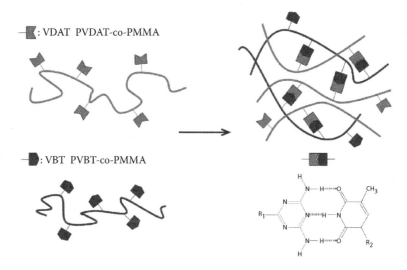

FIGURE 2.16
The supramolecular network structure formed from polymethyl methacrylate (PMMA) copolymer mixtures featuring specific multiple hydrogen-bonding interactions. (Reprinted with permission from Kuo, S. W., and Tsai, S. T. 2009. Complementary multiple hydrogen-bonding interactions increase the glass transition temperatures to PMMA copolymer mixtures. *Macromolecules* 42:4701–4711. Copyright 2009, American Chemical Society, USA.)

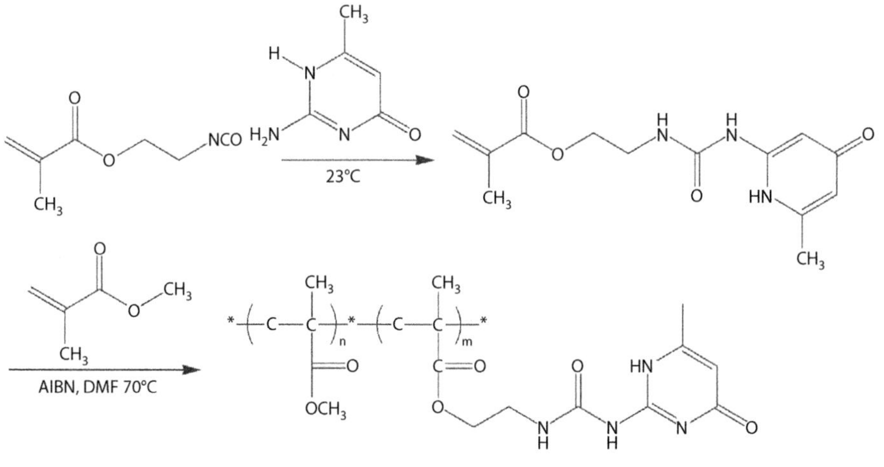

FIGURE 2.17
Synthesis of 2-ureido-4[1*H*]-pyrimidinone methyl methacrylate (UPyMA) monomer and poly(methyl methacrylate-co-2-ureido-4[1*H*]-pyrimidinone methyl methacrylate) (P(MMA-*co*-UPyMA) copolymers.

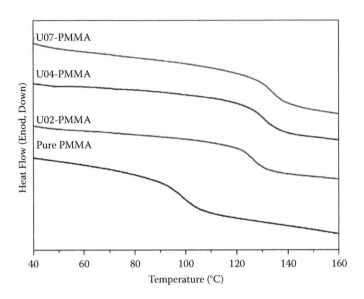

FIGURE 2.18
Differential scanning calorimetry (DSC) thermograms of P(MMA-*co*-UPyMA) copolymers.

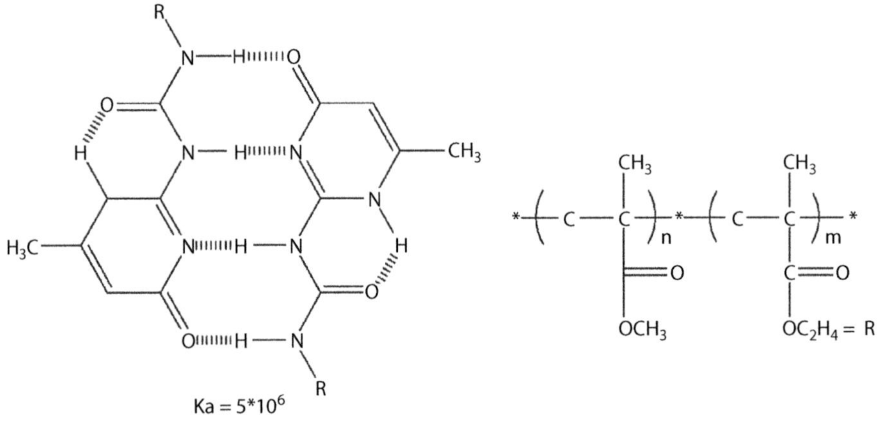

FIGURE 2.19
Self-complementary multiple hydrogen bonding of P(MMA-*co*-UPyMA) copolymers.

which was significantly less than Tg for U07-PMMA copolymer (135°C) at a similar molar incorporation of hydrogen-bonding sites. This is due to stronger self-complementary multiple hydrogen-bonding interaction (as shown in Figure 2.19) in P(MMA-*co*-UPyMA) copolymers than single hydrogen-bonding site in P(MMA-*co*-MAAM) copolymers.

2.4 Conclusions

This chapter has assembled recent studies on multiple hydrogen-bonding interactions in polymer blend systems. Research in this field has become quite well developed in recent years and offers various concepts to miscibility, thermal, rheology, and mechanically behaviors. Due to the diverse nature of polymers with specific functionality, the way is open to construct function into the newly formed materials for polymer blend systems.

References

1. Paul, D. R. 1978. *Polymer Blends*. D. R. Paul and S. Newman, Eds., Academic Press, New York.
2. Utracki, L. A. 1989. *Polymer Alloys and Blends*. Munich, Vienna, New York: Hanser.
3. Litmanovich, A. D., Plate, N. A., and Kudryavtsev, Y. V. 2002. Reactions in polymer blends: Interchain effects and theoretical problems. *Prog. Polym. Sci.* 27:915–970.
4. Macosko, C. W., Jeon, H. K., and Hoye, T. R. 2005. Reactions at polymer–polymer interfaces for blend compatibilization. *Prog. Polym. Sci.* 30:939–947.
5. Coleman, M. M., Pehlert, G. J., Yang, X., Stallman, J. B., and Painter, P. C. 1996. Self-association *versus* interassociation in hydrogen bonded polymer blends: 1. Determination of equilibrium constants from miscible poly(2,6-dialkyl-4-vinyl phenol) blends. *Polymer* 37:4753–4761.
6. Wang, J., Cheung, M. K., and Mi, Y. 2001. Miscibility of poly(ethyl oxazoline)/poly(4-vinylphenol) blends as investigated by the high-resolution solid-state ^{13}C NMR. *Polymer* 42:2077–2083.
7. Li, X., Goh, S. H., Lai, Y. H., and Wee, A. T. S. 2000. Miscibility of carboxyl-containing polysiloxane/poly(vinylpyridine) blends. *Polymer* 41:6563–6571.
8. Sawateri, C., and Kondo, T. 1999. Interchain hydrogen bonds in blend films of poly(vinyl alcohol) and its derivatives with poly(ethylene oxide). *Macromolecules* 32:1949–1955.
9. Cesteros, L. C., Isasi, J. R., and Katime, I. 1993. Hydrogen bonding in poly(4-vinylpyridine)/poly(vinyl acetate-*co*-vinyl alcohol) blends: An infrared study. *Macromolecules* 26:7256–7262.
10. Ma, C. C. M., Wu, H. D., and Lee, C. T. 1998. Strength of hydrogen bonding in the novolak-type phenolic resin blends. *J. Polym. Sci.: Polym. Phys.* 36:1721–1729.
11. Cassu, S. N., and Felisberti, M. I. 1999. Poly(vinyl alcohol) and poly(vinyl-pyrrolidone) blends: 2. Study of relaxations by dynamic mechanical analysis. *Polymer* 40:4845–4851.
12. Kuo, S. W., and Chang, F. C. 2001. Miscibility and hydrogen bonding in blends of poly(vinylphenol-*co*-methyl methacrylate) with poly(ethylene oxide). *Macromolecules* 34:4089–4097.

13. Huang, C. F., Kuo, S. W., Lin, F. J., Huang, W. J., Wang, C. F., Chen, W. Y., and Chang, F. C. 2006. Influence of PMMA-chain-end tethered polyhedral oligomeric silsesquioxanes on the miscibility and specific interaction with phenolic blends. *Macromolecules* 39:300–308.

14. Wang, C. F., Su, Y. C., Kuo, S. W., Huang, C. F., Sheen, Y. C., and Chang, F. C. 2006. Low-surface-free-energy materials based on polybenzoxazines. *Angew. Chem. Int. Ed.* 45:2248–2251.

15. Lin, H. C., Wang, C. F., Kuo, S. W., Tung, P. H., Huang, C. F., Lin, C. H., and Chang, F. C. 2007. Effect of intermolecular hydrogen bonding on low-surface-energy material of poly(vinylphenol). *J. Phys. Chem. B* 111:3404–3410.

16. Kuo, S. W., Lee, H. F., Huang, W. J., Jeong, K. U., and Chang, F. C. 2009. Solid state and solution self-assembly of helical polypeptides tethered to polyhedral oligomeric silsesquioxanes. *Macromolecules* 42:1619–1626.

17. Kuo, S. W., Wu, Y. C., and Wang, C. F. 2009. Preparing low-surface-energy polymer materials by minimizing intermolecular hydrogen-bonding interactions. *J. Phys. Chem. C* 113:20666–20673.

18. Lin, C. T., Kuo, S. W., and Chang, F. C. 2010. Glass transition temperature enhancement of PMMA through copolymerization with PMAAM and PTCM mediated by hydrogen bonding. *Polymer* 51:883–889.

19. Kuo, S. W. 2008. Hydrogen-bonding in polymer blends. *J. Polym. Res.* 15:459–486.

20. Coleman, M. M., and Painter, P. C. 1995. Hydrogen bonded polymer blends. *Prog. Polym. Sci.* 20:1–59.

21. Jiang, M., Mei, L., Xiang, M., and Zhou, H. 1999. Interpolymer complexation and miscibility enhancement by hydrogen bonding. *Adv. Polym. Sci.* 146:121–196.

22. He, Y., Zhu, B., and Inoue, Y. 2004. Hydrogen bonds in polymer blends. *Prog. Polym. Sci.* 29:1021–1051.

23. Binder, W. H., and Zirbs, R. 2007. Supramolecular polymers and networks with hydrogen bonds in the main-and side-chain. *Adv. Polym. Sci.* 207:1–78.

24. Bouteiller, L. 2007. Assembly via hydrogen bonds of low molar mass compounds into supramolecular polymer. *Adv. Polym. Sci.* 207:79–112.

25. Ten Brinke, G., Ruokolaine, J., and Ikkala, O. 2007. Supramolecular materials based on hydrogen-bonded polymers. *Adv. Polym. Sci.* 207:113–177.

26. Xu, H., Srivastava, S., and Rotello, V. M. 2007. Nanocomposites based on hydrogen bonds. *Adv. Polym. Sci.* 207:179–198.

27. Boreo, M., Ikeshoji, T., Liew, C. C., Terakura, K., and Parrinello, M. 2004. Hydrogen bond driven chemical reactions: Beckmann rearrangement of cyclohexanone oxime into ε-caprolactam in supercritical water. *J. Am. Chem. Soc.* 126:6280–6286.

28. Murata, T., Morita, Y., Yakiyama, Y., Fukui, K., Yamochi, H., Saito, G., and Nakasuji, K. 2007. Hydrogen-bond interaction in organic conductors: Redox activation, molecular recognition, structural regulation, and proton transfer in donor–acceptor charge-transfer complexes of TTF-imidazole. *J. Am. Chem. Soc.* 129:10837–10846.

29. Smith, J. D., Cappa, C. D., and Wilson, K. R. et al. 2004. Energetics of hydrogen bond network rearrangements in liquid water. *Science* 306:851–853.

30. Jones, W. D. 2000. Conquering the carbon-hydrogen bond. *Science* 287:1942–1943.

31. Deechongkit, S., Naguen, H., and Powers, E. T. et al. 2004. Context-dependent contributions of backbone hydrogen bonding to β-sheet folding energetics. *Nature* 430:101–105.

32. Mehta, R., and Dadmun, M. D. 2006. Small angle neutron scattering studies on miscible blends of poly(styrene-*ran*-vinyl phenol) with liquid crystalline polyurethane. *Macromolecules* 39:8799–8807.

33. Jeffrey, G. A., and Saenger, W. 1991. *Hydrogen Bonding in Biological Structures.* Berlin, Heidelberg, New York: Springer..

34. Calhorda, M. J. 2000. Weak hydrogen bonds: Theoretical studies. *Chem. Commun.* 10:801–809.

35. Goh, S. H., Lee, S. Y., Zhou, X., and Tan, K. L. 1999. X-ray photoelectron spectroscopic studies of interactions between styrenic polymers and poly(2,6-dimethyl-1,4-phenylene oxide). *Macromolecules* 32:942–944.

36. Goh, S. H., Lee, S. Y., Zhou, X., and Tan, K. L. 1998. X-ray photoelectron spectroscopic studies of interactions between poly(4-vinylpyridine) and poly(styrenesulfonate) salts. *Macromolecules* 31:4260–4264.

37. Jiao, H., Goh, S. H., and Valiyaveettil, S. 2001. Mesomorphic interpolymer complexes and blends based on poly(4-vinylpyridine)–dodecylbenzenesulfonic acid complex and poly(acrylic acid) or poly(*p*-vinylphenol). *Macromolecules* 34:7162–7165.

38. Coleman, M. M., Gref, J. F., and Painter, P. C. 1991. *Specific Interactions and the Miscibility of Polymer Blends.* Lancaster, PA: Technomic.

39. Coleman, M. M., and Painter, P. C. 2000. *Polymer Blend.* D. R. Paul, C. B. Bucknall, Eds. New York: Wiley.

40. Coleman, M. M., and Painter, P. C. 2006. *Miscible Polymer Blends: Background and Guide for Calculations and Design.* DEStech, Lancaster, PA.

41. Kou, S. W., and Chang, F. C. 2001. Studies of miscibility behavior and hydrogen bonding in blends of poly(vinylphenol) and poly(vinylpyrrolidone). *Macromolecules* 34:5224–5228.

42. Kou, S. W., Xu, H., Huang, C. F., and Chang, F. C. 2002. Significant glass-transition-temperature increase through hydrogen-bonded copolymers. *J. Polym. Sci.: Polym. Phys.* 40:2313–2323.

43. Kou, S. W., and Chang, F. C. 2002. Effect of inert diluent segment on the miscibility behavior of poly(vinylphenol) with poly(acetoxystyrene) blends. *J. Polym. Sci.: Polym. Phys.* 40:1661–1672.

44. Paul, D. R., and Barlow, J. W. 1984. A binary interaction model for miscibility of copolymers in blends. *Polymer* 25:487–494.

45. Krause S. 1991. On the intramolecular repulsion effect in random copolymer solubility. *Macromolecules* 24:2108.

46. Paul, D. R., and Merfeld, G. D. 2000. *Polymer Blend.* D. R. Paul, C. B. Bucknall, Eds. New York: Wiley.

47. Kuo, S. W., and Chang, F. C. 2001. Effect of copolymer composition on the miscibility of poly(styrene-*co*-acetoxystyrene) with phenolic resin. *Polymer* 42:9843–9848.

48. Kuo, S. W., and Chang, F. C. 2001. Effects of copolymer composition and free volume change on the miscibility of poly(styrene-*co*-vinylphenol) with poly(ε-caprolactone). *Macromolecules* 34:7737–7743.

49. de Mefathi, M. V., and Frechet, J. M. 1988. Study of the compatibility of blends of polymers and copolymers containing styrene, 4-hydroxystyrene and 4-vinylpyridine. *Polymer* 29:477–482.

50. Prinos, A., Dompros, A., and Panayiotou, C. 1998. Thermoanalytical and spectroscopic study of poly(vinyl pyrrolidone)/poly(styrene-*co*-vinyl phenol) blends. *Polymer* 39:3011–3016.

51. Zhu, K. J., Wang, L. Q., and Yang, S. L. 1994. Study of the miscibility of poly(*N*-vinyl-2-pyrrolidone) with poly[styrene-*co*-(4-hydroxystyrene)]. *Macromol. Chem. Phys.* 195:1965–1972.
52. Zhuang, H. F., Pearce, E. M., and Kwei, T. K. 1994. Miscibility studies of poly(styrene-*co*-4-vinylbenzenephosphonic acid diethyl ester) with poly(p-vinylphenol). *Macromolecules* 27:6398–6403.
53. Smith, J. R. 1996. Nucleic acid models. *Prog. Polym. Sci.* 21:209–253.
54. Watson, J. D., and Berry, A. 2003. *DNA: The Secret of Life*. Knopf: New York.
55. Binder, W. H. 2005 Polymeric ordering by H-bonds. Mimicking nature by smart building blocks. *Monatshefte Chemie* 136:1–19.
56. Liu, G., and Zhou, J. 2003. First- and zero-order kinetics of porogen release from the cross-linked cores of diblock nanospheres. *Macromolecules* 36:5279–5284.
57. Lutz, J. F., Thunemann, A. F., and Rurack K. 2005. DNA-like "melting" of adenine- and thymine-functionalized synthetic copolymers. *Macromolecules* 38:8124–8126.
58. Lutz, J. F., Thunemann, A. F., and Nehbring, R. 2005. Preparation by controlled radical polymerization and self-assembly via base-recognition of synthetic polymers bearing complementary nucleobases. *J. Polym. Sci., Polym. Chem.* 43:4805–4818.
59. Lutz, J. F., Pfeifer, S., Chanana, M., Thunemann, A. F., and Bienert, R. 2006. H-Bonding-directed self-assembly of synthetic copolymers containing nucleobases: Organization and colloidal fusion in a noncompetitive solvent. *Langmuir* 22:7411–7415.
60. Thibault, R. J., Hotchkiss, P. J., Gray, M., and Rotello, V. M. 2003. Thermally reversible formation of microspheres through non-covalent polymer cross-linking. *J. Am. Chem. Soc.* 125:11249–11252.
61. Uzun, O., Sanyal, A., Nakade, H., Thibault, R. J., and Rotello, V. M. 2004. Recognition-induced transformation of microspheres into vesicles: Morphology and size control. *J. Am. Chem. Soc.* 126:14773–14777.
62. Drechsler, U., Thibault, R. J., and Rotello, V. M. 2002. Formation of recognition-induced polymersomes using complementary rigid random copolymers. *Macromolecules* 35:9621–9623.
63. Mather, B. D., Baker, M. B., and Beyer, F. L. 2007. Supramolecular triblock copolymers containing complementary nucleobase molecular recognition. *Macromolecules* 40:6834–6845.
64. Mather, B. D., Lizotte, J. R., and Long, T. E. 2004. Synthesis of chain end functionalized multiple hydrogen bonded polystyrenes and poly(alkyl acrylates) using controlled radical polymerization. *Macromolecules* 37:9331–9337.
65. Yamauchi, K., Kanomata, A., Inoue, T., and Long, T. E. 2004. Thermoreversible polyesters consisting of multiple hydrogen bonding (MHB). *Macromolecules* 37:3519–3522.
66. Bazzi, H. S., and Sleiman, H. F. 2002. Adenine-containing block copolymers via ring-opening metathesis polymerization: Synthesis and self-assembly into rod morphologies. *Macromolecules* 35:9617–9620.
67. Park, T., and Zimmerman, S. C. 2006. Formation of a miscible supramolecular polymer blend through self-assembly mediated by a quadruply hydrogen-bonded heterocomplex. *J. Am. Chem. Soc.* 128:11582–11590.
68. Kuo, S. W., and Cheng, R. S. 2009. DNA-like interactions enhance the miscibility of supramolecular polymer blends. *Polymer* 50:177–188.

69. Benesi, H. A., and Hildebrand, J. H. 1949. A spectrophotometric investigation of the interaction of iodine with aromatic hydrocarbons. *J. Am. Chem. Soc.* 71:2703–2707.

70. Kyogoku, Y., Lord, R. C., and Rich, A. 1967. An infrared study of hydrogen bonding between adenine and uracil derivatives in chloroform solution. *J. Am. Chem. Soc.* 89:496–504.

71. Asanuma, H., Ban, T., Gotoh, S., Hishiya, T., and Komiyama, M. 1998. Hydrogen bonding in water by poly(vinyldiaminotriazine) for the molecular recognition of nucleic acid bases and their derivatives. *Macromolecules* 31:371–377.

72. Asanuma, H., Ban, T. S., Gotoh, S., Hishiya, T., and Komiyama, M. 1998. Precise recognition of nucleotides and their derivatives through hydrogen bonding in water by poly(vinyldiaminotriazine). *Supramol. Sci.* 5:405–410.

73. Asanuma, H., Ban, T., Gotoh, S., Hishiya, T., and Komiyama, M. 1998. Precise recognition of nucleic acid bases by polymeric receptors in methanol. Predominance of hydrogen bonding over apolar interactions. *J. Chem. Soc. Perkin Trans.* 2:1915–1918.

74. Beijer, F. H., Sijbesma, R. P., Vekemans, J. A. J. M., Meijer, E. M., Kooijman, H., and Spek, A. L. 1996. Hydrogen-bonded complexes of diaminopyridines and diaminotriazines: Opposite effect of acylation on complex stabilities. *J. Org. Chem.* 61:6371–6380.

75. Sherrington, D. C., and Taskinen, K. A. 2001. Self-assembly in synthetic macromolecular systems via multiple hydrogen bonding interactions. *Chem. Soc. Rev.* 30:83–93.

76. Kuo, S. W., and Hsu, C. H. 2010. Miscibility enhancement of supramolecular polymer blends through complementary multiple hydrogen bonding interactions. *Polymer International* 59:998–1005.

77. Yuichi, K., Kazuo, K., and Koichi, N. 1997. Synthesis of *N*-cyclohexylmaleimide for heat-resistant transparent methacrylic resin. *J. Appl. Polym. Sci.* 63:363–368.

78. Kine, B. B., and Novak, R. W. 1986. *Encyclopedia of Polymer Science and Engineering*, 2nd ed., J. L. Kroschwitz, Ed. New York: Wiley, 1, 234.

79. Otsu, T., Motsumoto, T., Kubota, T., and Mori, S. 1990. Reactivity in radical polymerization of *N*-substituted maleimides and thermal stability of the resulting polymers. *Polym. Bull.* 23:43–50.

80. Braun, D., and Czerwinski, W. K. 1987. Recent investigations on the rate of binary and ternary copolymerizations. *Makromol. Chem.* 10:415–439.

81. Mishra, A., Sinha, T. M. J., and Choudhary, V. 1998. Methyl methacrylate–*N*-chlorophenyl maleimide copolymers: Effect of structure on properties. *J. Appl. Polym. Sci.* 68:527–534.

82. Dong, S., Wang, Q., Wei, Y., and Zhang, Z. 1999. Study on the synthesis of heat-resistant PMMA. *J. Appl. Polym. Sci.* 72:1335–1339.

83. Tagaya, A., Harada, T., and Koike, K. et al. 2007. Improvement of the physical properties of poly(methyl methacrylate) by copolymerization with pentafluorophenyl methacrylate. *J. Appl. Polym. Sci.* 106:4219–4224.

84. Cornejo-Barvo, J. M., and Siegel, R. A. 1996. Water vapour sorption behaviour of copolymers of *N,N*-diethylaminoethyl methacrylate and methyl methacrylate. *Biomaterials* 17:1187–1193.

85. Chauhan, R., and Choudhary, V. 2009. Thermal and mechanical properties of copolymers of methyl methacrylate with N-aryl itaconimides. *J. Appl. Polym. Sci.* 112:1088–1095.

86. Teng, H., Koike, K., Zhou, D., Satoh, Z., and Koike, Y. 2009. High glass transition temperatures of poly(methyl methacrylate) prepared by free radical initiators. *J. Polym. Sci. Polym. Chem.* 47:315–317.
87. Kuo, S. W., Kao, H. C., and Chang, F. C. 2003. Thermal behavior and specific interaction in high glass transition temperature PMMA copolymer. *Polymer* 44:6873–6882.
88. Chen, J. K., Kuo, S. W., Kao, H. C., and Chang, F. C. 2005. Thermal properties, specific interactions, and surface energies of PMMA terpolymers having high glass transition temperatures and low moisture absorptions. *Polymer* 46:2354–2364.
89. Coleman, M. M., Xu, Y., and Painter, P. C. 1994. Compositional heterogeneities in hydrogen-bonded polymer blends: Infrared spectroscopic results. *Macromolecules* 27:127–134.
90. Kuo, S. W., Xu, H., Huang, C. F., and Chang, F. C. 2002. Significant glass-transition-temperature increase through hydrogen-bonded copolymers. *J. Polym. Sci. Polym. Phys.* 40:2313–2323.
91. Kuo, S. W., and Tsai, S. T. 2009. Complementary multiple hydrogen-bonding interactions increase the glass transition temperatures to PMMA copolymer mixtures. *Macromolecules* 42:4701–4711.
92. Nair, K. P., Breedveld, V., and Weck, M. 2008. Complementary hydrogen-bonded thermoreversible polymer networks with tunable properties. *Macromolecules* 41:3429–3438.
93. Nair, K. P., Pollino, J. M., and Weck, M. 2006. Noncovalently functionalized block copolymers possessing both hydrogen bonding and metal coordination centers. *Macromolecules* 39:931–940.
94. Burd, C., and Weck, M. 2005. Self-sorting in polymers. *Macromolecules* 38:7225–7230.
95. Celiz, A. D., and Scherman, O. A. 2008. Controlled ring-opening polymerization initiated via self-complementary hydrogen-bonding units. *Macromolecules* 41:4115–4119.
96. Kuo, S. W., and Tsai, H. T. 2012. Self-complementary multiple hydrogen bonding interactions increase the glass transition temperatures to PMMA copolymers. *J. Appl. Polym. Sci.* 123:3275–3282.

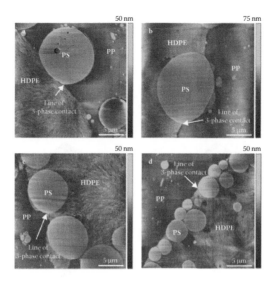

COLOR FIGURE 1.14
Focused ion beam–atomic force microscopy (FIB-AFM) micrographs of polystyrene/polypropylene/high-density polyethylene (PS/PP/HDPE) 10/45/45 blends after 30 min of quiescent annealing showing the gradual migration of the PS droplets toward the HDPE phase as a function of the increasing concentration of SEB. (a) 0.1% SEB, (b) 0.5% SEB, (c) 1% SEB, and (d) 2% SEB. (Reproduced from Virgilio, N., Desjardins, P., L'Esperance, G., and Favis, B. D. 2010. Modified interfacial tensions measured in situ in ternary polymer blends demonstrating partial wetting. *Polymer* 51:1472–1484 with permission from Elsevier.)

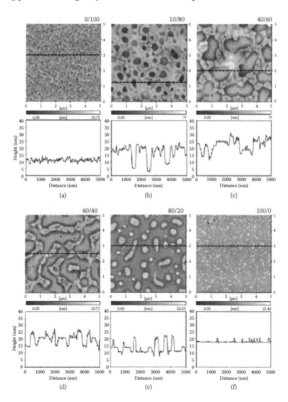

COLOR FIGURE 1.15
AFM images and line profiles of the surface topography of (a) CH$_3$-10SE, (b) 20/80 blend, (c) 40/60 blend, (d) 60/40 blend, (e) 80/20 blend, and (f) CF$_3$-10SE. (Reproduced from Sohn, E.-H., Kim, B. G., Chung, J.-S., and Lee, J.-C. 2010. Comb-like polymer blends of poly(oxyethylene)s with CH$_3$-terminated and CF$_3$-terminated alkylsulfonylmethyl side chains: Effect of terminal CF$_3$ moiety on the surface properties of the blends. *Journal of Colloid and Interface Science* 343:115–124 with permission from Elsevier.)

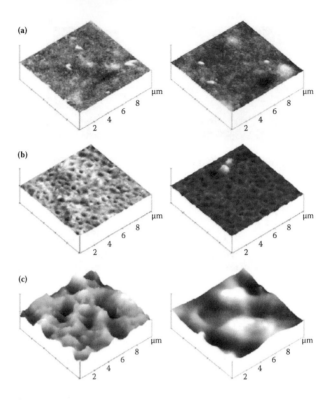

COLOR FIGURE 1.19
Atomic force microscopy height images of samples before (left) and after 30 h ultraviolet irradiation (right): (a) pectin (PEC), (b) PEC/polyvinylpyrrolidone (PVP) (80/20), and (c) PEC/PVP (60/40). (Reproduced from Kowalonek, J., and Kaczmarek, H. 2010. Studies of pectin/polyvinylpyrrolidone blends exposed to ultraviolet radiation. *European Polymer Journal* 46:345–353 with permission from Elsevier.)

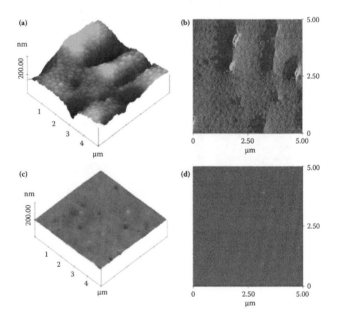

COLOR FIGURE 2.8
Tapping mode atomic force microscopy (AFM) images of polystyrene/poly(butyl methacrylate) (PS/PBMA) blend (a) height and (b) phase images, and A11-PS/T24-PBMA blend (c) height and (d) phase images. (Reprinted with permission from Kuo, S. W., and Cheng, R. S. 2009. DNA-like interactions enhance the miscibility of supramolecular polymer blends. *Polymer* 50:177–188. Copyright 2009, Elsevier Science Ltd., UK.)

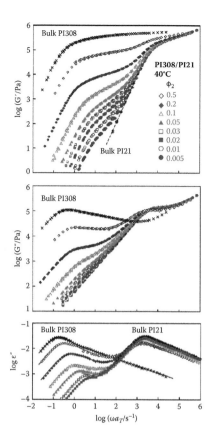

COLOR FIGURE 3.5

Viscoelastic and dielectric data of binary blends of monodisperse linear *cis*-polyisoprene samples with $M_1 = 2.1 \times 10^4$ (PI21) and $M_2 = 3.1 \times 10^5$ (PI308) at 40°C. The volume fraction ϕ_2 of the high-M component (PI308) is varied from 0.005 to 0.5. (Data taken, with permission, from Watanabe, H., S. Ishida, Y. Matsumiya, and T. Inoue. 2004a. Viscoelastic and dielectric behavior of entangled blends of linear polyisoprenes having widely separated molecular weights: Test of tube dilation picture. *Macromolecules* 37:1937–1951.)

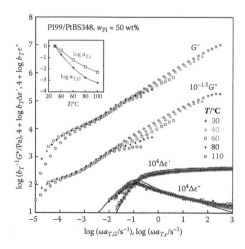

COLOR FIGURE 3.17

Test of time-temperature superposability for the PI99/PtBS348 miscible blend examined in Figure 3.16. The viscoelastic and dielectric data of the blend were separately shifted by different factors shown in the inset. The superposition is valid for the dielectric data but not for the viscoelastic data. (Data taken, with permission, from Watanabe, H., Q. Chen, Y. Kawasaki, Y. Matsumiya, T. Inoue, and O. Urakawa. 2011. Entanglement dynamics in miscible polyisoprene/poly(*p-tert*-butylstyrene) blends. *Macromolecules* 44:1570–1584.)

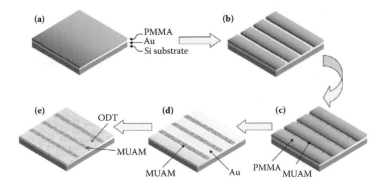

COLOR FIGURE 10.1
Patterning of hydrophilic/hydrophobic alkanethiols combining electron beam lithography (EBL) and self-assembly of alkanethiol molecules. (a) 150 nm thick polymethyl methacrylate (PMMA) resist spin coated onto the gold deposited on a silicon wafer; (b) Patterned PMMA trenches were defined by electron beam and development; (c) a hydrophilic 11-amino-1-undecanethiol hydrochloride (MUAM) assembled in the PMMA trench area; (d) PMMA resists were removed by acetone to produce patterned MUAM on gold; (e) backfilled by hydrophobic octadecanethiol (ODT) yielding the final chemical pattern. (Reprinted with permission from Wiley.)

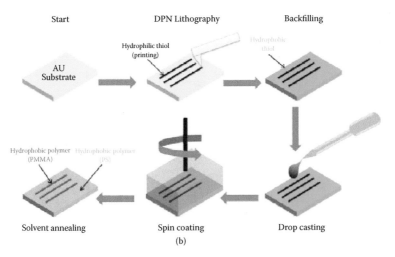

COLOR FIGURE 10.10
(a) Pattern geometries from the dip pen nanolithography (DPN) consisting of uniform and nonuniform features (template size 23 μm by 30 μm). (b) Fabrication method for the DPN templates and polymer assembly. (Reprinted with permission from Wiley.)

3

Component Dynamics in Miscible Polymer Blends

Hiroshi Watanabe

Kyoto University, Uji
Kyoto, Japan

Osamu Urakawa

Osaka University, Toyonaka
Osaka, Japan

CONTENTS

3.1 Introduction

In general, chemically different polymers are incompatible and exhibit macroscopic phase separation because of a vanishingly small mixing entropy of high molecular weight polymer chains, and the phase structure strongly influences the slow dynamics of blends (Utracki, 1989). Nevertheless, a considerable number of polymer pairs are known to be miscible in given ranges of temperature and composition. In early studies, blends of such miscible polymers were considered to be more or less similar to blends of chemically identical polymers and were subjected to just a vague focus of research. However, at temperatures above the glass transition temperature, the local concentration of the components in miscible blends fluctuates with time. In other words, the miscible blends are statically homogeneous but dynamically heterogeneous. This dynamic heterogeneity provides the miscible blends with a variety of unique features such as broad segmental relaxation/glass transition and significant thermorheological complexity (failure of time-temperature superposition) for both local and global relaxation processes (see, e.g., Alegria et al., 1994; Chen et al., 2008, 2011, 2012; Chung et al., 1994; Colby and Lipson, 2005; Ediger et al., 2006; Haley and Lodge, 2005; Haley et al., 2003; Hirose et al., 2003, 2004; Kumar et al., 1996; Lodge and McLeish, 2000; Lutz et al., 2004; Miller et al., 1990; Miura et al., 2001; Pathak et al., 1998, 1999, 2004; Sakaguchi et al., 2005; Takada et al., 2008; Urakawa, 2004; Urakawa et al., 2001; Watanabe and Urakawa, 2009; Watanabe et al., 2007, 2011; Zhao et al., 2008, 2009).

The above unique features, not observed for blends of chemically identical polymers, result from an interesting correlation between the length/time scales of the heterogeneity and the dynamic processes. This chapter briefly summarizes those features of the miscible blends in the linear response regime where the equilibrium chain motion determines the dynamic properties of the blends. The nonlinear features reflecting the nonequilibrium chain motion (under flow, for example) are also very interesting but are not included in this chapter. Readers interested in the nonlinear features are guided to recent papers by Endoh et al. (2008) and by Zhang et al. (2008) and the references therein.

3.2 Basics of Relaxation of Polymer Liquids

3.2.1 Viscoelastic Relaxation

At temperatures T above the glass transition temperature T_g, amorphous flexible polymers behave as liquids. Such polymeric liquids exhibit a mechanical stress when subjected to an instantaneous step shear strain γ at time 0; see Figure 3.1a. This stress essentially reflects the strain-induced *isochronal orientational anisotropy* of the chain conformation at various length scales ranging from a very local monomeric scale to a global scale spanning the ends of the chain (Inoue et al., 1997; McLeish, 2002; Watanabe, 1999). The anisotropy decays with time t because of the thermal motion of the polymer chains, and the stress $\sigma(t)$ relaxes accordingly. This stress relaxation is characterized by the relaxation modulus, $G(t) = \sigma(t)/\gamma$. The local anisotropy decays through the local, fast motion of the chain to govern the short time part of the stress relaxation, whereas the decay of the large-scale anisotropy, occurring through the slow global chain motion, determines the long time part of the stress relaxation. (Strain-induced distortion of atomic packing also contributes to the stress, but the stress having this structural origin relaxes almost instantaneously.) Consequently, the relaxation modulus $G(t)$ reflects the chain motion in various length scales to exhibit a relaxation mode distribution.

Under *infinitesimal* strain, the thermal motion of the chain activating the relaxation coincides with the motion at equilibrium. For this case, the stress $\sigma(t)$ under the step strain γ ($\ll 1$) is proportional to γ, and $G(t)$ is independent of γ and depends only on t. In this linear viscoelastic regime, the stress $\sigma(t)$ due to a strain $\gamma(t')$ of arbitrary history at $t' < t$ is expressed as a convolution of the relaxation modulus $G(t)$ and the strain rate $d\gamma/dt$ (Ferry, 1980):

$$\sigma(t) = \int_{-\infty}^{t} G(t-t')\, \frac{d\gamma(t')}{dt'}\, dt' \tag{3.1}$$

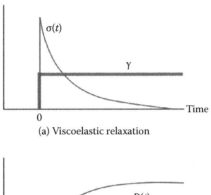

(a) Viscoelastic relaxation

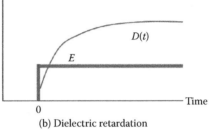

(b) Dielectric retardation

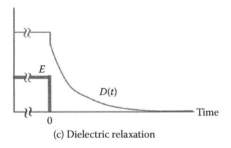

(c) Dielectric relaxation

FIGURE 3.1
(a) Viscoelastic relaxation, (b) dielectric retardation, and (c) dielectric relaxation.

Specifically, for sinusoidal strain $\gamma(t) = \gamma_0 \sin \omega t$ oscillating with a small amplitude γ_0 at an angular frequency ω (= $2\pi f$ with f being the frequency in Hz), Equation (3.1) gives

$$\sigma(t) = \gamma_0 \left[G'(\omega) \sin \omega t + G''(\omega) \cos \omega t \right] \qquad (3.2)$$

with

$$\text{dynamic storage modulus: } G'(\omega) = \omega \int_0^\infty G(t'') \sin \omega t'' \, dt''. \qquad (3.3)$$

$$\text{dynamic loss modulus: } G''(\omega) = \omega \int_0^\infty G(t'') \cos \omega t'' \, dt'' \qquad (3.4)$$

$G'(\omega)$ represents the elastic energy stored and released during one cycle of oscillation, while $G''(\omega)$ is related to the energy W dissipated as a heat during one cycle (per unit volume); $W = \pi \gamma_0^2 G''(\omega)$.

The relaxation mode distribution of $G(t)$ can be conveniently represented as a superposition of exponential decay modes with the intensity (= initial modulus) h_q and the relaxation time $\tau_{G,q}$:

$$G(t) = \sum_{q \geq 1} h_q \exp\left(-\frac{t}{\tau_{G,q}}\right) \text{ with } \tau_{G,1} > \tau_{G,2} > \tau_{G,3} > \dots \tag{3.5}$$

$\tau_{G,1}$ is the relaxation time of the slowest mode. From Equations (3.3), (3.4), and (3.5), the storage and loss moduli are expressed in terms of the viscoelastic relaxation spectrum $\{h_q, \tau_{G,q}\}$ as

$$G'(\omega) = \omega^2 \sum_{q \geq 1} h_q \frac{\tau_{G,q}^2}{1 + \omega^2 \tau_{G,q}^2}, \quad G''(\omega) = \omega \sum_{q \geq 1} h_q \frac{\tau_{G,q}}{1 + \omega^2 \tau_{G,q}^2} \tag{3.6}$$

As noted from Equation (3.6), the ω dependence of $G'(\omega)$ and $G''(\omega)$ reflects the relaxation mode distribution. In particular, these moduli exhibit the power-law-type terminal tails at low ω,

$$G'(\omega) = \omega^2 \sum_{q \geq 1} h_q \tau_{G,q}^2 \propto \omega^2 \tag{3.7}$$

and

$$G''(\omega) = \omega \sum_{q \geq 1} h_q \tau_{G,q} \propto \omega \quad \text{at } \omega \ll 1/\tau_{G,1}$$

These tails characterize completion of the relaxation at low ω. The second-moment average relaxation time, well defined in terms of the spectrum $\{h_q, \tau_{G,q}\}$, can be easily evaluated from the moduli data in this terminal zone as (Graessley, 1974; Watanabe, 1999)

$$\tau_G \equiv \frac{\displaystyle\sum_{q \geq 1} h_q \tau_{G,q}^2}{\displaystyle\sum_{q \geq 1} h_q \tau_{G,q}} = \left[\frac{G'}{\omega G''}\right]_{\omega \to 0} \tag{3.8}$$

This τ_G is an average of $\tau_{G,q}$ for all modes with the weighing factor of $h_q \tau_{G,q}$, and this factor is magnified for slow modes (having large $\tau_{G,q}$). Thus, τ_G is close to $\tau_{G,1}$ of the slowest relaxation mode and often utilized as the terminal relaxation time. Plots of $G''(\omega)$ against ω exhibit a peak if the relaxation intensity is considerably larger for a given, narrowly distributed group of modes compared to the remaining faster or slower modes. For this case, a reciprocal of the angular frequency for the peak, $\tau_{G,peak} = 1/\omega_{peak}$, can be utilized as a characteristic relaxation time for the given group of modes.

3.2.2 Dielectric Relaxation

Most organic molecules have electrical dipole moments. In the liquid state at $T > T_g$ (or at $T >$ melting temperature), these dipoles fluctuate due to the thermal motion of the molecules. When a constant electric field E is applied to such a liquid material placed in a capacitor, the capacitor instantaneously stores its bare charge, and the electron cloud and asymmetric (polarized) chemical bonds in the molecules are distorted by the electric field almost immediately. Then, the dipoles attached to the molecules are oriented in the direction of the electric field at a rate determined by the molecular motion. (No electronic/ionic direct current conductance is considered here.) These dynamic changes result in an evolution of the macroscopic electric displacement $D(t)$ defined as a sum of the instantaneously stored charge density on the capacitor and the extra charge density due to the distortion of the electron cloud/chemical bonds and the dipole orientation, as shown in Figure 3.1b. In the time scale of the molecular motion, this evolution can be expressed as (Kremer and Schönhals, 2003)

$$D(t) = \left[P_\infty + \Delta P \{1 - \Phi(t)\} \right] E \ (t > 0) \tag{3.9}$$

Here, P_∞ is the rapidly stored charge density normalized by the field intensity E; P_∞ is contributed from the bare capacitor and the distortion of the electron cloud/chemical bonds. ΔP corresponds to the saturated orientation of dipoles at $t = \infty$, and the factor $1 - \Phi(t)$ (= 0 and 1 at $t = 0$ and ∞) is the normalized dielectric retardation function reflecting the molecular motion. The function $\Phi(t)$ (= 1 and 0 at $t = 0$ and ∞), referred to as the normalized *dielectric relaxation function*, is most straightforwardly related to the molecular motion, as explained later for Equation (3.21): $\Phi(t)$ specifies the relaxation (or decay) of $D(t)$ on removal of the electric field E after full saturation of the charge density; see Figure 3.1c:

$$D(t) = \Delta P \Phi(t) E \ (t > 0) \tag{3.10}$$

In Equation (3.10), the origin of the time ($t = 0$) is set at the time for the removal of the field.

Under a weak electric field, the molecular motion activating the retarded growth (Equation 3.9) and relaxation (Equation 3.10) of $D(t)$ coincides with the thermal motion at equilibrium. Then, the linear dielectric response is observed, and the electric displacement $D(t)$ for an electric field $E(t')$ of arbitrary history at $t' < t$ is expressed as a convolution similar to Equation (3.1):

$$D(t) = P_\infty E(t) + \Delta P \int_{-\infty}^{t} \{1 - \Phi(t - t')\} \frac{dE(t')}{dt'} \, dt' \tag{3.11}$$

For a weak sinusoidal electric field $E(t) = E_0 \sin \omega t$, Equation (3.11) gives

$$D(t) = \varepsilon_{vac} E_0[\varepsilon'(\omega) \sin \omega t - \varepsilon''(\omega) \cos \omega t] \tag{3.12}$$

The dynamic dielectric constant $\varepsilon'(\omega)$ and the dielectric loss $\varepsilon''(\omega)$ appearing in Equation (3.12) are expressed in terms of the normalized dielectric relaxation function $\Phi(t)$ as

$$\varepsilon'(\omega) = \varepsilon_0 - \Delta\varepsilon \left\{ \omega \int_0^\infty \Phi(t'') \sin \omega t'' \, dt'' \right\} \tag{3.13}$$

$$\varepsilon''(\omega) = \Delta\varepsilon \left\{ \omega \int_0^\infty \Phi(t'') \cos \omega t'' \, dt'' \right\} \tag{3.14}$$

with

$$\varepsilon_0 = \frac{1}{\varepsilon_{vac}} (P_\infty + \Delta P) = \text{static dielectric constant} \tag{3.15}$$

and

$$\Delta\varepsilon = \frac{\Delta P}{\varepsilon_{vac}} = \text{dielectric relaxation intensity (due to dipole orientation)} \tag{3.16}$$

Here, ε_{vac} (= 8.85×10^{-12} C^2 J^{-1} m^{-1}) is the absolute permittivity of vacuum.

In general, the molecular motion allowing the dipole orientation has the intra- and intermolecular correlation. Thus, $\Phi(t)$ usually has a relaxation mode distribution and can be conveniently expressed as (Watanabe, 1999, 2001)

$$\Phi(t) = \sum_{q \geq 1} g_q \exp\left(-\frac{t}{\tau_{\varepsilon,q}}\right) \quad \text{with } \tau_{\varepsilon,1} > \tau_{\varepsilon,2} > \tau_{\varepsilon,3} > \dots \tag{3.17}$$

Here, g_q and $\tau_{\varepsilon,q}$ indicate the normalized intensity ($\Sigma_q g_q = 1$) and relaxation time of qth dielectric mode. From Equations (3.13) through (3.17), $\varepsilon'(\omega)$ and $\varepsilon''(\omega)$ are related to the dielectric relaxation spectrum $\{g_q, \tau_{\varepsilon,q}\}$ as

$$\Delta\varepsilon'(\omega) \equiv \varepsilon_0 - \varepsilon'(\omega) = \omega^2 \Delta\varepsilon \sum_{q \geq 1} g_q \frac{\tau_{\varepsilon,q}^2}{1 + \omega^2 \tau_{\varepsilon,q}^2},$$

$$\varepsilon''(\omega) = \omega\Delta\varepsilon \sum_{q \geq 1} g_q \frac{\tau_{\varepsilon,q}}{1 + \omega^2 \tau_{\varepsilon,q}^2} \tag{3.18}$$

As can be noted from Equation (3.18), the dielectric relaxation intensity $\Delta\varepsilon$ is experimentally evaluated from the data for the decrease of the dynamic dielectric constant, $\Delta\varepsilon'(\omega)$, and the dielectric loss, $\varepsilon''(\omega)$, as $\Delta\varepsilon = \Delta\varepsilon'(\omega \to \infty) = (2/\pi) \int_{-\infty}^{\infty} \varepsilon''(\omega)\, d \ln \omega$.

The expression of $\Delta\varepsilon'(\omega)$ and $\varepsilon''(\omega)$, Equation (3.18), is formally identical to that of the storage and loss moduli, Equation (3.6). Thus, all phenomenological relationships for those moduli are valid for $\Delta\varepsilon'(\omega)$ and $\varepsilon''(\omega)$. Specifically, $\Delta\varepsilon'(\omega)$ and $\varepsilon'(\omega)$ exhibit the power-law-type terminal tails at low ω where the relaxation has completed (cf. Equation 3.7):

$$\Delta\varepsilon'(\omega) = \omega^2 \Delta\varepsilon \sum_{q \geq 1} g_q \tau_{\varepsilon,q}^2 \propto \omega^2 \tag{3.19}$$

and

$$\varepsilon''(\omega) = \omega\Delta\varepsilon \sum_{q \geq 1} g_q \tau_{\varepsilon,q} \propto \omega \quad \text{at } \omega \ll 1/\tau_{\varepsilon,1}$$

The terminal (second-moment average) dielectric relaxation time can be evaluated from these tails as (cf. Equation 3.8)

$$\tau_\varepsilon \equiv \frac{\displaystyle\sum_{q \geq 1} g_q \tau_{\varepsilon,q}^2}{\displaystyle\sum_{q \geq 1} g_q \tau_{\varepsilon,q}} = \left[\frac{\Delta\varepsilon'}{\omega\varepsilon''} \right]_{\omega \to 0} \tag{3.20}$$

In addition, a reciprocal of the angular frequency for the peak of $\varepsilon''(\omega)$, $\tau_{\varepsilon,peak} = 1/\omega_{peak}$, can be utilized as a characteristic relaxation time for the most intensive group of modes.

Rigorous statistical mechanical analysis indicates that the dielectric relaxation function $\Phi(t)$ of an isotropic system in the linear response regime is equivalent to an autocorrelation function of a microscopic polarization $p(t)$ fluctuating through the molecular motion at equilibrium (Cole, 1967; Kubo, 1957):

$$\Phi(t) = \frac{\langle \mathbf{p}(t) \bullet \mathbf{p}(0) \rangle_{eq}}{\langle \mathbf{p}^2(0) \rangle_{eq}} \tag{3.21}$$

In Equation (3.21), the origin of the time ($t = 0$) is taken arbitrarily in the stationary equilibrium state, and $\langle ... \rangle_{eq}$ indicates the ensemble average at equilibrium. The microscopic polarization $\mathbf{p}$ is given by a sum of molecular dipoles. Thus, given that the dipole direction relative to the molecular axis is known, Equation (3.21) serves as a key for relating the macroscopic dielectric phenomenon to the molecular motion. The dielectric intensity $\Delta\varepsilon$ also includes the information for the molecular motion. Specifically, $\Delta\varepsilon$ (in meter-kilogram-second-ampere (MKSA) unit) is expressed as (cf. Kremer and Schönhals, 2003)

$$\Delta\varepsilon = \frac{4\pi\mu^2}{3k_BT} vFg \tag{3.22}$$

Here, μ and v denote the magnitude of molecular dipole and the number density of those dipoles, respectively; k_B is the Boltzmann constant; and T is the absolute temperature. F is a correction factor for a difference between the applied and internal electric fields ($F = (\varepsilon_0 + 2)^2/9$ in the Onsager form for nonpolar low-M molecules), and g is the Kirkwood–Fröhlich factor that represents the magnitude of the motional correlation of the dipoles (i.e., of the dipole-carrying molecules).

Now, we focus on polymeric materials. Most of polymers have monomeric dipoles classified as type-A, type-B, and type-C. The type-A and type-B dipoles are directly attached to the chain backbone in the directions parallel and perpendicular to the backbone, respectively; cf. Figure 3.2. The dielectrically observed fluctuation of these dipoles is activated only by the motion of the chain backbone. In contrast, the type-C dipole (not shown in Figure 3.2) is attached to the side groups and fluctuates through the side group motion even in the absence of the backbone motion.

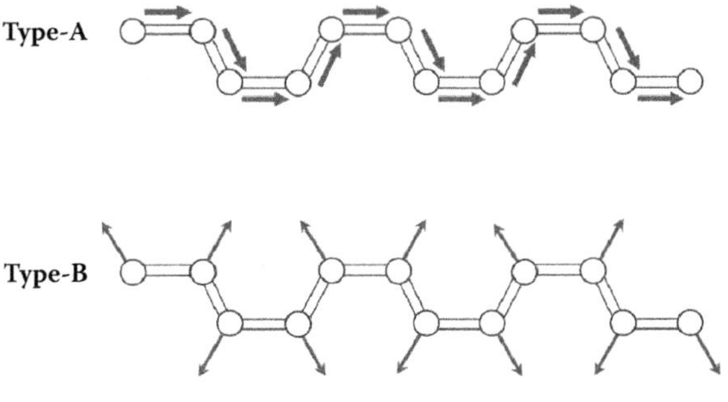

FIGURE 3.2
Type-A and type-B dipoles of a polymer chain.

For a linear chain having the type-A monomeric dipoles *without* inversion, the total dipole is proportional to the end-to-end vector **R**. Then, the polarization of the system of such type-A polymers is expressed as $\mathbf{p}(t) \propto \Sigma_\alpha \mathbf{R}_\alpha(t)$ with α being the index specifying the chain, and Equation (3.21) is rewritten as (Adachi and Kotaka, 1993; Watanabe, 2001)

$$\Phi(t) = \frac{\langle \mathbf{R}(t) \bullet \mathbf{R}(0) \rangle_{eq}}{\langle R^2 \rangle_{eq}} \quad \text{for type-A chain} \tag{3.23}$$

(A relationship valid for Gaussian chains, $\langle \mathbf{R}_\alpha(t) \bullet \mathbf{R}_\beta(0) \rangle_{eq} = \langle \mathbf{R}_\alpha(0) \bullet \mathbf{R}_\beta(0) \rangle_{eq} = 0$ for $\alpha \neq \beta$, has been utilized in the derivation of Equation 3.23.) Thus, the dielectric relaxation of type-A polymer systems detects the end-to-end vector fluctuation (global motion) of *respective* chains therein. Correspondingly, Equation (3.22) reduces to

$$\Delta\varepsilon = \frac{4\pi\tilde{\mu}^2}{3k_BT} \nu \langle R^2 \rangle_{eq} \quad \text{for type-A chain} \tag{3.24}$$

where ν is the number density of the chains, and $\tilde{\mu}$ denotes the magnitude of the type-A dipole per unit backbone length of the chain ($F \cong 1$ for the global relaxation) (Adachi and Kotaka, 1993). The Kirkwood–Fröhlich factor g appearing in Equation (3.22) is set to be unity in Equation (3.24) because of the Gaussian relationship explained above (not because of lack of interchain correlation of the global motion; this correlation does exist for entangled polymers, as explained later). As noted from Equation (3.24), the dielectric intensity $\Delta\varepsilon$ serves as a measure of the mean-square end-to-end distance, $\langle R^2 \rangle_{eq}$.

For a system of type-B chains, the polarization can be expressed as $\mathbf{p}(t) \propto \Sigma_{j,\alpha} \mathbf{n}_\alpha^j(t)$ with $\mathbf{n}_\alpha^j$ being the unit normal vector of jth monomeric unit of αth chain defined with respect to the backbone of this chain. The monomeric units belonging to either the same or different chains exhibit some local packing correlation, and sequences of a few units in the same chain always have the orientational correlation of $\mathbf{n}_\alpha^j$. Thus, for the type-B polymers, Equation (3.21) can be rewritten as

$$\Phi(t) = \frac{\sum_{corr} \langle \mathbf{n}_\alpha^j(t) \bullet \mathbf{n}_\beta^i(0) \rangle_{eq}}{\sum_{corr} \langle \mathbf{n}_\alpha^j(0) \bullet \mathbf{n}_\beta^i(0) \rangle_{eq}} \tag{3.25}$$

where the summation Σ_{corr} is taken for the monomeric units (either in the same or different chains) having the orientational correlation. The corresponding dielectric relaxation intensity is given by Equation (3.22), and the Kirkwood–Fröhlich factor g appearing therein reflects this correlation. Thus, the dielectric relaxation of type-B chains provides us with the information related to the correlated motion of the monomeric units (or, monomeric segments).

3.2.3 Thermorheological Behavior

The storage and loss moduli, G' and G'', exhibit the viscoelastic relaxation mode distribution (cf. Equation 3.6). The relaxation times and intensities of those modes, $\tau_{G,q}$ and h_q, are determined by the molecular motion and thus change with the temperature T. In general, the molecular motion is accelerated with increasing T so that $\tau_{G,q}$ decreases while h_q increases on the increase of T. If all modes exhibit the same fractional changes of their $\tau_{G,q}$ and h_q on a change of temperature from T_r (reference temperature) to T, $\tau_{G,q}$ and h_q satisfy simple relationships:

$$\tau_{G,q}(T) = a_{T,G}\tau_{G,q}(T_r), \; h_q(T) = b_T h_q(T_r) \tag{3.26}$$

Here, the factors $a_{T,G}$ and b_T, referred to as the viscoelastic shift factor and the intensity factor, respectively, are functions of T and T_r but are independent of the mode index q. From Equations (3.26) and (3.6), the moduli at temperatures T and T_r are mutually related as

$$G'(\omega;T) = b_T G'(\omega a_{T,G};T_r), \; G''(\omega;T) = b_T G''(\omega a_{T,G};T_r) \tag{3.27}$$

If the modulus data of a material satisfy Equation (3.27), this material is referred to as thermorheologically simple and obeys the (viscoelastic) time–temperature superposition.

Equation (3.26), the prerequisite of this simplicity, is satisfied if the slow relaxation modes result from accumulation of fast modes. This is the case for the slow, global viscoelastic relaxation of homopolymer liquids (that corresponds to the accumulated motion of the Rouse segment) (Inoue et al., 2002; Watanabe, 2009). The fast viscoelastic relaxation process of homopolymers, being related to the glass transition and often referred to as the segmental relaxation, is also thermorheologically simple, except in a close vicinity of the glass transition temperature T_g. For both processes, $a_{T,G}$ are well described by the William–Landell–Ferry (WLF) equation (Ferry, 1980):

$$\log a_T = -\frac{C_1^{WLF}(T - T_r)}{C_2^{WLF} + (T - T_r)} \tag{3.28}$$

The corresponding intensity factor b_T is essentially given by T/T_r (which reflects the entropy elasticity), although the thermal expansion of the material and a small change of the chain dimension with T have a minor contribution to this factor (Graessley, 2008).

The WLF coefficients C_1^{WLF} and C_2^{WLF} appearing in Equation (3.28) are common and thus the shift factor $a_{T,G}$ is the same for the global and local relaxation processes at T well above T_g. The segmental friction can be obtained from analysis of the $a_{T,G}$ data at such high T (Ferry, 1980). However, in the close vicinity of T_g, the segmental relaxation process becomes thermorheologically complex to violate Equations (3.26) and (3.27) because the motional correlation of the monomeric units is enhanced at $T \sim T_g$. For this case, the segmental relaxation mode distribution broadens with decreasing T, and the segmental relaxation time increases more strongly compared to the global relaxation time.

Another example of the thermorheological complexity is found for multicomponent polymer systems. If the motion of different component chains contributes to the viscoelastic moduli of the system, the thermorheological complexity prevails even for the case that respective components exhibit the simplicity but have different shift factors. In more general cases, each component exhibits the complexity, thereby providing the systems with very strong complexity. This strong complexity of multicomponent systems (miscible blends) is explained in Section 3.5.2.5.

Now, we turn our attention to the dielectric relaxation. The decrease of dynamic dielectric constant $\Delta\varepsilon'$ and the dielectric loss ε'' have the mode distribution (cf. Equation 3.18) formally identical to the viscoelastic mode distribution, and the above explanation for the viscoelastic relaxation applies also to the dielectric relaxation: For the simplest case, the relaxation times $\tau_{\varepsilon,q}$ and nonnormalized intensities $\Delta\varepsilon g_q$ of the dielectric modes exhibit the same change with T irrespective of the mode index q as

$$\tau_{\varepsilon,q}(T) = a_{T,\varepsilon}\tau_{\varepsilon,q}(T_r), \ \Delta\varepsilon(T)g_q(T) = b_T^{-1}\Delta\varepsilon(T_r)g_q(T_r) \tag{3.29}$$

For polymers, the intensity factor for the dielectric quantities often (though not always) coincides with reciprocal of b_T ($\sim T/T_r$) for the viscoelastic properties. This feature has been incorporated in Equation (3.29). Corresponding to Equation (3.29), $\Delta\varepsilon'$ and ε'' obeyed the time-temperature superposition:

$$\Delta\varepsilon'(\omega;T) = b_T^{-1}\Delta\varepsilon'(\omega a_{T,\varepsilon};T_r), \ \varepsilon''(\omega;T) = b_T^{-1}\varepsilon''(\omega a_{T,\varepsilon};T_r) \tag{3.30}$$

This superposition is experimentally known to be valid for the segmental relaxation of type-B homopolymers except in a close vicinity of T_g, which suggests that the magnitude of correlation between the monomeric units hardly changes with T (in a range well above T_g). The superposition is also valid for the global dielectric relaxation of type-A homopolymers because the global motion results from accumulation of local motion of the Rouse segments. The shift factors $a_{T,\varepsilon}$ for these relaxation processes are described by the WLF equation having a functional form shown in Equation (3.28). However, for miscible blends of chemically different chains, the dielectric superposition usually fails for the segmental relaxation process because of the difference in $a_{T,\varepsilon}$ of the component chains. This failure, together with a delicate failure noted for the global relaxation of the components in those blends, is explained later in more detail.

3.3 Relaxation in Homopolymer Systems (Reference for Miscible Blends)

3.3.1 Overview for Monodisperse Systems

Dynamic behavior of homopolymers serves as a reference for understanding the uniqueness of the behavior of miscible blends of chemically different chain. From this point, this section is devoted to briefly summarizing the dynamic behavior of homopolymers.

As an example of the behavior for monodisperse homopolymers, Figure 3.3 shows the storage and loss moduli, G' and G'', measured for monodisperse linear polystyrene (PS) samples having the molecular weights M as indicated (data taken from Chen et al., 2010; Matsumiya et al., 2011; Schausberger et al., 1985; Uno, 2009). For comparison, the decrease of dynamic dielectric constant from the static constant, $\Delta\varepsilon'(\omega)$ $(= \varepsilon_0 - \varepsilon'(\omega))$, and the dielectric loss, $\varepsilon''(\omega)$, measured for a high-M sample (with $10^{-3}M = 96$) are multiplied by a factor of 10^7 and shown with the large circles (data taken from Matsumiya et al., 2011; Uno, 2009). All these data were subjected to the time-temperature superposition (cf. Equations 3.27 and 3.30) and reduced at 180°C. For the lowest-M sample with $10^{-3}M = 9.5$, a minor correction for the T_g decrease (that resulted in a decrease of the segmental relaxation time) was made on the basis of the WLF analysis. Clearly, the moduli data show two distinct relaxation processes, the segmental relaxation process at high $\omega > 10^6$ s^{-1} and the global relaxation process at lower ω. The terminal tails of the moduli (Equation 3.7) are clearly noted for the latter process. In contrast, the dielectric data exhibit just the segmental relaxation at high ω followed by the terminal tails (cf. Equation 3.19) because PS has only type-B dipole and is dielectrically inert in the time scale of global motion.

For the moduli data, the time-temperature superposition fails at intermediate ω between the segmental and global relaxation processes because these processes exhibit different $a_{T,G}$ at low $T \sim T_g$ (see, e.g., Adachi and Kotaka, 1993; Inoue et al., 1991, 1996; Kremer and Schönhals, 2003). (This failure is not well resolved in the compressed scale of the plots shown in Figure 3.3.) The superposition works separately at high and low ω where the viscoelastic data are dominated by one of these processes. In contrast, the dielectric data satisfy the superposition in the entire range of ω because those data detect just the segmental relaxation process, although it fails in a close vicinity of

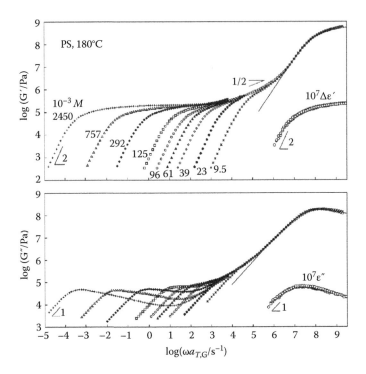

FIGURE 3.3
Storage and loss moduli, G' and G'', measured for monodisperse linear polystyrene samples at 180°C. The numbers indicate $10^{-3}M$. For comparison, the decrease of the dynamic
dielectric constant from the static constant, $\Delta\varepsilon'$, and the dielectric loss, ε'', of a high-M
sample ($10^{-3}M = 96$) are also shown. (Data taken, with permission, from Chen, Q., A.
Uno, Y. Matsumiya, and H. Watanabe. 2010. Viscoelastic mode distribution of moderately entangled linear polymers. *J. Soc. Rheol. Japan (Nihon Reoroji Gakkaishi)* 38:187–193;
Matsumiya, Y., A. Uno, H. Watanabe, T. Inoue, and O. Urakawa. 2011. Dielectric and viscoelastic investigation of segmental dynamics of polystyrene above glass transition temperature: Cooperative sequence length and relaxation mode distribution. *Macromolecules*,
44:4355–4363; Schausberger, A., G. Schindlauer, and H. Janeschitz-Kriegl. 1985. Linear
elastico-viscous properties of molten standard polystyrenes I. Presentation of complex
moduli; role of short range structural parameters. *Rheol. Acta*, 24:220–227; and Uno, A. 2009.
Dielectric Determination of Segmental Size of Polystyrene, MS Dissertation, Graduate School of
Engineering, Kyoto University.)

T_g (not examined in Figure 3.3) (cf. Adachi and Kotaka, 1993; Kremer and
Schönhals, 2003).

The viscoelastic and dielectric data in the segmental relaxation process are
insensitive to the molecular weight M and the molecular weight distribution (either in the monodisperse systems or blends). In fact, the viscoelastic
and dielectric data for this process are indistinguishable for PS samples with
$M \geq 10^4$ (cf. Matsumiya et al., 2011; Uno, 2009). This insensitivity naturally
emerges because the mechanism of the local motion of polymer chains is
independent of M. The M-insensitive viscoelastic and dielectric data for the

segmental relaxation process of PS are associated with the same shift factors ($a_{T,G} = a_{T,\varepsilon}$), but the relaxation time is larger for the latter by a factor $\cong 6$, as can be noted in Figure 3.3. This difference reflects a difference of the molecular origins of the viscoelastic and dielectric relaxation processes. Both processes detect the same local chain dynamics, but the viscoelastic process averages this dynamics as the isochronal orientational anisotropy, while the dielectric process averages the same dynamics as the decay of orientational memory. Extensive studies are still being made for this difference and for the segmental relaxation time and mode distribution in the close vicinity of T_g, the latter being frequently analyzed on the basis of the Kohlrausch–Williams–Watts function, $\Phi(t) = \exp\{(-t/\tau^*)^\alpha\}$ with $0 < \alpha \leq 1$ (see, e.g., Kremer and Schönhals, 2003). Some detail of those studies is explained later in relation to the segmental relaxation behavior in miscible blends.

Now, we turn our focus on the global dynamics. For monodisperse linear homopolymers, the terminal relaxation behavior reflecting the global dynamics has been extensively studied from both experimental and theoretical aspects (see, e.g., Doi and Edwards, 1986; Graessley, 2008; McLeish, 2002; Rubinstein and Colby, 2003; Watanabe, 1999). The terminal behavior is associated with universal features that depend only on the molecular weight and its distribution as well as the topological structure (e.g., branching) of the chains. Specifically, for *monodisperse* linear chains including PS examined in Figure 3.3, the M dependence of the viscoelastic terminal relaxation time τ_G (defined by Equation 3.8) is known to be described by empirical equations:

$$\text{For low-}M \text{ polymers: } \tau \propto \zeta_s M^2 \tag{3.31}$$

$$\text{For high-}M \text{ polymers: } \tau \propto \zeta_s M^{3.5(\pm 0.2)}/M_{e,\text{bulk}}^{1.5(\pm 0.2)} \tag{3.32}$$

where ζ_s denotes the friction coefficient of the Rouse segment, the smallest motional unit during the global relaxation process as discussed by Inoue et al. (1997, 2002). The Rouse segment is not identical to the monomeric segment intimately related to the fast glassy relaxation and glass transition, but the friction coefficients of these two type of segments exhibit the same T dependence at high T well above T_g (see, e.g., Adachi and Kotaka, 1993; Inoue et al., 1997, 2002; Kremer and Schönhals, 2003). The dielectric terminal relaxation time of monodisperse type-A chains, τ_ε (defined by Equation 3.20), is close to $2\tau_G$ and is also described by Equations (3.31) and (3.32), as explained later in more detail (see Watanabe, 2009; Watanabe et al., 2002).

The crossover from Equation (3.31) to Equation (3.32) is attributed to the topological constraint for the global motion referred to as *entanglement*. The polymer chains are physically uncrossable to each other so that the high-M chains mutually constrain their global motion. A density of this topological constraint in bulk systems is characterized by the entanglement molecular weight $M_{e,\text{bulk}}$ appearing in Equation (3.32): High-M polymers exhibit a plateau in their G'

data (cf. Figure 3.3), and $M_{e,bulk}$ is evaluated from the height of this plateau, G_N, as $M_{e,bulk} = cRT/G_N$ (c = mass concentration). Some models introduce a prefactor <1 in this expression of M_e (see, e.g., Graessley, 2008; Rubinstein and Colby, 2003). However, this prefactor is not important in this chapter.

3.3.2 Overview for Binary Blends

Binary blends of chemically identical polymers of different molecular weights M_1 and M_2 ($>M_1$) are the simplest polydisperse systems, and extensive studies have been made for the effects of blending on global relaxation (see, e.g., McLeish, 2002; Watanabe, 1999). The local segmental relaxation does not change on blending (unless M_1 is very small to decrease T_g of the blend). In contrast, the global relaxation behavior strongly changes on blending. Specifically, the behavior of binary blends is classified into several cases according to M_1 and M_2 of the short and long components and the volume fraction ϕ_2 of the latter, as schematically shown in Figure 3.4. In zone **I** where the short chains (with $M_1 < M_{e,bulk}$) are in the intrinsically nonentangled state and the long chains are too dilute to be mutually entangled ($\phi_2 < \{M_{e,bulk}/M_2\}^{1/d}$ with the dilution exponent $d \cong 1.3$), both long and short chains behave as if they were in the nonentangled monodisperse bulk state to exhibit the relaxation time specified by Equation (3.31). In zone **II** where the short chains are in the nonentangled state but the long chains are concentrated ($\phi_2 > \{M_{e,bulk}/M_2\}^{1/d}$) to exhibit their mutual entanglements, the short chains behave as a simple solvent for the long chains, and in long time scales the blend behaves as an entangled solution of the long chains. For this case, the terminal relaxation time of the long chain is described by

$$\tau^{[long]} \propto \zeta_s M_2^{3.5(\pm 0.2)} / \{M_{e,bulk} / \phi_2^d\}^{1.5(\pm 0.2)} \text{ with } d \cong 1.3 \qquad (3.33)$$

The factor $\{M_{e,bulk}/\phi_2^d\}$ coincides with the entanglement molecular weight $M_{e,soln}$ in the solution.

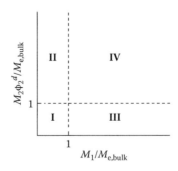

FIGURE 3.4
Zones for different types of relaxation behavior in binary blends of chemically identical polymers.

Interesting entanglement relaxation behavior is observed in zones **III** and **IV**. As an example of this behavior, Figure 3.5 shows the G' and G'' data for binary blends of monodisperse linear *cis*-polyisoprene (PI) samples with $M_1 = 2.1 \times 10^4$ (PI21) and $M_2 = 3.1 \times 10^5$ (PI308) reported by Watanabe et al. (2004a). The data for the dielectric loss ε'' are also shown (bottom panel). Those M_i are well above $M_{e,bulk}$ ($= 5.0 \times 10^3$ for bulk PI), and the components are mutually entangled in the blends.

Differing from PS examined in Figure 3.3, PI has the type-A dipole so that the global motion activates both viscoelastic and dielectric relaxation having the same shift factor, $a_{T,G} = a_{T,\varepsilon}$. This feature can be confirmed in Figure 3.5. (PI also has the type-B dipole and the segmental relaxation is dielectrically active. However, the segmental relaxation of PI occurs at high ω not covered in Figure 3.5 and has no contribution to the data shown therein.) From the G'' and ε'' data of the components in bulk (cross and plus symbols in Figure 3.5), we note that the dielectric τ_ε and the viscoelastic τ_G are close to each other (actually $\tau_G \cong \tau_\varepsilon/2$), and thus the global motion is dielectrically active. These τ data of the components in bulk are described by Equation (3.32). For nonentangled monodisperse linear PI with $M \le M_{e,bulk}$, τ_ε is again close to τ_G ($\cong \tau_\varepsilon/2$) and described by Equation (3.31) (see, e.g., Watanabe, 2009; Watanabe et al., 2002, 2004a, 2004b).

In Figure 3.5, the PI308/PI21 binary blends clearly exhibit two-step relaxation. Because $M_2 \gg M_1$, the terminal relaxation is much slower for the long PI308 chain than for the short PI21 chain irrespective of the ϕ_2 value. Thus, the slow step is attributed to the global relaxation of the long PI308 chains, and the fast step, to the global relaxation of the short PI21 chain (and a partial relaxation of the PI308 chain activated by the PI21 relaxation). The high-ω plateau modulus sustained by all these component chains does *not* change with ϕ_2. Correspondingly, the relaxation time of the short chain (seen for the fast step) increases with ϕ_2 just moderately (by a factor less than 3 even for $\phi_2 = 0.5$). However, the relaxation behavior of the long PI308 chain strongly changes with ϕ_2, as summarized below.

For small $\phi_2 < \phi_{2,ent} = \{M_{e,bulk}/M_2\}^{1/d}$ ($= 0.042$) (i.e., in zone **III** shown in Figure 3.4), the long chains are not mutually entangled but with the short chains. In the *dilute limit* at $\phi_2 \le 0.01$ ($\ll \phi_{2,ent}$), the viscoelastic and dielectric terminal relaxation times of the long chain (evaluated from the data shown in Figure 3.5) were found to be independent of ϕ_2 and described by an empirical equation (Sawada et al., 2007; Watanabe, 1999; Watanabe et al., 2004a):

$$\tau^{[long]} \propto \zeta_s M_1^\alpha M_2^2 / M_{e,bulk}^\alpha \text{ with } \alpha \cong 3 \text{ (for } M_2 \gg M_1 > M_{e,bulk}) \qquad (3.34)$$

The strong M_1 dependence of $\tau^{[long]}$ shown by Equation (3.34) suggests that the global motion of the short chains activates the relaxation of the long chain. In contrast, for large $\phi_2 \gg \phi_{2,ent}$ (zone **IV** of Figure 3.4), the long chains are mutually entangled with each other to exhibit $\tau^{[long]}$ being dependent on ϕ_2

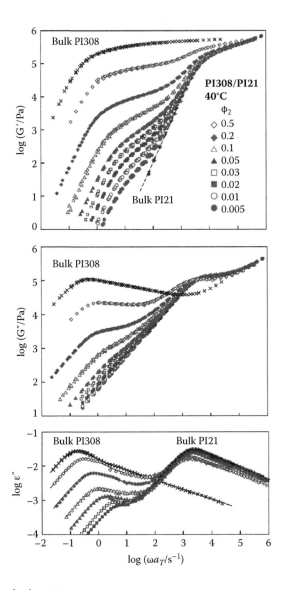

FIGURE 3.5 (See color insert.)
Viscoelastic and dielectric data of binary blends of monodisperse linear *cis*-polyisoprene samples with $M_1 = 2.1 \times 10^4$ (PI21) and $M_2 = 3.1 \times 10^5$ (PI308) at 40°C. The volume fraction ϕ_2 of the high-M component (PI308) is varied from 0.005 to 0.5. (Data taken, with permission, from Watanabe, H., S. Ishida, Y. Matsumiya, and T. Inoue. 2004a. Viscoelastic and dielectric behavior of entangled blends of linear polyisoprenes having widely separated molecular weights: Test of tube dilation picture. *Macromolecules* 37:1937–1951.)

but insensitive to M_1. This M_1-insensitivity suggests that $\tau^{[long]}$ is determined by the motion of the long chains by themselves, irrespective of the motion of the short chains. In this limit of large ϕ_2 (and $M_2 \gg M_1$) the $\tau^{[long]}$ data are described by Equation (3.33), indicating that the short chains behave as the simple solvent in the time scale of the terminal relaxation of the long chains (see, e.g., Watanabe, 1999; Watanabe et al., 2004a).

Corresponding to the above change of $\tau^{[long]}$ with ϕ_2, the terminal visco-elastic mode distribution of the long chains changes from the broad, power-law-type distribution without being associated by the plateau of G' at low ω (for small ϕ_2) to the narrow distribution associated with this plateau (for large ϕ_2), as noted in Figure 3.5. It should be emphasized that the $\tau_G^{[long]}$ data are described by Equations (3.34) and (3.33) only in the small-ϕ_2 and large-ϕ_2 asymptotes and only for the blends having $M_2 \gg M_1$ ($>M_{e,bulk}$) and exhibiting $\tau^{[long]} \gg \tau^{[short]}$. (Specifically, the asymptotic behavior can be clearly observed only for cases of $\tau^{[long]} > 10^4 \tau^{[short]}$; cf. Watanabe, 1999.) The blends exhibit broad crossover of $\tau_G^{[long]}$ and of the relaxation mode distribution at intermediate ϕ_2, as noted in Figure 3.5. For blends with M_2 not much larger than M_1 ($>M_{e,bulk}$), this crossover zone is spread over the entire range of ϕ_2, and neither Equation (3.44) nor Equation (3.43) is valid. Thus, zones **III** and **IV** shown in Figure 3.4 are further divided into subzones according to the M_2 and M_1 values (according to the $\tau^{[long]}/\tau^{[short]}$ ratio) (cf. Rubinstein and Colby, 2003; Watanabe, 1999).

In Figure 3.5, it should also be noted that the changes of the mode distribution with ϕ_2 are different for the dielectric and viscoelastic data. The visco-elastic data exhibit the crossover of this distribution, as explained above. Correspondingly, the low-ω plateau height of G' and the corresponding peak height of G'' rapidly decrease with decreasing ϕ_2 from 1 to the threshold value for the long-long entanglement, $\phi_{2,ent} = \{M_{e,bulk}/M_2\}^{1/d}$. For the data shown in the top and middle panel of Figure 3.5, these plateau and peak heights scale roughly as $\phi_2^{2.3}$. In contrast, in the bottom panel of Figure 3.5, no significant mode broadening is observed for the low-ω ε'' data detecting the global motion of the long chain, and the low-ω peak height of ε'' scales as ϕ_2. This difference emerges because the viscoelastic and dielectric relaxation processes average the same chain dynamics in different ways. Specifically, the slow viscoelastic relaxation detects the isochronal orientational anisotropy of the *entanglement segments* having the molecular weight $M_{e,bulk}$. The anisotropy of the long chains decays with the global motion of the short chains because this motion allows successive entanglement segments in the long chain to exhibit cooperatively motion and mutual equilibration, as fully discussed by Watanabe (1999). In contrast, the slow dielectric relaxation of the long PI chain reflects fluctuation of the end-to-end vector of this chain (cf. Equation 3.23). This fluctuation is hardly activated by the global motion of the short chains as long as the entanglements among the long chains remain effective, indicating that the short chain motion hardly affects the dielectric mode distribution/peak height of the long chains. This difference between the viscoelastic and dielectric relaxation has been utilized in analysis of the dynamic behavior of the PI/PI blends (Watanabe, 2009), as explained in the following section.

3.3.3 Molecular Models for Global Dynamics

Extensive theoretical studies have been made for the mechanisms of the global chain motion that leads to the viscoelastic (and dielectric) terminal relaxation of the homopolymer systems explained in the previous section. Several molecular models have been proposed, as briefly explained below.

3.3.3.1 Model for Nonentangled Chains

For nonentangled polymers, the most widely accepted model is the Rouse model (see, e.g., Doi and Edwards, 1986; McLeish, 2002; Rubinstein and Colby, 2003; Watanabe, 1999). This model coarse-grains the actual linear chain into a sequence of N submolecules, each having a friction coefficient ζ_s and the exhibiting the entropic elasticity, as schematically illustrated in Figure 3.6a. The strength of the entropic spring is given by $3k_BT/b^2$ with b^2 representing the mean-square end-to-end distance of the submolecule at equilibrium. The dynamics of the Rouse chain is determined by a balance of the frictional force, elastic spring force, and thermal Brownian force acting on each submolecule. The viscoelastic relaxation of the Rouse chain corresponds to the relaxation of the orientational anisotropy of respective submolecules determined by this force balance. For monodisperse linear chains of the molecular weight M and mass concentration c at the absolute temperature T, the viscoelastic $G(t)$ and dielectric $\Phi(t)$ calculated from the Rouse model in the continuous limit (for $N \to \infty$) can be compactly expressed as (Doi and Edwards, 1986; Watanabe, 1999)

$$G(t) = \frac{cRT}{M} \sum_{q=1}^{N} \exp\left(-\frac{q^2 t}{\tau_{R,G}}\right), \quad \Phi(t) = \sum_{q=\text{odd}}^{N} \frac{8}{q^2 \pi^2} \exp\left(-\frac{q^2 t}{\tau_{R,\varepsilon}}\right) \quad (3.35)$$

with

$$\tau_{R,G} = \frac{\zeta_s b^2 N^2}{6\pi^2 k_B T} = \frac{\zeta_{\text{chain}} \langle R^2 \rangle_{\text{eq}}}{6\pi^2 k_B T} \propto M^2, \quad \tau_{R,\varepsilon} = 2\tau_{R,G} \quad (3.36)$$

In the calculation of $\Phi(t)$, the chain is assumed to have the type-A dipoles without inversion. In Equations (3.35) and (3.36), $\tau_{R,G}$ and $\tau_{R,\varepsilon}$, respectively, indicate the viscoelastic and dielectric relaxation times of the slowest Rouse mode. These τ_R can be expressed in terms of the directly measurable parameters, the total friction of the chain ζ_{chain} $(= N\zeta_s)$, and the mean-square end-to-end distance at equilibrium $\langle R^2 \rangle_{\text{eq}} = Nb^2$, as shown in Equation (3.36). Thus,

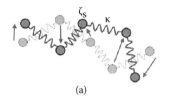

(a)

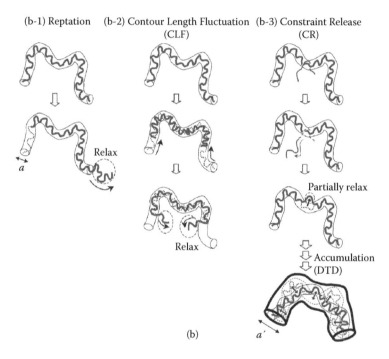

(b-1) Reptation   (b-2) Contour Length Fluctuation   (b-3) Constraint Release
(CLF)   (CR)

(b)

FIGURE 3.6
(a) Rouse model and (b) tube model.

the predictions of the Rouse model in long time scales are not affected by our choice of the submolecule number N, which demonstrates the physical soundness of the model. In relation to this point, it should be emphasized that the Rouse model considers *internally equilibrated* submolecules (that behave as the entropic springs explained above) and is valid in any time scale longer than the equilibration time of the submolecule. The minimum possible size of the submolecule being consistent with the viscoelastic data of homopolymer systems has been examined through rheo-optical experiments by Inoue et al. (1991, 1996, 2002). These experiments suggested that the smallest possible submolecule, often referred to as the *Rouse segment*, has the molecular weight close to that of the statically well-defined Kuhn segment.

The Rouse predictions are considerably close to the data for nonentangled monodisperse polymers, for example, the viscoelastic moduli data of low-M

PS shown in Figure 3.3 and the relaxation time data described by the empirical equation ($\tau_G \cong \tau_\varepsilon/2 \propto \zeta_S M^2$) (Equation 3.31). The Rouse prediction for blends, given by averages of Equation (3.35) over the component fractions, is also close to the data.

3.3.3.2 Model for Entangled Chains

Because of the success of the Rouse model for nonentangled polymers, this model has been utilized also as a starting model for entangled polymers. The entanglement results from the topological constraint for deeply interpenetrating coils of high-M chains and thus reflects, by nature, the interchain motional correlation. Nevertheless, the model most widely utilized so far is the tube model that treats these constraints as a dynamic mean field and represents them as an impenetrable tube surrounding a chain of our focus, as schematically illustrated in Figure 3.6b (see, e.g., Doi and Edwards, 1986; McLeish, 2002; Rubinstein and Colby, 2003; Watanabe, 1999, 2009).

The early versions of the tube model considered a Rouse chain trapped in a *spatially fixed* tube having a diameter $a = b_R n_e^{1/2}$, where n_e is the number of the Rouse segments per entanglement segment (of the molecular weight $M_{e,bulk}$), and b_R is the step length of the Rouse segment. In those fixed-tube models, the large-scale motion of the chain is allowed only in the direction of the tube axis and the relaxation occurs only for the entanglement segments that have escaped the initial tube (defined at time 0). For linear chains, this motion occurs only through two representative mechanisms illustrated in parts (b-1) and (b-2), the reptation equivalent to the curvilinear diffusion along the tube axis and the contour length fluctuation (CLF) equivalent to breathing of the chain contour through the Rouse modes along the tube axis. The viscoelastic data of entangled monodisperse linear chains, including the moduli data (such as those shown in Figure 3.3) and empirical $M^{3.5}$ dependence of the relaxation time, are reasonably described by the fixed-tube models combining the reptation and CLF mechanisms. However, these models cannot describe the behavior of the binary blends; for example, the low-ω plateau height for the blends examined in Figure 3.5 scales roughly as $\phi_2^{2.3}$ (for $\phi_2 > \phi_{2,ent}$), while the fixed-tube model predicts this height to scale as ϕ_2. In addition, for star branched chains, the fixed tube model gives an unrealistically long relaxation time, as pointed out by Ball and McLeish (1989).

Thus, the model has been refined through incorporation of the constraint release (CR) mechanism schematically illustrated in part (b-3) of Figure 3.6. In the CR process, a topological constraint for a given chain is removed transiently on the global motion of the tube-forming (entangling) chains, and the entanglement segment of the given chain at the release point is allowed to jump over a distance $>a$ in the direction lateral to the tube axis. This jump is equivalent to the local motion of the tube, and the CR mechanism introduces the interchain motional correlation into the tube model, though in a crude way. Because the local CR jump would happen with the same probability for all entanglement segments, the CR mechanism basically (though not exactly) accelerates the relaxation of the chain *uniformly* along the backbone. The global CR relaxation occurs through accumulation of such local CR-jumps and can be approximately modeled as the retarded Rouse relaxation. For the *dilute* long chains entangled only with the short matrix chains in the blends, the corresponding CR relaxation functions $G_{CR,2}(t)$ and $\Phi_{CR,2}(t)$ are given by Equation (3.35) with N therein being replaced by the entanglement number per long chain ($= M_2/M_{e,bulk}$), and $\tau_{R,G}$ ($= \tau_{R,\varepsilon}/2$), by the terminal viscoelastic CR time given below (Graessley, 1982; Watanabe, 1999):

$$\tau_{CR,G}^{[long]} = K(z)\left(\frac{M_2}{M_{e,bulk}}\right)^2 \tau_G^{[short]}M_1^{-0.5} \propto \zeta_s M_1^\alpha M_2^2 / M_{e,bulk}^{\alpha+0.5} \text{ with } \alpha \cong 3 \quad (3.37)$$

Here, $K(z)$ is a numerical factor determined by the local gate number z ($=$ number of chains sustaining each entanglement), as introduced by Graessley (1982). The relaxation time of the short chain, $\tau_G^{[short]}$ ($\propto M_1^{3.5(\pm 0.2)}/M_{e,bulk}^{1.5(\pm 0.2)}$) (cf. Equation 3.32), naturally appears in Equation (3.37) because the global motion of the short chain activates the local CR-jump of the dilute long chain. The number of short chains penetrating an entanglement segment decreases as $M_1^{-0.5}$, and the $M_1^{-0.5}$ factor in Equation (3.37) accounts for this decrease (cf. Klein, 1986; Watanabe, 1999). The original CR model accounted for neither this decrease nor CLF of the short chain (Graessley, 1982). The M_1 and M_2 dependence of τ deduced from the Rouse-CR model (Equation 3.37) agrees with the empirical observation (Equation 3.34) for the dilute long chains entangled with much shorter matrix chains. (Because $M_{e,bulk}$ does not change on blending, a delicate difference of the $M_{e,bulk}$ dependence seen between Equations (3.37) and (3.34) cannot be tested from the data for bulk blends. The data for blend solutions are still limited to make a conclusive test.) Furthermore, the Rouse mode distribution of $G_{CR,2}(t)$ is close (though not identical) to the distribution observed for those dilute long chains (see Watanabe, 1999, 2009; Watanabe et al., 2004b). Thus, the Rouse-like CR relaxation has been experimentally confirmed for the blends. This relaxation should occur also in monodisperse systems where all chains are equivalent to each other and the tube-forming chains are unequivocally mobile to the same extent as the chain trapped in the tube.

The accumulation of CR-jumps results in the chain motion in the direction lateral to the tube axis over a distance well above the tube diameter a ($\propto M_{e,bulk}^{1/2}$), as schematically shown at the bottom of part (b-3) of Figure 3.6. Thus, the CR mechanism dynamically dilates an effective tube diameter defined in a *coarse-grained* time scale, and the chain is regarded to be constrained in the dilated tube (supertube) in that time scale, as first pointed out by Marrucci (1985). A model based on this molecular picture was first proposed by Marrucci (1985) and later refined by Ball and McLeish (1989) and by Milner and McLeish (1998). Because this coarse-graining molecular picture is easily applicable to binary blends containing concentrated long chains (that are entangled among themselves as well as with the short chains), the CR process has been treated as the dynamic tube dilation (DTD) process in most of the molecular models so far proposed. The basic parameters for this DTD picture are the dilated tube diameter $a'(t)$ in a given time scale t and the fraction $\varphi'(t)$ of the *unrelaxed* entanglement segments (= the segments at the time t that are still trapped in a hypothetical dilated tube of the diameter $a'(t)$ defined at time 0). For binary blends, $\varphi'(t)$ is given by an average of the unrelaxed fractions $\varphi_j'(t)$ of the components therein: $\varphi'(t) = \phi_1\varphi_1'(t) + \phi_2\varphi_2'(t)$ with ϕ_j being the volume fraction of the component j.

The dilated tube diameter $a'(t)$ gives the number $\beta(t)$ of successive entanglement segments that are mutually equilibrated through their CR motion over the distance $a'(t)$: $\beta(t) = \{a'(t)/a\}^2$ (see, e.g., Watanabe, 1999). These $\beta(t)$ entanglement segments, each behaving as the independent entropic unit to sustain the stress at $t = 0$, are merged into an dilated segment on the mutual CR-equilibration to behave as one stress-sustaining unit in the time scale of t. Thus, the CR-equilibration results in a decrease of the number of the stress-sustaining unit by a factor of $1/\beta(t)$. Correspondingly, the DTD molecular picture leads to an expression of the relaxation modulus:

$$\text{DTD expression: } G(t) = G_N \frac{\varphi'(t)}{\beta(t)} = G_N\varphi'(t)\{a/a'(t)\}^2 \qquad (3.38)$$

where G_N is the entanglement plateau modulus (sustained by the undilated entanglement segments).

Most of the available CR-DTD models assume that the relaxed portions of the chains behave as a simple solvent (diluent) for the unrelaxed portions having the fraction $\varphi'(t)$, and a bulk system having this fraction is equivalent to a polymer solution of the concentration $\varphi'(t)$. If this assumption is valid, the entanglement mesh (tube) in the bulk system is *fully dilated* to that in the solution. This argument leads to the replacement of $a'(t)$ by the entanglement mesh size in the solution, $a_{sol} = a/\{\varphi'(t)\}^{d/2}$ with $d \cong 1.3$ (dilation exponent). For this *full-DTD* molecular picture, Equation (3.38) is simplified as

$$\text{full-DTD: } G(t) = G_N\{\varphi'(t)\}^{1+d} \qquad (3.39)$$

The time evolution (decay) of $\varphi'(t)$ has been calculated in several full-DTD models, and the corresponding $G(t)$ and complex modulus $G^*(\omega) = G'(\omega) + iG''(\omega)$ have been compared with the data (see Ball and McLeish, 1989; Marrucci, 1985; Milner and McLeish, 1998). Agreement of the model prediction and data was reported for monodisperse linear and star-branched polymer.

Nevertheless, we should remember that the viscoelastic relaxation captures just one aspect of the polymer dynamics—that is, the relaxation of isochronal orientational anisotropy of the entanglement segments (= sub-molecules adopted in long time scales). The DTD picture is established for the chain motion, and is to be tested also for nonviscoelastic relaxation processes. For this purpose, the dielectric data of type-A chains such as PI are very useful. The DTD process hardly changes the end-to-end vector of the chain, as can be easily noted from the illustration at the bottom of part (b-3) of Figure 3.6. Thus, the normalized dielectric relaxation function $\Phi(t)$ of type-A chains, detecting the end-to-end vector fluctuation (Equation 3.23), is essentially identical to the unrelaxed fraction of the entanglement segments $\varphi'(t)$, as pointed out/formulated by Watanabe et al. (2004a). This feature allows us to experimentally determine $\varphi'(t)$ from the $\Phi(t)$ data (after a very minor correction for the chain motion in the dilated tube edges) and test the full-DTD relationship, Equation (3.39), in a purely empirical way without relying on any model for calculation of $\varphi'(t)$. For monodisperse linear PI, this test has been made, and the validity of the full-DTD picture was confirmed (cf. Watanabe, 2009; Watanabe et al., 2004a). However, for PI/PI binary blends with small ϕ_2, Equation (3.39) does not describe the viscoelastic data: As an example, Figure 3.7 compares the prediction of (3.39) (dotted curves) with the data of the normalized relaxation modulus $G(t)/G_N$ (circles) for the PI308/PI21

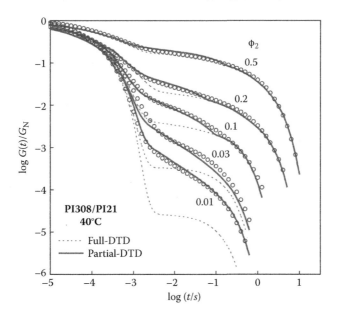

FIGURE 3.7

Comparison of the normalized relaxation modulus of PI308/PI21 blends at 40°C examined in Figure 3.5 with the prediction of full dynamic tube dilation (DTD) model (dotted curves) and partial-DTD model (solid curve). (Data taken, with permission, from Watanabe, H., S. Ishida, Y. Matsumiya, and T. Inoue. 2004b. Test of full and partial tube dilation pictures in entangled blends of linear polyisoprenes. *Macromolecules* 37:6619–6631.)

blends (cf. Watanabe et al., 2004b). (The G', G'', and ε'' data of these blends, summarized in Figure 3.5, were utilized to evaluate the $G(t)/G_N$ data shown in Figure 3.7 and $\varphi'(t)$ appearing in Equation 3.39.) Clearly, Equation (3.39) significantly underestimates the modulus data at intermediate to long time scales. This result, obtained without relying on any model calculation of $\varphi'(t)$, allows us to conclude that the full-DTD picture fails to consistently describe the viscoelastic and dielectric data of the PI/PI blends. The failure of the full-DTD picture was found also for monodisperse star-branched PI and Cayley-tree-type branched PI, as reported by Watanabe et al. (2002, 2008).

The failure of the full-DTD picture corresponds to the failure of the underlying assumption that the relaxed portions of the chains behave as a simple solvent and the relaxing bulk system is equivalent to a polymer solution of the concentration $\varphi'(t)$. For the full-DTD picture to be valid, the chain should quickly explore all available conformations over the distance $a'(t) = a\{\varphi'(t)\}^{-d/2}$ perpendicular to the tube axis, thereby allowing a sequence of $\beta(t)$ ($= \{a'(t)/a\}^2$) entanglement segments to behave as an enlarged stress-sustaining unit as a whole. However, this exploration requires a time for activating the CR jumps for all these $\beta(t)$ entanglement segments. Thus, if this CR time is longer than the time scale of our focus, t, the chain cannot explore all conformations in time, and the size of the effectively dilated segment $a_{eff}(t)$ becomes smaller than the size $a\{\varphi'(t)\}^{-d/2}$ assumed in the full-DTD picture. Watanabe et al. (2002, 2004a, 2008) found that $a_{eff}(t)$ evaluated from the available CR time data is smaller than $a\{\varphi'(t)\}^{-d/2}$ for the PI/PI blends and star/Cayley-tree PI in the range of t where the full-DTD prediction fails. (For monodisperse linear PI, $a_{eff}(t)$ was comparable/larger than $a\{\varphi'(t)\}^{-d/2}$ in the entire range of t, which resulted in the validity of the full-DTD picture.)

The DTD picture has been refined by Watanabe et al. (2004b) on the basis of the above results. This refined picture allows the tube to dilate only partially up to a diameter $a_{CR}(t)$ determined by the CR mechanism for the case of $a_{CR}(t) < a\{\varphi'(t)\}^{-d/2}$, and up to $a\{\varphi'(t)\}^{-d/2}$ in the opposite case. Within this *partial-DTD* picture, the dilated tube diameter $a'(t)$ appearing in the general DTD relationship (Equation 3.38) is given by

$$\text{partial-DTD: } a'(t) = \mathbf{min}[a\{\varphi'(t)\}^{d/2}, a_{CR}(t)] \ (d \cong 1.3) \tag{3.40}$$

The dielectrically obtained $\varphi'(t)$ data and the CR time data were utilized in Equation (3.40) to evaluate $a'(t)$, and the normalized modulus $G(t)/G_N$ deduced from the partial-DTD picture was calculated by utilizing this $a'(t)$ in Equation (3.38). The modulus calculated for the PI/PI blends (Watanabe et al., 2004b) is shown with the solid curves in Figure 3.7. These curves are in agreement with the data (circles), demonstrating the validity of the partial-DTD picture. The validity was confirmed also for the monodisperse star/Cayley-tree PI (see Watanabe et al., 2006, 2008).

The success of the partial-DTD picture demonstrates the importance of the consistency in the coarse-graining of the time and length scales. Namely, for a given time scale t, there is a maximum length scale for coarse-graining the segments into an *internally equilibrated* larger segment. For the DTD process, this equilibration occurs through the CR-Rouse mechanism. However, the importance of the equilibration is not limited to the DTD process but also noted for the Rouse relaxation process within the entanglement segment. The entanglement segment behaves as the entropic stress-sustaining unit only in a time scale where all monomeric units therein are mutually equilibrated through the Rouse mechanism. This point becomes key in the later discussion for the miscible blends of chemically different chains.

3.4 Local Dynamics in Miscible Blends

3.4.1 Self-Concentration

The glass transition of polymers occurs on thermal activation of local segmental motion of polymer chains. As a part of the study for the glass transition, the local dynamics in polymer blends has been extensively investigated with various experimental methods that include nuclear magnetic resonance, thermal, viscoelastic, and dielectric methods. In early studies, a blend was judged to be uniform and phase-separated, respectively, if it exhibited single/sharp and multiple/broad glass transition. This judgment is based on a classical molecular picture that the dynamic environment is the same for the monomeric segments of uniformly mixed polymer chains. In fact, this molecular picture is valid for many blend systems (Utracki, 1989).

Nevertheless, the above molecular picture fails for some miscible blends, as revealed from recent experiments. For example, blends of *cis*-polyisoprene (PI) and poly(vinyl ethylene) (PVE) shows the lower critical solution temperature (LCST) phase behavior, as confirmed from scattering experiments by Tomlin and Roland (1992). The segmental mobilities of PI and PVE in the blends, detected with ^{13}C-NMR (nuclear magnetic resonance) and ^{2}H-exchange NMR spectroscopy by Miller et al. (1990) and Chung et al. (1994), are different even in the miscible state at temperatures $T < T_{LCST}$. Correspondingly, a very broad, almost two-step glass transition emerges for the thermal data of the PI/PVE miscible blends, as reported by Chung et al. (1994) and Sakaguchi et al. (2005). (An example of differential scanning calorimetry [DSC] data for such broad glass transition is later shown in Figure 3.24.) These results demonstrate a difference of the dynamic frictional environment for the segments of different component chains in miscible blends. In other words, the component chains have different effective glass transition temperatures $T_{g,eff}$ in the miscible blends, as pointed out by Chung et al. (1994).

Because the torsional barrier for the segmental motion around the chain backbone is different for the chemically different component chains, these chains naturally exhibit a difference of their $T_{\mathrm{g,eff}}$ in the miscible blends. (This barrier should be determined by the chemical structure of the chain and also by the packing state of the segments in the blends.) However, $T_{\mathrm{g,eff}}$ is affected more significantly by the connectivity of the monomeric segments along the chain backbone. Chung et al. (1994) and Lodge and McLeish (2000) focused on the self-concentration that exclusively results from this connectivity: The segments neighboring along the chain backbone always occupy neighboring volume elements in the blends, thereby leading to the self-concentration (enrichment of their own concentration). Thus, in blends of polymers A and B, an *effective* volume fraction $\phi_{\mathrm{eff}}^{[X]}$ of the segment of component X (= A or B) is larger than a nominal volume fraction ϕ_X (macroscopic average in the whole blend), and the effective $T_{\mathrm{g,eff}}^{[X]}$ of the component X is determined by the effective $\phi_{\mathrm{eff}}^{[X]}$. Consequently, the components A and B exhibit respective glass transitions in A- and B-rich environments (characterized by the effective $\phi_{\mathrm{eff}}^{[A]}$ and $\phi_{\mathrm{eff}}^{[B]}$), thereby exhibiting different $T_{\mathrm{g,eff}}$ in the miscible blends.

The self-concentration, never occurring in homopolymer systems, is one of the most unique features of the miscible blends in the length scale of the monomeric units. The Lodge–McLeish model utilizes the self-concentrated volume fraction $\phi_{\mathrm{self}}^{[X]}$ and the nominal volume fraction ϕ_X of the segment X to express the effective $\phi_{\mathrm{eff}}^{[X]}$ of this segment as (Lodge and McLeish, 2000)

$$\phi_{\mathrm{eff}}^{[X]} = \phi_{\mathrm{self}}^{[X]} + \left(1 - \phi_{\mathrm{self}}^{[X]}\right)\phi_X \ \text{ with X = A, B} \tag{3.41}$$

$\phi_{\mathrm{self}}^{[X]}$ is expressed in terms of the characteristic ratio $C_{\infty}^{[X]}$ of the component chain X, the molecular weight $M_0^{[X]}$ and the number $k^{[X]}$ of the backbone bonds per repeating unit of this chain, a volume $V_K^{[X]}$ corresponding to the Kuhn length b_K ($V_K^{[X]} \sim b_K^3$), the mass density of the blend ρ, and the Avogadro constant N_A as

$$\phi_{\mathrm{self}}^{[X]} = \frac{C_{\infty}^{[X]} M_0^{[X]}}{k^{[X]} V_K^{[X]} \rho N_A} \ \text{ with X = A, B} \tag{3.42}$$

Combining the above expression of $\phi_{\mathrm{eff}}^{[X]}$ with an empirical equation that describes the *single* T_g° in hypothetical (or idealized) miscible blends exhibiting no self-concentration, we can express the effective $T_{\mathrm{g,eff}}^{[X]}$ as a function of the nominal volume fraction ϕ_X. The Fox equation given below is frequently utilized as this empirical equation.

$$\frac{1}{T_g^{\circ}} = \frac{\phi_A^{\circ}}{T_g^{\text{pure A}}} + \frac{\phi_B^{\circ}}{T_g^{\text{pure B}}} \tag{3.43}$$

Here, ϕ_A° and ϕ_B° (= $1 - \phi_A^{\circ}$) represent the volume fractions of the components A and B in the hypothetical blend and are not identical to the macroscopic ϕ_A and ϕ_B. From Equations (3.41) through (3.43), $T_{g,\text{eff}}^{[A]}$ and $T_{g,\text{eff}}^{[B]}$ of the components A and B subjected to the self-concentration effect are given by (Lodge and McLeish, 2000)

$$T_{g,\text{eff}}^{[A]} = T_g^\circ\left(\phi_A^\circ = \phi_{\text{eff}}^{[A]}\right), \quad T_{g,\text{eff}}^{[B]} = T_g^\circ\left(\phi_B^\circ = \phi_{\text{eff}}^{[B]}\right) \tag{3.44}$$

Equations (3.41) through (3.44) allow us to utilize the $T_g^{\text{pure X}}$ and ϕ_X data to evaluate $T_{g,\text{eff}}^{[X]}$ on the basis of the Lodge–McLeish model.

Experimentally, the $T_{g,\text{eff}}^{[X]}$ value can be estimated from the data of the relaxation time τ_X^* of the monomeric segment of the component X without relying on the Lodge–McLeish (or other) model but with an assumption for the temperature dependence of τ_X^* in the miscible blends: τ_X^* in the miscible blends is frequently assumed to obey the WLF equation (equivalent to the Vogel–Fulture–Tamman [VFT] equation; cf. Kremer and Schönhals, 2003) and related to $\tau_{X,r}^*$ at a reference state as $\log \tau_X^* = \log \tau_{X,r}^* + \log a_T$, with the shift factor a_T being described by Equation (3.28). Utilizing the WLF coefficients C_1^{WLF} and C_2^{WLF} of the bulk component X in Equation (3.28) and replacing the reference temperature T_r therein by the effective $T_{g,\text{eff}}^{[X]}(\phi)$ of this component in the blend, we may relate τ_X^* at two different temperatures as

$$\log \tau_X^*(T,\phi) = \log \tau_X^*\left(T_{g,\text{eff}}^{[X]}(\phi)\right) - \frac{C_1^{\text{WLF}}\left(T - T_{g,\text{eff}}^{[X]}(\phi)\right)}{C_2^{\text{WLF}} + T - T_{g,\text{eff}}^{[X]}(\phi)} \tag{3.45}$$

(Equation 3.45 reduces to the WLF equation for bulk if $T_{g,\text{eff}}^{[X]}$ is replaced by bulk $T_g^{[X]}$.)

As an example of the segmental relaxation behavior, Figure 3.8 shows the dielectric loss ε'' of miscible blends of poly(2-chlorostyrene) (P2CS450; $M_{\text{P2CS}} = 4.5 \times 10^5$) and poly(vinyl methyl ether) (PVME96; $M_{\text{PVME}} = 9.6 \times 10^4$) samples measured by Urakawa et al. (2001). The measurements were conducted at a fixed frequency of $f = 1$ kHz ($\omega = 6.28 \times 10^3$ s^{-1}) in a range of T as indicated. The monomeric units of P2CS and PVME have the type-B dipole perpendicular to the chain backbone (cf. Figure 3.2). Because P2CS and PVME have comparable magnitudes of their dipoles, their segmental relaxation (often referred to as the α-relaxation) is clearly detected as the ε'' peaks of comparable heights. The high-T and low-T peaks of the blends are assigned as the segmental relaxation of P2CS and PVME, respectively. It should be emphasized that P2CS and PVME exhibit separate segmental relaxation processes in the miscible state because of the self-concentration and intrachain torsional barrier explained earlier.

From the data shown in Figure 3.8 (and the other sets of data), the segmental relaxation time can be determined as $\tau_X^*(T) = \tau_{\varepsilon,\text{peak}} = 1/\omega_{\text{peak}}(T)$, where $\omega_{\text{peak}}(T)$ is the angular frequency for the ε'' peak at a given temperature T. Figure 3.9 shows T dependence of the τ_{P2CS}^* and τ_{PVME}^* data thus obtained.

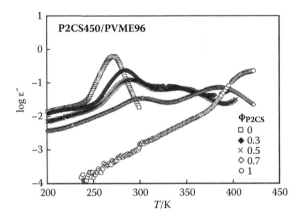

FIGURE 3.8
Dielectric loss ε'' of miscible blends of poly(2-chlorostyrene) (P2CS450; $M_{P2CS} = 4.5 \times 10^5$) and poly(vinyl methyl ether) (PVME96; $M_{PVME} = 9.6 \times 10^4$) with various P2CS volume fractions ϕ_{P2CS} measured at 1 kHz at various temperatures. (Data taken, with permission, from Urakawa, O., Y. Fuse, H. Hori, Q. Tran-Cong, and O. Yano. 2001. A dielectric study on the local dynamics of miscible polymer blends: Poly(2-chlorostyrene)/poly(vinyl methyl ether). *Polymer* 42:765–773.)

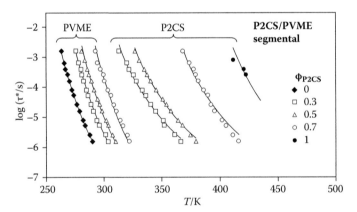

FIGURE 3.9
Temperature dependence of dielectrically determined relaxation times τ^* of monomeric segments of poly(2-chlorostyrene) (P2CS) and poly(vinyl methyl ether) (PVME) in P2CS/PVME miscible blends with various P2CS volume fractions ϕ_{P2CS} as indicated. (Data taken, with permission, from Urakawa, O., Y. Fuse, H. Hori, Q. Tran-Cong, and O. Yano. 2001. A dielectric study on the local dynamics of miscible polymer blends: Poly(2-chlorostyrene)/poly(vinyl methyl ether). *Polymer* 42:765–773.)

The solid curves indicate the results of fitting those data by the WLF equation, Equation (3.45), with the coefficients C_1^{WLF} and C_2^{WLF} being fixed at the bulk values ($C_1^{WLF} = 8.17$ and $C_2^{WLF} = 62$ K for P2CS; $C_1^{WLF} = 12.4$ and $C_2^{WLF} = 40.3$ K for PVME) and the effective $T_{g,eff}^{[X]}$ being treated as a fitting parameter. Reasonable fitting is achieved to give an empirical estimate of $T_{g,eff}^{[X]}$.

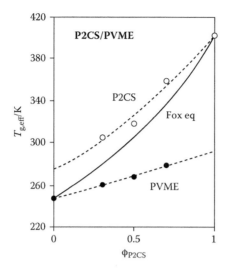

FIGURE 3.10
Changes of effective glass transition temperature of components in poly(2-chlorostyrene)/
poly(vinyl methyl ether) (P2CS/PVME) miscible blends with P2CS volume fraction ϕ_{P2CS}. (Data
taken, with permission, from Urakawa, O., Y. Fuse, H. Hori, Q. Tran-Cong, and O. Yano. 2001.
A dielectric study on the local dynamics of miscible polymer blends: Poly(2-chlorostyrene)/
poly(vinyl methyl ether). *Polymer* 42:765–773.)

Figure 3.10 shows plots of the estimated $T_{g,eff}^{[X]}$ against the (macroscopic)
volume fraction of P2CS, ϕ_{P2CS}. The solid curve represents the Fox equation,
Equation (3.43), with ϕ_X^o being replaced by the nominal ϕ_X (i.e., for the case of
the absence of the self-concentration effect). The $T_{g,eff}^{[X]}$ data (circles) deviate
from this curve significantly, demonstrating the importance of the self-concentration effect. These $T_{g,eff}^{[X]}$ data were fitted with Equation (3.44) combined
with Equations (3.41) and (3.43), with the self-concentrated volume fraction
$\phi_{self}^{[X]}$ being utilized as a fitting parameter. Good fitting was achieved with the
parameter values of $\phi_{self}^{[P2CS]} = 0.25$ and $\phi_{self}^{[PVME]} = 0.6$, as shown with the dashed
curves. This empirical value of $\phi_{self}^{[P2CS]}$ for P2CS is close to the theoretical
value, $\phi_{self}^{[P2CS]} = 0.22$, evaluated from Equation (3.42) utilizing separately deter-
mined/known molecular parameters, while the empirical $\phi_{self}^{[PVME]}$ for PVME
is considerably larger than the theoretical $\phi_{self}^{[PVME]}$ ($= 0.25$).

Thus, for the P2CS/PVME blends, the above result suggests not only the
basic validity of the concept of self-concentration incorporated in the Lodge–
McLeish model but also the limitation of the model (i.e., the necessity of
empirical adjustment of $\phi_{self}^{[X]}$). The basic validity and the limitation of this
model have been confirmed also for the other miscible blend systems (see,
e.g., Chung et al., 1994; Ediger et al., 2006; Hirose et al., 2003, 2004; Lutz et al.,
2004; Pathak et al., 1998, 1999; Urakawa, 2004; Zhao et al., 2008). In particular,
^{13}C-NMR experiments by Lutz et al. (2004) demonstrated that dilute low-M

cis-polyisoprene (PI) chains have different $\phi_{\text{self}}^{[\text{PI}]}$ values in miscible matrices of high-M poly(vinyl ethylene) (PVE), poly(1,4-butadiene) (PB), and polystyrene (PS), $\phi_{\text{self}}^{[\text{PI}]} = 0.41$ (in PVE), 0.85 (in PB), and 0.20 (in PS). The empirical $\phi_{\text{self}}^{[\text{PI}]}$ value in the PVE matrix is close to the prediction of Equation (3.42) ($\phi_{\text{self}}^{[\text{PI}]} = 0.45$), but those in the PB and PS matrices are considerably different from the prediction.

For an accurate description of the local dynamics of the monomeric segments and the corresponding glass transition, factors other than the self-concentration need to be considered. One of those factors is the concentration fluctuation explained later in detail (see Section 3.4.2.2). The remaining factors include the intrachain torsional barrier (as explained earlier) and the motional cooperativity of neighboring segments, the latter being discussed extensively for homopolymer systems (see, e.g., Kanaya and Kaji, 2001). Recently, this motional cooperativity was modeled for binary blends by Cangialosi et al. (2006) on the basis of the Adam-Gibbs theory (Adam and Gibbs, 1965). In addition, the size of the segment could change if the system density considerably changes on blending (Inoue et al., 2002), which may also enhance the deviation from the prediction of the Lodge–McLeish model. These factors deserve further investigation.

3.4.2 Length Scale of Segmental Dynamics

Motion of the monomeric segments unequivocally governs the glass transition phenomenon, but the size of the segment has been rarely specified. Hirose et al. (2003, 2004) and Urakawa (2004) conducted extensive dielectric studies for miscible blends of linear PI and PVE to challenge this problem. Features of the segmental motion revealed in their work are summarized below.

3.4.2.1 Analysis Utilizing Internal Reference of Length Scale

Both *cis*-polyisoprene (PI) and poly(vinyl ethylene) (PVE) have the type-B dipoles perpendicular to the chain backbone, and PI also has the type-A dipoles parallel along the backbone (cf. Figure 3.2). The dielectric relaxation detects the fluctuation of these dipoles, as explained in Section 3.2.2. The fluctuation of the type-B dipoles is activated by the fast, local motion of the monomeric segments, which enables the dielectric investigation of this motion. In contrast, the slow dielectric relaxation of PI due to the type-A dipoles exclusively detects the fluctuation of the end-to-end-vector **R** (see Equation 3.23). These dielectric features of PI and PVE are clearly noted in Figure 3.11, where the ε'' data are shown for a PI/PVE blend with the component molecular weights $M_{\text{PI}} = 1.2 \times 10^4$ and $M_{\text{PVE}} = 6 \times 10^4$ and the PI content $w_{\text{PI}} = 75$ wt% (Hirose et al., 2003). The data measured at different temperatures are converted to the master curve after the time-temperature superposition with the reference temperature of $T_r = -20°C$, as explained later in more detail. The three distinct dispersions seen at high, middle, and low

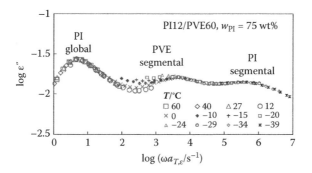

FIGURE 3.11

Test of time-temperature superposability for dielectric loss data of a miscible blend of *cis*-polyisoprene (PI12; $M_{PI} = 1.2 \times 10^4$) and poly(vinyl ethylene) (PVE60; $M_{PVE} = 6 \times 10^4$) with the PI content $w_{PI} = 75$ wt%. (Data taken, with permission, from Hirose, Y., O. Urakawa, and K. Adachi. 2003. Dielectric study on the heterogeneous dynamics of miscible polyisoprene/poly(vinyl ethylene) blends: Estimation of the relevant length scales for the segmental relaxation dynamics. *Macromolecules* 36:3699–3708.)

angular frequencies ω are assigned as the segmental relaxation of PI, the segmental relaxation of PVE, and the global relaxation of PI, respectively. The PVE chains have no type-A dipole and thus exhibit no slow dielectric dispersion reflecting their global motion. (However, the global motion of PVE was slower than that of PI and observed as the terminal viscoelastic relaxation of the blend as a whole.)

In the master curve shown in Figure 3.11, the ε'' data at various T (between −39°C and 60°C) were shifted along the ω axis by an appropriate factor $a_{T,\varepsilon}$ to achieve the best superposition in the segmental and global relaxation regimes for PI. This shift (for PI) gives very poor superposition in the segmental relaxation regime for PVE, which results from a difference of the effective $T_{g,eff}$ of the components in the blends: The WLF-type shift factor for the segmental relaxation, $a_{T,\varepsilon} = \tau^*(T)/\tau^*(T_r)$ with τ^* being the segmental relaxation time, is more strongly dependent on T for PVE (having high $T_{g,eff}$) than for PI, thereby violating the superposition for the blend as a whole.

The global and segmental relaxation processes of PI in the PI/PVE blends are simultaneously observed dielectrically (see Figure 3.11). The global relaxation of PI chains corresponds to their motion over the average end-to-end distance, $\langle R_{PI}^2 \rangle^{1/2}$ ($\propto M_{PI}^{1/2}$). Thus, with the aid of this $\langle R_{PI}^2 \rangle^{1/2}$ serving as an internal reference length, a length scale for the segmental relaxation of PI can be estimated from data of the segmental and global dielectric relaxation times, τ^* and τ. Specifically, for nonentangled PI chains, the global dynamics can be satisfactorily described by the Rouse model (cf. Section 3.3.3.1), and a length scale ξ of relaxation in a given time scale t can be specified by the Rouse relationship (in a continuous limit), $\xi^4 \propto t$. For several PI/PVE blends containing nonentangled PI, Hirose et al.

(2003) assumed the validity of this relationship in a range of ξ from the PI chain dimension $\langle R_{PI}^2 \rangle^{1/2}$ down to the segmental length scale ξ^* and estimated the ξ^* value from the τ^* and τ data and the $\langle R_{PI}^2 \rangle^{1/2}$ data as

$$\xi^* = \left(\frac{\tau^*}{\tau} \right)^{1/4} \langle R_{PI}^2 \rangle^{1/2} \qquad (3.46)$$

The resulting ξ^* ($= 0.9 \pm 0.04$ nm) was considerably close to the Kuhn segment length of PI, $b_K = 0.68$ nm.

For estimation of ξ^* for PVE in the PI/PVE blends, Hirose et al. (2003) utilized the effective $T_{g,eff}^{[PI]}$ and $T_{g,eff}^{[PVE]}$ of PI and PVE in the blends to reduce the segmental τ^* data of PI and PVE in the *isofrictional* state (where the PI and PVE segments have the same friction coefficient). They applied Equation (3.46) to those τ^* data to estimate $\xi^* = 1.8 \pm 0.3$ nm for PVE. This estimate is reasonably close to the Kuhn segment length of PVE, $b_K = 1.16$ nm.

Although the continuous Rouse relationship, $\xi^4 \propto t$, does not hold accurately at small length scales, it seems reasonable to conclude, from the above results based on this relationship, that the dynamically defined segmental length scale ξ^* is reasonably close to the statistically well-defined Kuhn segment length b_K, and the relaxation of the monomeric segments of PI and PVE (related to their separate glass transitions) occurs over different length scales.

3.4.2.2 Analysis Based on Mode Broadening

For further examination of the segmental length scale for PVE in the PI/PVE blends, Hirose et al. (2004) analyzed the broadness of the dielectric mode distribution observed as the ω dependence of the ε'' data. As an example, Figure 3.12 shows ω dependence of the ε'' data at 10°C measured for a PI/PVE blend with $M_{PI} = 1.2 \times 10^4$, $M_{PVE} = 6 \times 10^4$, and $w_{PI} = 17$ wt% (circles). Because of this small w_{PI} value, the relaxation seen in Figure 3.12 is almost exclusively attributed to the segmental motion of PVE in the blend. The dielectric segmental relaxation of bulk PVE is satisfactorily described by the Havriliak–Negami (HN) empirical equation (Hirose et al., 2004):

$$\varepsilon_{HN}''(\omega; \tau_{HN}) = -\mathrm{Im} \left\{ \frac{\Delta\varepsilon}{\left[1 + (i\omega\tau_{HN})^\alpha \right]^\beta} \right\} \text{ with } i = \sqrt{-1} \qquad (3.47)$$

where $\Delta\varepsilon$ ($= 0.108$) is the dielectric relaxation intensity, τ_{HN} is the HN-type nominal relaxation time, and α ($= 0.721$) and β ($= 0.391$) are the HN parameters for PVE. The HN equation does not describe the real terminal behavior, $\varepsilon'' \propto \omega$ at low ω (cf. Equation 3.19) but can be utilized for phenomenological

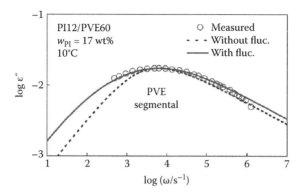

FIGURE 3.12
Dielectric loss ε'' measured at 10°C for a PI12/PVE60 miscible blend with $M_{PI} = 1.2 \times 10^4$, $M_{PVE} = 6 \times 10^4$, and $w_{PI} = 17$ wt% (circles). Dotted and solid curves, respectively, indicate the results of fitting with Havriliak–Negami empirical equation without and with the mode broadening due to the concentration fluctuation. (Data taken, with permission, from Hirose, Y., O. Urakawa, and K. Adachi. 2004. Dynamics in disordered block copolymers and miscible blends composed of poly(vinyl ethylene) and polyisoprene. *J. Polym. Sci. Part B: Polym. Phys.* 42:4084-4094.)

description of the intensive part of the dielectric relaxation around the ε'' peak. This ε_{HN}'' for bulk PVE, shifted in the double-logarithmic scale to match the ε'' peak frequency and height with the ε'' data of the blend, is shown in Figure 3.12 with the dotted curve. The dielectric relaxation mode distribution of the blend is broader than this HN curve, in particular at low ω.

Hirose et al. (2004) attributed this broadening to the concentration fluctuation, the characteristic dynamic feature of miscible blends (not observed for homopolymer systems). Namely, in the time scale of the segmental relaxation, the fluctuation of the local PVE volume fraction ϕ_{local} does not vanish, and ϕ_{local} varies in space around its average, ϕ_0. This distribution of ϕ_{local}, resulting in a spatial variation of the dynamic (frictional) environment for the PVE segments to broaden the mode distribution, may be described by the Zetsche–Fischer (ZF) model (1994) giving a Gaussian distribution function of ϕ_{local} around the average:

$$\psi(\phi_{local}) = \frac{1}{\sqrt{2\pi\sigma^2}} \exp\left(-\frac{\{\phi_{local} - \phi_0\}^2}{2\sigma^2}\right) \tag{3.48}$$

Here, σ^2 is the variance; $\sigma^2 = \langle(\phi_{local} - \phi_0)^2\rangle$. Considering this distribution, Hirose et al. (2004) expressed ε'' for the segmental relaxation of PVE averaged over the distribution function $\psi(\phi_{local})$:

$$\varepsilon''(\omega) = \frac{\phi_{PVE} \int_0^1 \varepsilon_{HN}''\left(\omega; \tau_{HN}(\phi_{local})\right) \psi(\phi_{local}) \, d\phi_{local}}{\int_0^1 \psi(\phi_{local}) \, d\phi_{local}} \qquad (3.49)$$

In Equation (3.49), $\varepsilon_{HN}''(\omega; \tau_{HN})$ represents the HN function (Equation 3.47) describing the intensive part of the ε'' data of bulk PVE. Namely, Equation (3.49) assumes that the relaxation mode distribution for the PVE segments in a given local frictional environment (characterized by ϕ_{local}) is identical to that for the segments in bulk. In addition, the characteristic time $\tau_{HN}(\phi_{local})$ in this environment is assumed to be dependent only on ϕ_{local}.

For further analysis, Hirose et al. (2004) replaced this ϕ_{local} dependence of τ_{HN} in the blend by the dependence on the macroscopic ϕ_{PVE} observed for the PVE segmental relaxation time τ_{PVE}^* ($= 1/\omega_{peak}$) in PVE-rich PI/PVE blends. The local volume fraction ϕ_{local} determining this τ_{PVE}^* does not rigorously coincide with the macroscopic ϕ_{PVE} but should be close to the effective $\phi_{eff}^{[PVE]}$ discussed earlier. However, a difference between ϕ_{PVE} and $\phi_{eff}^{[PVE]}$ was rather small in the PVE-rich blends. For this reason, the above replacement can be made as the first approximation. For the same reason, the average ϕ_o appearing in Equation (3.48) can be approximated as the macroscopic ϕ_{PVE}.

Under these assumptions and approximations, Hirose et al. (2004) fitted the ε'' data of the PI/PVE blends with Equations (3.48) and (3.49). (Actually, the fitting was made after conversion of ψ appearing in Equation 3.49 into a distribution function of τ_{HN}.) In Figure 3.12, the solid curve indicates the best-fit result obtained with the variance value of $\sigma^2 = 0.013$. Hirose et al. (2004) converted this variance value into a characteristic length of the concentration fluctuation, r_a, through a general expression for scattering from Gaussian chains exhibiting the concentration fluctuation (Zetsche and Fischer, 1994):

$$\sigma^2 = \left\langle \left(\phi_{local} - \phi_o \right)^2 \right\rangle = \frac{b^3}{2\pi^2 r_a^2} \int_0^\infty \left\{ \frac{3(\sin \, qr_a - qr_a \cos \, qr_a)}{q^2 r_a^2} \right\}^2 S(q) \, dq \quad (3.50)$$

Here, b, q, and $S(q)$ denote the monomer bond length, the magnitude of scattering vector, and the structural factor, respectively. Utilizing an expression of $S(q)$ based on the random phase approximation (known to be valid for miscible blends) (de Gennes, 1979), Hirose et al. (2004) evaluated r_a from the σ^2 value. The result, $r_a = 1.6$ to 1.7 nm, agreed with the segmental length scale dielectrically estimated with the aid of Equation (3.46), $\xi^* = 1.8 \pm 0.3$ nm for PVE, and was reasonably close to the Kuhn segment length of PVE, $b_K = 1.16$ nm. This result lends further support to the molecular picture that the dynamically defined segmental length scale ξ^* is reasonably close to the statistical Kuhn segment length b_K.

Here, some comments need to be added for the results of the above analysis. Zetsche and Fischer (1994) examined the segmental relaxation of the PVME component in PS/PVME blends inside the miscible region. The PVME segmental relaxation was easily detected with the dielectric method (because the dipole is much larger for polar PVME than for nonpolar PS). They analyzed the data on the basis of Equation (3.50) to obtain the characteristic length of the concentration fluctuation $r_a \cong 7$ nm (at T_g). This r_a is much larger than the Kuhn segment length, $b_K = 1.16$ nm for PVME. This result differs from the results by Hirose et al. ($r_a \cong b_K$) explained above. This discrepancy could be partly due to the assumption of a common, single T_g for the PS and PVME components in the PS/PVME blends incorporated in the analysis by Zetsche and Fischer (1994). This assumption broadens the distribution function $\psi(\phi_{local})$ that matches the ε'' data and might have resulted in overestimation of r_a.

For quantitative description of the segmental dynamics of the components in the miscible blends, both concentration fluctuation and self-concentration effects should be consistently incorporated in the model. This approach has already been examined by Kumar et al. (1996), Kamath et al. (1999), and Colby and Lipson (2005). In particular, the model by Colby and Lipson adopted the r_a value (~1 nm) *comparable to* the Kuhn segment length to reasonably describe the segmental relaxation time and mode distribution for both PI and PVE components in PI/PVE blends.

3.5 Global Dynamics in Miscible Blends

For homopolymer systems, the global motion of polymer chains and the resulting terminal relaxation and flow behavior have been studied extensively, as explained in earlier sections. In particular, for entangled linear homopolymers, the mechanisms of the chain motion/relaxation such as reptation, contour length fluctuation (CLF), constraint release (CR), and dynamic tube dilation (DTD) have been established and an effect of combination of these mechanisms has been understood to a considerable depth. The global dynamics in miscible blends of chemically different polymers would be basically understood within the framework of this knowledge for homopolymers. Nevertheless, there are quite important differences between the homopolymer systems and blends. First, the effective $T_{g,eff}$ determining the frictional environment is different for different components in the blends, as explained earlier. In addition, the entanglement length a (= tube diameter discussed earlier) could be different for the components in their bulk state and in the blend. These differences would affect the global dynamics of the component chains in the blends. In particular, the frictional difference of the component chains could result in an interesting situation that one component being less entangled but having a larger friction relaxes slower than the other component. Interesting dynamic coupling of the component chains is expected for such cases.

This section is devoted to discussion of the characteristic features of the miscible blends that include the coupling of the global dynamics of the dynamically asymmetric component chains. We first focus on miscible blends of *cis*-polyisoprene (PI) and poly(vinyl ethylene) (PVE) having a moderate dynamic asymmetry of the components. The PI/PVE blends demonstrate basic similarities and differences between the miscible blends and homopolymer systems. Then, we turn our attention to miscible blends of PI and poly(*p-tert*-butyl styrene) (PtBS) exhibiting the strongest dynamic asymmetry among the miscible blends so far investigated. The effect of the dynamic asymmetry on the global relaxation of the components is most clearly examined for these blends.

The segmental dynamics is strongly affected by the local heterogeneity of the composition in the length scale of the monomeric segment, and the effective volume fraction $\phi_{eff}^{[X]}$, not the macroscopically averaged volume fraction ϕ_X, determines the segmental dynamics of the component X, as explained in the previous section. In contrast, such very local heterogeneity is smeared for the global dynamics. Thus, the global dynamics is determined by the friction ζ_s of the Rouse segment (= smallest motional unit for the global dynamics), the molecular weight M_X of the component X, and the macroscopic ϕ_X of the component. (ζ_s is basically determined by $T_{g,eff}$.) For this reason, the effective $\phi_{eff}^{[X]}$ defined for the monomeric segment is not explicitly incorporated in the explanation/discussion below.

3.5.1 Global Relaxation in Polyisoprene/Poly(vinyl ethylene) (PI/PVE) Blend

3.5.1.1 Thermorheological Behavior of PI/PVE Blend

The PI/PVE blend is one of the most extensively studied miscible blends (see, e.g., Arendt et al., 1997; Chung et al., 1994; Haley and Lodge, 2005; Haley et al., 2003; Hirose et al., 2003, 2004; Miller et al., 1990; Pathak et al., 2004; Sakaguchi et al., 2005; Tomlin and Roland, 1992; Urakawa, 2004; Zawada et al., 1994a, 1994b). The entanglement lengths a and T_g of PI and PVE in respective bulk states are

$$\text{bulk PI: } a \cong 5.8 \text{ nm}, T_g \cong -70°C \tag{3.51}$$

$$\text{bulk PVE: } a \cong 5.1 \text{ nm}, T_g \cong 0°C \tag{3.52}$$

The entanglement length is similar for bulk PI and PVE and thus remains similar also in PI/PVE blends (as explained later in more detail). However, the effective $T_{g,eff}$ is different for PI and PVE in the blends (higher for PVE)

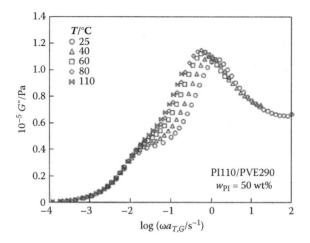

FIGURE 3.13

Loss modulus G'' of a PI110/PVE290 miscible blend with $M_{PI} = 1.1 \times 10^5$, $M_{PVE} = 2.9 \times 10^5$, and polyisoprene (PI) content of $w_{PI} = 50$ wt% measured at various temperatures as indicated. The data are shifted along ω axis to test the time-temperature superposability. (Data taken, with permission, from Arendt, B. H., R.. Krishnamoorti, J. A. Kornfield, and S. D. Smith. 1997. Component dynamics in miscible blends: Equally and unequally entangled polyisoprene/polyvinylethylene. *Macromolecules* 30:1127–1137.)

mostly because of the self-concentration effect explained earlier. (PI is anti-plasticized by PVE while PVE is plasticized by PI on blending.) This difference of $T_{g,eff}$ affects the global relaxation of the PI/PVE blends.

As an example demonstrating this effect, Figure 3.13 shows the loss modulus G'' of a blend of PI110 ($M_{PI} = 1.1 \times 10^5$) and PVE290 ($M_{PVE} = 2.9 \times 10^5$) samples with the PI content of $w_{PI} = 50$ wt% reported by Arendt et al. (1997). In the blend, PI110 and PVE290 behave as the fast and slow components entangled with each other, as can be expected from the difference of $T_{g,eff}$ and the similarity of a explained above and also from the M_{PI} and M_{PVE} values. The G'' data measured at various temperatures $T = 25$ to $110°C$ are semilogarithmically plotted against the angular frequency ω and further shifted along the ω axis to achieve the best superposition at low ω. This superposition is equivalent to the superposition of the modulus of the slow component, PVE290, and the shift factor $a_{T,G}$ reflects a change of the friction coefficient of PVE290 with T. The high-ω peak of G'', reflecting the global relaxation of PI110 (and a partial relaxation of PVE290 activated by the PI relaxation through the CR/DTD mechanism), is not superposed with the $a_{T,G}$ factor for PVE but shifts to lower $\omega a_{T,G}$ with increasing T. Thus, the increase of T accelerates the global relaxation less significantly for PI110 than for PVE290, which can be naturally understood from the friction coefficient of the Rouse segment that exhibits the WLF-type T dependence (cf. Equation 3.28) with $T_{g,eff}$ being utilized as the reference temperature T_r.

3.5.1.2 Component Dynamics in PI/PVE Blend

The complex modulus G^* ($= G' + iG''$) of the binary blends can be analyzed with the aid of the rheo-optical data. For homopolymers, the stress-optical coefficient C does not depend on the molecular weight and the molecular weight distribution. Thus, G^* and the complex birefringence coefficient K^* ($= K' + iK''$) of the binary blend of *chemically identical* polymers always satisfy the well-known stress-optical rule in the rubbery-to-flow zone:

$$K^*(\omega) = CG^*(\omega) \qquad (3.53)$$

Those $G^*(\omega)$ and $K^*(\omega)$ are contributed from the components in the blend and can be expressed as

$$G^*(\omega) = G_1^*(\omega) + G_2^*(\omega) \qquad (3.54)$$

$$K^*(\omega) = K_1^*(\omega) + K_2^*(\omega) \qquad (3.55)$$

Here, $G_i^*(\omega)$ and $K_i^*(\omega)$ are the *bare* modulus and birefringence coefficient of the component i not normalized by the volume fraction ϕ_i of this component, with the subscripts $i = 1$ and 2 standing for the short- and long-chain components, respectively. As clearly noted from Equation (3.53), we cannot resolve the component moduli just from the $G^*(\omega)$ and $K^*(\omega)$ data of the blend. For this resolution, one component needs to be labeled for rheo-optical distinction from the other component. In fact, rheodichroism tests have been made for blends of chemically identical but deuterated and protonated component polymers to determine $G_i^*(\omega)$ and $K_i^*(\omega)$, the latter being evaluated from the dynamic dichroism signal (see, e.g., Ekanayake et al., 2002; Kornfield et al., 1989; Ylitalo et al., 1991a, 1991b). These tests revealed that Equation (3.53) is not valid for $G_i^*(\omega)$ and $K_i^*(\omega)$ of respective components because the monomeric segments of the components are orientationally coupled with each other. Specifically, $K_1^*(\omega)$ of the short-chain component was found to just partially relax in a range of ω where its $G_1^*(\omega)$ has fully relaxed. These results indicate that a monomeric segment tends to be locally oriented in the same direction as the surrounding segments because of the nematic coupling due to the entropic packing effect, as first pointed out by Kornfield et al. (1989).

The birefringence and dichroism detect the orientational anisotropy summed for all monomeric segments. In contrast, the modulus in the rubbery relaxation/flow zone reflects the orientational anisotropy of much larger units such as the entanglement segments. The nematic coupling between the monomeric segments is smeared in the length scale of those large segments to negligibly change the modulus and hardly affect the global dynamics of the chains, as deduced from theoretical analysis by Watanabe et al. (1991) and by Doi and Watanabe (1991). The corresponding relationship between $G_i^*(\omega)$ and $K_i^*(\omega)$ was theoretically derived by Doi et al. (1989):

$$K_2^*(\omega) = C\{(1 - \theta\phi_1)G_2^*(\omega) + \theta\phi_2 G_1^*(\omega)\} \tag{3.56}$$

$$K_1^*(\omega) = C\{\theta\phi_1 G_2^*(\omega) + (1 - \theta\phi_2)G_1^*(\omega)\} \tag{3.57}$$

Here, θ is the nematic coupling constant between the monomeric segments of the components. From Equation (3.57), we note that $K_1^*(\omega)$ of the short-chain component does not fully relax even in the range of ω where its modulus $G_1^*(\omega)$ has fully relaxed but the modulus $G_2^*(\omega)$ of the long-chain component remains unrelaxed. This theoretical prediction is consistent with experiments. We can also confirm that Equations (3.56) and (3.57) give $K^*(\omega) = K_1^*(\omega) + K_2^*(\omega) = C\{G_1^*(\omega) + G_2^*(\omega)\}$ and are consistent with the stress-optical rule, Equation (3.53).

For miscible blends of chemically different chains A and B, the above formulation for $K^*(\omega)$ is modified by Zawada et al. (1994b) and Arendt et al. (1997):

$$K^*(\omega) = C_A[G_A^*(\omega) - \theta\{\phi_A G_B^*(\omega) - \phi_B G_A^*(\omega)\}]$$

$$+ C_B[G_B^*(\omega) - \theta\{\phi_B G_A^*(\omega) - \phi_A G_B^*(\omega)\}] \tag{3.58}$$

Here, C_A and C_B are the stress-optical coefficients of the components A and B in respective bulk systems, and θ denotes the coupling constant averaged for all monomeric units of the components. In general, C_A is different from C_B so that $K^*(\omega)$ and $G^*(\omega)$ $(= G_A^*(\omega) + G_B^*(\omega))$ of the A/B miscible blends are not proportional to each other. Thus, for those blends, no specific optical labeling is required for determination of the component moduli $G_i^*(\omega)$: We can just apply Equations (3.58) and (3.54) (with 1 = A and 2 = B) to the $K^*(\omega)$ and $G^*(\omega)$ data of the A/B blend to determine $G_i^*(\omega)$, given that the average coupling coefficient θ is known. In general, θ is unknown a priori. However, for the PI110/PVE290 blend examined in Figure 3.13, Arendt et al. (1997) analyzed the $K^*(\omega)$ and $G^*(\omega)$ data to demonstrate that the θ value should be in a very narrow range for reproducing the full relaxation of $G_{PI}^*(\omega)$ of PI110 (fast component) in the high-ω zone around the G'' peak frequency, ω_{peak}. As an example, Figure 3.14 shows the results of their analysis for the blend at 60°C (Arendt et al., 1997). The normalized storage modulus of PI110, $\phi_{PI}^{-1}G_{PI}'$, evaluated for $\theta = 0.34$ exhibits the terminal tail ($G' \propto \omega^2$; cf. Equation 3.7) in the expected range of ω, but a slightly smaller θ (= 0.33) gives an artificial relaxation tail at low ω ($\ll \omega_{peak}$), whereas a slightly larger θ (= 0.35) results in physically unreasonable ω dependence of the modulus (stronger than the asymptotic proportionality to ω^2). The normalized modulus of PVE290 (slow component) hardly changes in this narrow range of θ. Thus, the θ value and the component moduli are simultaneously determined with negligibly small uncertainties.

Figure 3.15 examines the time-temperature superposability of the component moduli in the PI110/PVE290 blend determined with the above method (Arendt et al. 1997). The reference temperature is $T_r = 25°C$. The shift factor $a_{T,G}$ was separately chosen for PI110 and PVE290 to achieve the best superposition

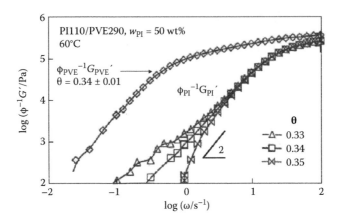

FIGURE 3.14

Normalized storage moduli of polyisoprene (PI) and poly(vinyl ethylene) (PVE) components in the PI110/PVE290 miscible blend (w_{PI} = 50 wt%) obtained from rheo-optical analysis with the nematic coupling constant θ being varied in a narrow range between 0.33 and 0.35. Physically reasonable moduli were obtained for θ = 0.34 but not for θ = 0.33 and 0.35. (Data taken, with permission, from Arendt, B. H., R.. Krishnamoorti, J. A. Kornfield, and S. D. Smith. 1997. Component dynamics in miscible blends: Equally and unequally entangled polyisoprene/polyvinylethylene. *Macromolecules* 30:1127–1137.)

for respective components. This factor was of the WLF type and more strongly dependent on T for PVE290 than for PI110, which reflects a difference of the effective $T_{g,eff}$ (higher for PVE290) (see Equation 3.28 with $T_r = T_{g,eff}$).

As noted in the top panel of Figure 3.15, the normalized $\phi_{PI}^{-1} G_{PI}^*$ data of PI110 are well superposed within the accuracy of the above analysis. This result suggests that the relaxation mechanism of PI110 hardly changes with T. Because PI110 is the fast component in the blend and entangled with the slow component, PVE290, the reptation and CLF mechanism explained earlier (cf. Figure 3.6) would have dominated the relaxation of PI110 to give the validity of the superposition. (For the PI110 chains entangling with the slow PVE290 chains, the CR mechanism should have suppressed compared to that in bulk PI110 system.) In fact, a good superposition is achieved also for the dielectrically detected end-to-end vector fluctuation process of the other PI chains entangled with slower PVE chains; see the low-ω dielectric data in Figure 3.11. Thus, the relaxation behavior of fast PI entangled with slow PVE can be understood within the molecular picture established for the chemically uniform PI/PI blends.

In contrast, for PVE290 in the blend examined in the bottom panel of Figure 3.15, the superposition of $\phi_{PVE}^{-1} G_{PVE}^*$ is valid at low ω (where the fast PI110 has relaxed) but not at high ω (where PI110 has not relaxed). Specifically, the fast relaxation of PVE290 is noted as the up-turn of the $\phi_{PVE}^{-1} G_{PVE}''$ curves at high ω, and this fast relaxation shifts toward the terminal relaxation of PVE290 with increasing T to violate the superposition. This behavior can

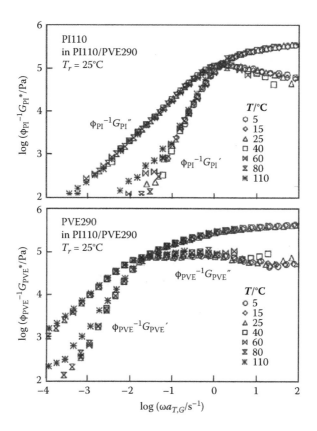

FIGURE 3.15

Test of the time-temperature superposability for the component moduli in the PI110/PVE290 miscible blend (w_{PI} = 50 wt%). The reference temperature is T_r = 25°C, and the rheo-optically obtained normalized moduli of the components are shifted along the ω axis to achieve the best superposition. (Data taken, with permission, from Arendt, B. H., R.. Krishnamoorti, J. A. Kornfield, and S. D. Smith. 1997. Component dynamics in miscible blends: Equally and unequally entangled polyisoprene/polyvinylethylene. *Macromolecules* 30:1127–1137.)

be again understood with the aid of the molecular picture for the chemically uniform PI/PI blends: The PVE290 chains would have relaxed partly at high ω through the CR mechanism activated by the global motion of PI110, as similar to the behavior of a high-M component in PI/PI blends. Thus, the relaxation time of this high-ω process is dominated by the motion of the PI110 chain, and the T dependence of this time is essentially determined by $T_{g,eff}$ of PI110. In contrast, the T dependence of the terminal relaxation time of PVE290 (possibly occurring through the reptation and CLF mechanisms) is determined by $T_{g,eff}$ of PVE290. Because $T_{g,eff}$ is higher for PVE290 than for PI110, the spacing between these two times decreases with increasing T, which naturally results in the observed failure of the superposition. It should be noted that the difference of $T_{g,eff}$ is a unique feature of the chemically heterogeneous miscible blends, but the relaxation mechanisms of the components therein are still similar to those in the chemically uniform PI/PI blends. In this sense, the dynamic asymmetry in the PI/PVE blends is not strong enough to activate a relaxation mechanism absent in the PI/PI blends.

3.5.2 Global Relaxation in Polyisoprene/Poly(*p-tert* butyl styrene) (PI/PtBS) Blends

Poly(*p-tert* butyl styrene) (PtBS) and *cis*-polyisoprene (PI) are miscible in surprisingly wide ranges of temperature and composition despite a huge difference in their chemical structures, as found by Yurekli and Krishnamoorti (2004). Because PI has the type-A dipole while PtBS does not, the slow dielectric response of PI/PtBS blends exclusively detects the global dynamics of the PI chains therein, which enables us to examine the dynamics of the PI component without theoretical assumptions.

The entanglement length a and the glass transition temperature T_g of bulk PtBS, summarized in Equation (3.59), are significantly larger and higher, respectively, than those of PI (Equation 3.51):

$$\text{bulk PtBS: } a = 11.7 \text{ nm}, \ T_g \cong 150°C \tag{3.59}$$

Specifically, the large difference of the entanglement lengths of PI and PtBS allows us to examine the effect of blending on this length, and the large difference of bulk T_g results in a large dynamic asymmetry of PI and PtBS. The large asymmetry allows the PtBS chains to behave as effectively immobilized obstacles during the global relaxation of PI, thereby strongly affecting (retarding) the PI relaxation. The following sections explain the component dynamics in the PI/PtBS blends and examine these effects.

3.5.2.1 Overview of Entanglement Relaxation in High-M PI/PtBS Blend

As an example of the dynamic behavior of PI/PtBS blends having a large dynamic asymmetry, Figure 3.16 shows the data of the storage and loss moduli, G' and G'', and the dielectric loss, ε'', measured for a blend of PI99 ($M_{PI} = 9.9 \times 10^4$) and PtBS348 ($M_{PtBS} = 3.5 \times 10^5$) with $w_{PI} = 50$ wt% (Watanabe et al., 2011). These high-M components, having $M_{PI}/M_{e,bulk}^{[PI]} \cong 20$ and $M_{PtBS}/M_{e,bulk}^{[PtBS]} \cong 9$ in respective bulk systems, are well entangled with each other also in the blend. Only the data at representative T are shown for clarity of the plots, and the ε'' data are multiplied by a factor of 10^4. At all T examined, the blends exhibit two-step viscoelastic relaxation, and the second step (terminal) relaxation is much slower than the dielectric relaxation exclusively detecting the global motion of the type-A PI99 chains. In addition, the first step viscoelastic relaxation occurs in the same range of ω as the dielectric relaxation. These results indicate that the PI99 and PtBS348 chains behave as the fast and slow components in the blend at the temperatures examined.

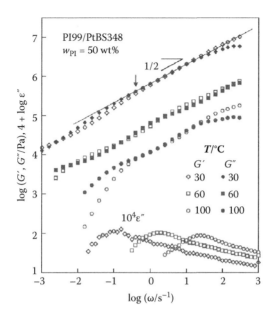

FIGURE 3.16
Storage and loss moduli, G' and G'', and the dielectric loss, ε'', of a high-M PI99/PtBS348 miscible blend ($M_{PI} = 9.9 \times 10^4$, $M_{PtBS} = 3.5 \times 10^5$, $w_{PI} = 50$ wt%) measured at temperatures as indicated. The ε'' data are multiplied by a factor of 10^4. (Data taken, with permission, from Watanabe, H., Q. Chen, Y. Kawasaki, Y. Matsumiya, T. Inoue, and O. Urakawa, 2011. Entanglement dynamics in miscible polyisoprene/poly(*p-tert*-butylstyrene) blends. *Macromolecules* 44:1570–1584.)

The two-step viscoelastic relaxation seen for the high-M PI99/PtBS348 blend is qualitatively similar to that of entangled PI/PI blends (cf. Figure 3.5). However, the viscoelastic data of the PI99/PtBS348 blend do not satisfy the time-temperature superposition, as demonstrated in Figure 3.17 where the data are subjected to a minor intensity correction (reduction by the factor of $b_T = T/T_r$; cf. Equation 3.27) and shifted along the ω axis by a factor $a_{T,G}$ to achieve the best superposition of the b_T^{-1} G'' data at $\omega a_{T,G} \cong 10^{-3}$ s^{-1}. This shift gives branches of the G' and G'' curves at high ω, clearly demonstrating the failure of the superposition. This failure is partly attributable to a difference of the shift factors for the components, PI99 and PtBS348: Both PI99 and PtBS348 contribute to the viscoelastic data at high ω, while PtBS348 dominates the data at low ω (where PI99 has already relaxed). In fact, the shift factor $a_{T,\varepsilon}$ for PI99 evaluated from the superposition of the dielectric $\Delta\varepsilon'$ and ε'' data is much less dependent on T compared to the viscoelastic $a_{T,G}$ that reflects, at low T, the behavior of PtBS348; see the inset in Figure 3.17.

The above mechanism of the failure of the superposition has also been noted for the entangled PI/PVE blends having just a moderate dynamic asymmetry of the components (cf. Figure 3.13). However, Figure 3.17 demonstrates that the superposition fails for the viscoelastic data of the PI99/PtBS348 blend not only at high ω but also at low ω. The failure at low ω

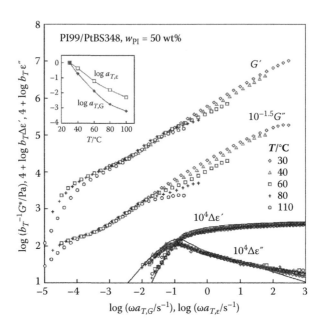

FIGURE 3.17 (See color insert.)
Test of time-temperature superposability for the PI99/PtBS348 miscible blend examined in Figure 3.16. The viscoelastic and dielectric data of the blend were separately shifted by different factors shown in the inset. The superposition is valid for the dielectric data but not for the viscoelastic data. (Data taken, with permission, from Watanabe, H., Q. Chen, Y. Kawasaki, Y. Matsumiya, T. Inoue, and O. Urakawa. 2011. Entanglement dynamics in miscible polyisoprene/poly(*p-tert*-butylstyrene) blends. *Macromolecules* 44:1570–1584.)

suggests that the global dynamics of the slow component (PtBS348) changes with T. This change, not seen for the PI/PVE blend (cf. low-ω data in the bottom panel of Figure 3.15), is the other mechanism for the failure of the superposition for the PI99/PtBS348 blend.

As noted in Figure 3.16, the PI99/PtBS348 blend exhibits a plateau of G' and a peak of G'' at high T (100°C) in a range of $\omega > 20$ s^{-1} where the global relaxation of PI is dielectrically observed. Similar high-ω plateau, being attributed to the entanglement among all component chains, has been noted for the PI/PVE and PI/PI blends. In contrast, at low T (30°C), neither G'-plateau nor G''-peak is observed for the PI99/PtBS348 blend in the PI relaxation zone at $\omega > 0.1$ s^{-1}. Instead, the G' and G'' data at low T exhibit the Rouse-type power-law behavior, $G' = G'' \propto \omega^{1/2}$ (solid line), and increase monotonically *beyond* the levels of G'-plateau/G''-peak seen at high T. (This power-law behavior is deduced from Equation 3.35 at t well below $\tau_{R,G}$, as explained later in more detail.) The disappearance of the high-ω entanglement plateau at low T, not seen for the PI/PVE and PI/PI blends, is a remarkable feature of the PI/PtBS blend having a very large dynamic asymmetry of the components.

The disappearance explained above suggests a significant change of the global relaxation mechanism of PI99 (fast component in the blend) with *T*. This change of the relaxation mechanism largely contributes to the failure of the time-temperature superposition of the viscoelastic data of the PI99/PtBS348 blend. In relation to this point, we note in Figure 3.17 that the superposition satisfactorily works for the dielectric $\Delta\varepsilon'$ and ε'' data of PI99 (with the shift factor $a_{T,\varepsilon}$ shown in the inset). Thus, the relaxation mechanisms of PI99 at high and low *T* are different but should be still associated with the same dielectric mode distribution (the mode distribution of the end-to-end vector fluctuation; cf. Equation 3.23). This point is discussed later in more detail.

3.5.2.2 Entanglement Length in High-M PI/PtBS Blend

Because the entanglement length *a* is considerably different for bulk PI (Equation 3.51) and PtBS (Equation 3.59), the effect of blending on this length can be clearly examined for the PI99/PtBS348 blend. The viscoelastic data of this blend in the high-ω plateau zone at high *T*, reflecting the entanglements among all component chains, enable a quantitative test of several mixing rules of *a* and the entanglement molecular weight M_e ($\propto a^2$) reported in literature:

$$\frac{1}{a} = \frac{\phi_A}{a_A} + \frac{\phi_B}{a_B} \quad \text{(Pathak et al., 2004)} \qquad (3.60)$$

$$a = n_A a_A + n_B a_B \quad \text{(Chen et al., 2008; Watanabe et al., 2011)} \qquad (3.61)$$

$$\frac{1}{M_e^{1/2}} = \frac{\phi_A}{\left\{M_{e,\text{bulk A}}\right\}^{1/2}} + \frac{\phi_B}{\left\{M_{e,\text{bulk B}}\right\}^{1/2}} \quad \text{(Haley and Lodge, 2005)} \qquad (3.62)$$

In Equations (3.60) and (3.61), a_X is the entanglement length in the *bulk system* of the component X (= A, B), ϕ_X is the volume fraction of the component X, and n_X denotes the number fraction of the Kuhn segments of the component X in the blend. $M_{e,\text{bulk }X}$ appearing in Equation (3.62) is the entanglement molecular weight of the component X in bulk. (Equation 3.61 was originally formulated for the packing length $p \cong a/20$; Chen et al., 2008. However, a small variation of the p/a ratio among polymer species can be neglected to rewrite the original expression in the form of Equation 3.61; Watanabe et al. 2011.)

Equations (3.60) through (3.62) can be tested on the basis of a formal blending law of the complex modulus, $G^*(\omega) = \Sigma_{X=A,B} G_X^*(\omega)$ with $G_X^*(\omega)$ being the bare (nonnormalized) complex modulus of the component X in *the blend*. (This definition of $G_X^*(\omega)$ is identical to that utilized in Equation 3.58.) As explained earlier, the CR effect on the viscoelastic relaxation changes on blending. Thus, in general, $G_X^*(\omega)$ differs from $G_{X,\text{bulk}}^*(\omega)$ of bulk component X, and this formal blending law just indicates the stress additivity of the components. However, in the *high-ω plateau zone* where Equations (3.60) through (3.62) are tested, the fast component has hardly relaxed and thus activates no significant CR relaxation for the fast and slow components. In this zone, $G_X^*(\omega)$ at a given T can be safely approximated to have the same relaxation mode distribution as $G_{X,\text{bulk}}^*(\omega)$ of the bulk at the same T, as suggested from the G^* data of the chemically uniform PI/PI blends (see the high-ω data in Figure 3.15) and of PS/PS blends (Watanabe, 1999). Then, the above blending law can be rewritten as

$$G^*(\omega) = \sum_{X=A,B} \phi_X I_X G_{X,\text{bulk}} {}^*(\omega\Lambda_X) \text{ in the high-}\omega \text{ plateau zone} \qquad (3.63)$$

with

$$\Lambda_X = \frac{\tau_{G,X}}{\tau_{G,X}^{[\text{bulk}]}} \qquad (3.64)$$

and

$$I_X = \left(\frac{a_X}{a}\right)^2 \text{ (when Equations 3.60 and 3.61 are utilized)} \qquad (3.65)$$

$$I_X = \frac{M_{e,\text{bulk}\,X}}{M_e} \text{ (when Equation 3.62 is utilized)} \qquad (3.66)$$

Here, Λ_X denotes a difference between the viscoelastic terminal relaxation times of the component X (= PI, PtBS) in bulk, $\tau_{G,X}^{[\text{bulk}]}$, and in the blend, $\tau_{G,X}$. I_X represents a difference of the entanglement plateau heights normalized to unit volume fraction of the component X in bulk and blend. I_X is determined according to the mixing rules, Equations (3.60) through (3.62).

Because the PI99 chains relax much faster than the entangling PtBS348 chains (cf. Figure 3.16), the constraint release (CR) mechanism hardly contributes to the PI99 relaxation. For this case, the terminal viscoelastic relaxation time $\tau_{G,\text{PI}}$ of PI99 appearing in Equation (3.64) can be safely replaced by the dielectric $\tau_{\varepsilon,\text{PI}}$, the latter being evaluated from ω_{peak} for the narrow ε'' peak in Figure 3.16 as $\tau_{\varepsilon,\text{PI}} = 1/\omega_{\text{peak}}$ (= 0.04 s at 100°C). Thus, Λ_{PI} is evaluated from this $\tau_{\varepsilon,\text{PI}}$ and the $\tau_{G,\text{PI}}^{[\text{bulk}]}$ data of bulk PI99 as $\Lambda_{\text{PI}} = \tau_{\varepsilon,\text{PI}}/\tau_{G,\text{PI}}^{[\text{bulk}]}$. We also note that the first step relaxation of the slow PtBS348 chains corresponds to their CR relaxation activated by the global motion of the PI99 chains. Thus, $\tau_{G,\text{PtBS}}$ appearing in Equation (3.64) can be safely replaced by $\tau_{G,\text{PI}}$ (= $\tau_{\varepsilon,\text{PI}}$) of PI99, and Λ_{PtBS} is evaluated from $\tau_{\varepsilon,\text{PI}}$ and the $\tau_{G,\text{PtBS}}^{[\text{bulk}]}$ data of bulk PtBS as $\Lambda_{\text{PtBS}} = \tau_{\varepsilon,\text{PI}}/\tau_{G,\text{PtBS}}^{[\text{bulk}]}$.

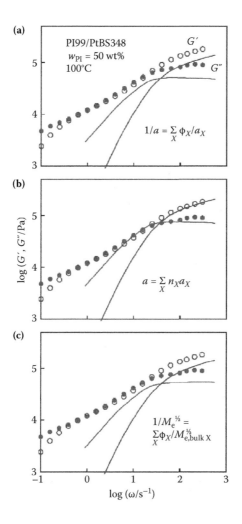

FIGURE 3.18

Test of the mixing rules of the entanglement length *a* for the PI99/PtBS348 miscible blend examined in Figure 3.16. In panels (a), (b), and (c), respectively, the blend moduli calculated from mixing rules, Equations (3.60), (3.61), and (3.62), are shown with the curves. Comparison of these curves with the data (circles) suggests the success of Equation (3.61) and failure of Equations (3.60) and (3.62). (Data taken, with permission, from Watanabe, H., Q. Chen, Y. Kawasaki, Y. Matsumiya, T. Inoue, and O. Urakawa. 2011. Entanglement dynamics in miscible polyisoprene/poly(*p-tert*-butylstyrene) blends. *Macromolecules* 44:1570–1584.)

Utilizing those Λ_X values together with the I_X values deduced from the mixing rules, Equations (3.60) through (3.62), Watanabe et al. (2011) calculated G^* from the $G_{\mathrm{PI,bulk}}^*$ and $G_{\mathrm{PtBS,bulk}}^*$ data with the aid of Equation (3.63). Figure 3.18 compares G^* (solid curves) calculated for the PI99/PtBS348 blend at 100°C with the data (circles). As noted for the curves in panels (a) and (c), Equations (3.60) and (3.62) considerably underestimate the modulus in the high-ω plateau zone at $\omega \geq 20$ s^{-1}, where the mixing rules are to be tested. In contrast, Equation (3.61) describes the data in this zone, as noted for the curves in panel (b). This validity of Equation (3.61) has been confirmed for other PI/PtBS blends having different compositions or component molecular weights (see Watanabe et al., 2011).

In the current molecular picture, the entanglement density in homopolymer systems is related to the packing length p, and p is proportional to the entanglement length a ($p \cong a/20$) (see Fetters et al., 1994, 1999, 2007). Thus, for the PI/PtBS blends, the high-ω plateau modulus is determined by an average of a_{PI} and a_{PtBS}. Because the Kuhn segment is the fundamental unit for description of the flexible polymer conformation, this average naturally includes the number fractions of these segments as the weighing factors to give Equation (3.61).

Note also that the above test of the mixing rule was successfully made because of the considerable difference of the a_{PI} and a_{PtBS} values (cf. Equations 3.51 and 3.59). For the other pair of components such as PI and PVE, the difference of the a_X values is much smaller so that Equations (3.60) through (3.62) give nearly the same a value and cannot be clearly distinguished experimentally.

3.5.2.3 Component Relaxation Mechanisms in High-M PI/PtBS Blend

The entangled PI99/PtBS348 blend examined in Figures 3.16 and 3.17 exhibits the two-step viscoelastic relaxation associated with the low-ω and high-ω plateaus of G' at high T. This behavior is similar to that of chemically uniform PI/PI blends with large ϕ_2 shown in Figure 3.5. Thus, the relaxation mechanisms of the components in the PI99/PtBS348 blend can be understood on the basis of the behavior of the PI/PI blends. Namely, the fast component (PI99) appears to fully relax through the reptation and CLF mechanisms in the entanglement network with the mesh size a specified by Equation (3.61). (Because PI99 relaxes much faster than PtBS348, the PI99 relaxation should be hardly contributed from the CR mechanism.) The slow component (PtBS348) appears to partially relax at high ω through the CR mechanism activated by the global motion of PI99. After this CR relaxation, PI99 basically behaves as a solvent, and the effective entanglement mesh size for PtBS348 increases from a (cf. Equation 3.61) to that in a corresponding solution, $a_{PtBS}/\phi_{PtBS}{}^{d/2}$ with a_{PtBS} and d ($\cong 1.3$) being the entanglement length in bulk PtBS and the dilation exponent, respectively. The terminal relaxation of PtBS348 should have occurred through the reptation, CLF, and CR mechanisms in this dilated entanglement mesh, as suggested from coincidence of the terminal relaxation mode distribution observed for the blend and a real PtBS348 solution in a low-M solvent (Watanabe et al., 2011). All these relaxation mechanisms are identical to those in the PI/PI blends, although the entanglement mesh size for PtBS348 before its CR relaxation (corresponding to the high-ω plateau of G') is dependent on ϕ_{PtBS} (on n_{PtBS}; cf. Equation 3.61) and is much smaller than

a_{PtBS} of bulk PtBS. Thus, no qualitative difference of the relaxation behavior is noted for the PI99/PtBS348 blend at high T and the PI110/PVE290 blend (cf. Figures 3.13 and 3.15), the latter having just a moderate dynamic asymmetry.

In contrast, at low T, the PI99/PtBS348 blend exhibits no high-ω plateau of G', but the Rouse-like power-law increases of G' and of G'' well beyond the level of the expected high-ω plateau (cf. Figures 3.16 and 3.17). This behavior, not observed for the PI/PI and PI/PVE blends, should reflect the very large dynamic asymmetry between PI99 and PtBS348, as discussed below.

For the high-ω plateau corresponding to the entanglement mesh size a (Equation 3.61) to be observed, the Rouse equilibration within this mesh should be much faster than the global motion at length scales greater than a. Because the effective $T_{g,eff}$ is considerably higher for PtBS than for PI, the WLF-type friction coefficient of the Rouse segment, ζ_s, is much larger for PtBS than for PI at low T. Correspondingly, the *intrinsic* Rouse equilibration time over the length a, $\tau_a° = \zeta_s a^2 N_R / 6\pi^2 k_B T \propto \zeta_s$ with N_R being the number of the Rouse segment per entanglement segment ($a = b_R N_R^{1/2}$ with $b_R =$ Rouse segment size; cf. Equation 3.36), is much longer for PtBS. Obviously, the PI chain cannot take a conformation overlapping the PtBS segments. Thus, the concentrated PtBS chains (with $w_{PtBS} = 50$ wt%) should behave as densely distributed obstacles to *topologically* hinder the PI chain from exploring, within the intrinsic $\tau_{a,PI}°$, all *local* conformations at a length scale of a. For this case, the PI and PtBS chains should be cooperatively Rouse equilibrated, and a time necessary for this equilibration should be essentially determined by the PtBS motion over the length a and given by $\tau_{a,PtBS}° = \zeta_{s,PtBS} a^2 N_R / 6\pi^2 k_B T$. Once this equilibration is completed, the PI chain can exhibit its global relaxation at a time $\tau_{G,PI}$ proportional to $\zeta_{s,PI}$ of its Rouse segment. Because $\zeta_{s,PI} \ll \zeta_{s,PtBS}$ at low T, this $\tau_{G,PI}$ can be rather close to $\tau_{a,PtBS}°$. For this case, the PI chains exhibit their global relaxation soon after the cooperative Rouse equilibration, and the CR relaxation of PtBS activated by this global relaxation immediately follows. Then, the high-ω plateau sustained by the PI and PtBS chains has a width too narrow to be resolved experimentally and is masked by the cooperative Rouse equilibration process, as noted in Figure 3.16.

The above argument can be examined through direct comparison of $\tau_{a,PtBS}°$ and $\tau_{G,PI}$. Because the PtBS chains are effectively immobilized over the length scale $> a$ in the time scale of the global relaxation of PI, $\tau_{G,PI}$ can be safely evaluated as the dielectric relaxation time of PI, $\tau_{\varepsilon,PI}$ ($= 10$ s at 30°C; cf. Figure 3.16). The time required for the cooperative Rouse equilibration process, τ_a ($= \tau_{a,PtBS}°$), can be evaluated on the basis of a power-law-type expression of the modulus of the Rouse model (equivalent to Equation 3.35) (Osaki et al., 2001):

$$G'(\omega) = G''(\omega) = 1.111 \frac{cRT}{M} (\omega \tau_{R,G})^{1/2} \quad \text{for } \omega \text{ well above } 1/\tau_{R,G} \quad (3.67)$$

For the evaluation of τ_a, M in Equation (3.67) is replaced by the entanglement molecular weight of the component X in the blend ($X = $ PI, PtBS), $M_{e,X} = \{a/a_X\}^2 M_{e,\text{bulk}X}$ with a being evaluated from Equation (3.61): $a = 6.3$ nm for the PI99/PtBS348 blend examined in Figure 3.16. After this replacement, Equation (3.67) gives the G' and G'' values corresponding to τ_a for this blend at 30°C as

$$G'(1/\tau_a) = G''(1/\tau_a) = 1.111 \left\{ \frac{c_{\text{PI}} RT}{M_{e,\text{PI}}} + \frac{c_{\text{PtBS}} RT}{M_{e,\text{PtBS}}} \right\} = 3.6 \times 10^5 \text{ Pa} \qquad (3.68)$$

These G' and G'' values are attained for the modulus data of that blend at $\omega_a = 0.4$ s^{-1} ($\tau_a = 2.5$ s), as shown with the arrow in Figure 3.16. (This τ_a value was in agreement with $\tau_{a,\text{PtBS}}{}^\circ$ ($= 2.4$ s) obtained from the WLF analysis of the shift factor for PtBS in the blend; see Watanabe et al., 2011.) The τ_a thus obtained is considerably close to $\tau_{e,\text{PI}}$ ($= 10$ s). In addition, the intrinsic Rouse equilibration time of PI in the blend obtained from the WLF analysis, $\tau_{a,\text{PI}}{}^\circ = 1.1 \times 10^{-4}$ s at 30°C, was orders of magnitude shorter than τ_a, and thus the equilibration for PI was strongly retarded by PtBS (cf. Watanabe et al., 2011). These results lend support to the above argument that relates the lack of the high-ω entanglement plateau at low T to the cooperative Rouse equilibration of PI strongly retarded by PtBS. It should also be noted that at 100°C the intrinsic $\tau_{a,\text{PtBS}}{}^\circ$ ($= 6.8 \times 10^{-5}$ s) of PtBS in the blend was much shorter than the global $\tau_{e,\text{PI}}$ of PI ($= 0.05$ s; cf. Figure 3.16) thereby allowing the high-ω plateau to be clearly observed at 100°C.

Watanabe et al. (2011) further examined the above molecular mechanism of the lack of the high-ω plateau through simple modeling. They modeled the modulus of PI corresponding to this mechanism as

$$G_{\text{PI}}^*(\omega) = \frac{c_{\text{PI}} RT}{M_{e,\text{PI}}} \sum_{q=1}^{N_R} \frac{i\omega\tau_a / r_q^2}{1 + i\omega\tau_a / r_q^2} + \phi_{\text{PI}} I_{\text{PI}} G_{\text{PI,bulk}}^*(\omega\Lambda_{\text{PI}}) \qquad (3.69)$$

with

$$r_q = \sin\left\{ \frac{q\pi}{2(N_R + 1)} \right\} \sin^{-1}\left\{ \frac{\pi}{2(N_R + 1)} \right\} \qquad (3.70)$$

The first summation term in Equation (3.69) indicates the modulus due to the cooperative Rouse equilibration of a sequence of $N_R + 1$ Rouse segments in each entanglement segment (with the molecular weight $M_{e,\text{PI}}$) occurring at time τ_a. This term is expressed in a form utilizing a ratio r_q of discrete eigenvalues shown in Equation (3.70). ($r_q = q$ for $N_R \gg 1$, which corresponds to the continuous Rouse expression of $G(t)$ given by Equation 3.35). Because the Rouse and Kuhn segments of flexible PI chains are similar in size, the sequence length N_R can be estimated from $M_{K,PI}$ ($\cong 130$) of the Kuhn segment as $N_R \cong M_{e,PI}/M_{K,PI}$ (cf. Watanabe et al., 2011).

The second term in Equation (3.69) represents the terminal entanglement relaxation of PI occurring at $\tau_{G,PI}$ ($= \tau_{\varepsilon,PI}$ for PI relaxing much faster than PtBS, as explained earlier). As explained for Equation (3.63), this term is approximately expressed in terms of the modulus $G_{PI,bulk}*$ of bulk PI, the PI volume fraction ϕ_{PI}, the relaxation time shift factor $\Lambda_{PI} = \tau_{G,PI}/\tau_{G,PI}^{[bulk]}$, and the intensity factor I_{PI} given by Equation (3.65).

For the PtBS chains in relatively short time scales *before the relaxation of their mutual entanglement*, Watanabe et al. (2011) formulated their modulus in the blend as

$$G_{PtBS}*(\omega) = \frac{c_{PtBS}RT}{M_{e,PtBS}} \sum_{q=1}^{N_R} \frac{i\omega\tau_a/r_q^2}{1+i\omega\tau_a/r_q^2}$$

$$+ \frac{c_{PtBS}RT}{M_{e,soln}^{[PtBS]}} \sum_{q=1}^{N_{CR}-1} \frac{i\omega\tau_{CR}/\rho_q^2}{1+i\omega\tau_{CR}/\rho_q^2} + \frac{c_{PtBS}RT}{M_{e,soln}^{[PtBS]}} \tag{3.71}$$

with

$$\rho_q = \sin\left\{\frac{q\pi}{2N_{CR}}\right\} \sin^{-1}\left\{\frac{\pi}{2N_{CR}}\right\}, \quad N_{CR} = \frac{M_{e,soln}^{[PtBS]}}{M_{e,PtBS}} \tag{3.72}$$

The first summation term in Equation (3.71) represents the modulus due to the cooperative Rouse equilibration. Because the PtBS and PI chains are equilibrated cooperatively to have the common τ_a, the onset time of this equilibration, $\tau_a/r_{N_R}^2$, and the number N_R determining this onset time were approximated to be common for these chains.

After this Rouse equilibration, the global motion of the PI chains activates the CR relaxation of the PtBS chains to increase the effective M_e for PtBS from $M_{e,PtBS}$ in the PI/PtBS blends ($= \{a/a_{PtBS}\}^2 M_{e,bulk\ PtBS}$) to $M_{e,soln}^{[PtBS]}$ in the solution having the same ϕ_{PtBS} as the blend ($= M_{e,bulk\ PtBS}/\phi_{PtBS}^d$ with $d \cong 1.3$). The second summation term in Equation (3.71) represents the modulus for this CR process occurring at the terminal CR time τ_{CR} for a sequence of N_{CR} ($= M_{e,soln}^{[PtBS]}/M_{e,PtBS}$) entanglement segments of PtBS in the blend. This CR term is expressed in the discrete Rouse form with the eigenvalue ratio ρ_q given by Equation (3.72). Because the local CR hopping of the PtBS chain is activated by the global motion of the PI chains, the onset time for the CR process, $\tau_{CR}/q_{N_{CR}-1}^2$, should be determined by $\tau_{\varepsilon,PI}$ of PI. Watanabe et al. (2011) utilized the Graessley model (Graessley, 1982) to relate $\tau_{CR}/q_{N_{CR}-1}^2$ and $\tau_{\varepsilon,PI}$ as

$$\tau_{CR}/q_{N_{CR}-1}^2 = \Lambda^{[CR]}(z)\,\tau_{\varepsilon,PI} \text{ with } \Lambda^{[CR]}(z) = \frac{1}{z}\left(\frac{\pi^2}{12}\right)^z \qquad (3.73)$$

Here, z is the local jump gate number (typically in a range of $z = 2$ to 4) treated as an adjustable parameter ($z = 2$ for the PI/PtBS blends, as shown later). Finally, the third term in Equation (3.71) corresponds to the plateau modulus sustained only by the PtBS chains before they exhibit the global relaxation.

Here, a comment needs to be added for the τ_{CR} value. Equation (3.73) assumes the proportionality between $\tau_{CR}/q_{N_{CR}-1}^2$ and $\tau_{\varepsilon,PI}(\propto M_{PI}^{3.5})$, which does not match the empirical equation for $\tau_{CR}^{[long]}$ ($\propto M_{PI}^{\alpha}$ with $\alpha \cong 3$; Equation 3.34) explained earlier. However, Equation (3.73) was applied to PI/PtBS blends containing two PI samples of rather close M_{PI} (= 9.9×10^4 and 1.3×10^5) (cf. Watanabe et al., 2011). The difference of the M_{PI} dependence in Equations (3.73) and (3.34) gave just a minor numerical difference for τ_{CR} in these blends (by factor of 14%) and negligibly affected the results of the test of the above molecular model. In other words, the application of Equation (3.73) to those blends effectively treated the CR onset time $\tau_{CR}/q_{N_{CR}-1}^2$ as an adjustable parameter in a range of $0.11\tau_{\varepsilon,PI} \le \tau_{CR}/q_{N_{CR}-1}^2 \le 0.34\tau_{\varepsilon,PI}$ (the range corresponding to $z = 2$ to 4), which is consistent with the $\tau_{CR}^{[long]}$ data for PI/PI blends explained earlier.

The basic parameters in the above model, τ_a and $\tau_{\varepsilon,PI}$, were determined experimentally, as explained in the previous section. The other parameters, $M_{e,X}$ (X = PI, PtBS), N_R, and N_{CR} were evaluated from a in the blend, a_X and $M_{e,bulk\,X}$ in bulk, $M_{K,PI}$ ($\cong 130$), and $M_{e,soln}^{[PtBS]}$. The values of these parameters are summarized in Table 3.1. The blend modulus $G^* = G_{PI}^* + G_{PtBS}^*$ was calculated from Equations (3.69) through (3.72) with these parameter values. Figure 3.19 compares the model calculation with the G^* data for several PI/PtBS blends as indicated (sample code numbers showing $10^{-3}M$; cf. Watanabe et al., 2011). Despite the approximate use of $G_{PI,bulk}^*$ in the model (Equation 3.69), the G^* calculated with a reasonable value of $z = 2$ ($\tau_{CR}/q_{N_{CR}-1}^2 = 0.34\tau_{\varepsilon,PI}$; solid curves) is surprisingly close to the data (symbols) for all blends examined. This result lends support to the molecular picture underlying the model, the cooperative Rouse equilibration of the PI and PtBS chains being slower than the intrinsic Rouse equilibration of PI and leading to the lack of high-ω plateau at low T.

Watanabe et al. (2011) utilized the dynamic birefringence data of the PI128/PtBS348 blend ($w_{PI} = 50$ wt%) to further examine the component dynamics therein. Figure 3.20a shows the complex shear birefringence coefficient $K^* (= K' + iK'')$ measured for this blend at 30°C. K' was negative in the entire range of ω examined, and K'' was also negative at $\omega < 100$ s^{-1} where the blend exhibited the Rouse-like power-law behavior. Thus, the plots are shown for their absolute values, $|K'|$ and $|K''|$. For general cases of the nematic coupling of the monomeric segments explained earlier, $K^* (= K_A^* + K_B^*)$ and $G^* (= G_A^* + G_B^*)$ of the A/B blend are related to each other through

TABLE 3.1

Parameter Values Utilized in the Model Calculation
for Figures 3.19 and 3.20

	PI99/PtBS348 w_{PI} = 50 wt% 30°C	PI128/PtBS348 w_{PI} = 50 wt% 30°C	PI128/PtBS348 w_{PI} = 40 wt% 60°C
a/nm^a	6.3	6.3	6.5
τ_a/s^b	2.5	2.5	1.0
$\tau_\varepsilon^{PI}/s^c$	10	20	2.8
$10^{-3} M_{e,PI}^d$	5.8	5.8	6.2
$10^{-3} M_{e,PtBS}^d$	10.8	10.8	11.5
N_R^e	44	44	47
N_{CR}^f	9	9	7
$10^{-2}\tau_{CR}/s^g$	2.7	5.5	0.47

[a] Determined from Equation (3.61).
[b] Evaluated from G^* data (cf. Equation 3.68).
[c] Evaluated from ε'' data.
[d] $M_{e,X} = \{a/a_X\}^2 M_{e,bulk X}$ (X = PI, PtBS).
[e] $N_R = M_{e,PI}/M_{K,PI}$.
[f] $N_{CR} = M_{e,soln}^{[PtBS]} / M_{e,PtBS}$.
[g] $\tau_{CR} = \Lambda^{[CR]}(z)\tau_{e,PI} q_{N_{CR}-1}^2$ (cf. Equation 3.73).

Equation (3.58) with a nonzero coupling constant θ. Watanabe et al. (2011) evaluated the stress-optical coefficients of PI and PtBS in the blend at 30°C, C_{PI} (= 9.7 × 10⁻¹⁰ Pa⁻¹) and C_{PtBS} (= −5 × 10⁻⁹ Pa⁻¹), from the C data in respective bulk systems after a minor correction of T. They utilized these C_{PI} and C_{PtBS} values in Equation (3.58) with θ = 0 (*no nematic coupling*) to evaluate the component moduli in the PI128/PtBS348 blend, G_{PI}^* and G_{PtBS}^*. In Figures 3.20b and 3.20c, the unfilled circles and squares show these component moduli for θ = 0. For comparison, the PI modulus G_{PI}' obtained with θ = 0.3 (close to the θ value for PI/PVE blends; cf. Figure 3.14) is shown with small filled circles in panel (b). The dielectric loss data detecting the end-to-end vector fluctuation of the PI chains are also shown in panel (b).

As noted in panel (b), G_{PI}^* obtained with θ = 0 exhibits the terminal viscoelastic relaxation behavior exactly in the range of ω where the terminal dielectric relaxation is experimentally observed. On the other hand, G_{PI}' evaluated with θ = 0.3 shows no terminal relaxation in this range of ω. (This lack of the terminal relaxation was found even for a smaller θ value ≤ 0.1.) Thus, the coupling constant in the PI/PtBS blends is quite small (and practically zero), probably because the PI and PtBS chains have quite different chemical structures (and cannot be very densely packed in the system).

In panels (b) and (c), we also note that G_{PI}^* and G_{PtBS}^* (evaluated with θ = 0) exhibit the Rouse-like power-law increases beyond the level of the high-ω entanglement plateau (3.6 × 10⁵ Pa; cf. Equation 3.68) expected for the 50 wt% PI/PtBS blend at 30°C. The viscoelastic modulus at ω well below the

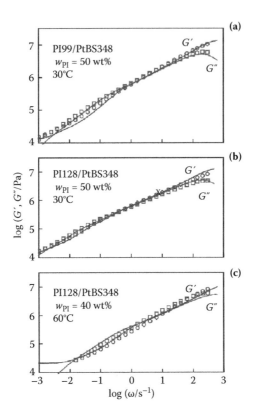

FIGURE 3.19

In panels (a)–(c), comparison of the blend moduli calculated from the model (curves; Equations 3.69 through 3.73) with the moduli data (symbols) for various high-M polyisoprene/ poly(p-$tert$ butyl styrene) (PI/PtBS) blends as indicated. The sample code numbers of the blends indicate $10^{-3}M$ of the components. The model considers the cooperative Rouse equilibration and successive constraint release (CR)/reptation relaxation of the component chains, and the model parameters summarized in Table 3.1 were determined experimentally. (Redrawn, with permission, from Watanabe, H., Q. Chen, Y. Kawasaki, Y. Matsumiya, T. Inoue, and O. Urakawa. 2011. Entanglement dynamics in miscible polyisoprene/poly(p-$tert$-butylstyrene) blends. *Macromolecules* 44:1570–1584).

segmental relaxation zone is free from the effect of the nematic coupling, as explained earlier. For this reason, this effect was not incorporated in the simple model, Equations (3.69) through (3.72). The component moduli calculated from this model (solid curves; with the model parameters summarized in Table 3.1) agree with the rheo-optically determined moduli (unfilled symbols for $\theta = 0$) surprisingly well. These results lend strong support to the mechanism of the lack of the high-ω plateau, the retarded cooperative Rouse equilibration of PI that masks this plateau.

It should be emphasized that the cooperative Rouse equilibration is intimately related to the fundamental aspect of polymer rheology explained

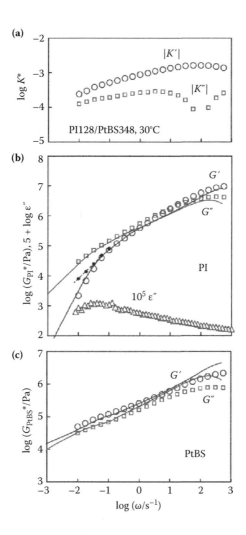

FIGURE 3.20

(a) Data of the dynamic birefringence coefficient K^* of the PI128/PtBS348 blend examined in the middle panel of Figure 3.19. These data were analyzed to give the moduli of the polyisoprene (PI) and poly(p-$tert$ butyl styrene) (PtBS) components shown in (b) and (c). In panels (b) and (c), the unfilled circles and squares indicate the component moduli obtained with the nematic coupling constant $\theta = 0$. In (b), the small filled circles indicate the storage modulus of PI obtained with $\theta = 0.3$, and the triangles show the dielectric loss data multiplied by a factor of 10^5. The curves in (b) and (c) indicate the moduli calculated from the model considering the cooperative Rouse equilibration and successive constraint release (CR)/reptation relaxation (Equations 3.69 through 3.73). (Data taken, with permission, from Watanabe, H., Q. Chen, Y. Kawasaki, T. Matsumiya, T. Inoue, and O. Urakawa. 2011. Entanglement dynamics in miscible polyisoprene/poly(p-$tert$-butylstyrene) blends. *Macromolecules* 44:1570–1584.)

earlier. That is, the mechanical stress reflects the orientational anisotropy of submolecules exploring *all* internal conformations in a given time scale. At low T, the PI chain *cannot* explore all conformations at length scales $\leq a$ within its intrinsic Rouse time because the slow PtBS chains behave as densely dispersed obstacles in this time scale, thereby hindering the PI chain from this exploration. This retardation mechanism in the length scale of a is never observed for homopolymer systems and is the most remarkable feature of the miscible PI/PtBS blends having the large dynamic asymmetry of the component chains.

3.5.2.4 Length Scale of Cooperativity

The cooperative Rouse equilibration of the PI and PtBS chains in the blends over the entanglement length a is physically reasonable, as discussed in the previous section. In fact, the experimentally proved mixing rule (Equation 3.61) is based on the molecular idea of the coincidence of a for chemically different components (that results in the balance of the local chain tension of those components), which is consistent with this mechanism. In relation to this point, it should also be pointed out that the model (Equations 3.69 through 3.72) modified for separate equilibration of those components cannot describe the G^* data of the blend (see Watanabe et al., 2011).

Thus, in the length scale of entanglement, the chemically different components exhibit dynamic cooperativity through the Rouse dynamics. Rouse dynamics is associated with the length scale of equilibration ξ that grows with time as $\xi(t) \sim a(t/\tau_a)^{1/4}$, where τ_a is the equilibration time. This ξ can also be expressed in terms of the step length b_R and the basic relaxation time τ_R^* of the Rouse segment, the smallest motional unit during the Rouse equilibration, as $\xi(t) \sim b_R(t/\tau_R^*)^{1/4}$. These expressions of $\xi(t)$ suggest that the strong cooperativity discussed in the previous section is limited to the length scale between a and b_R and the time scale between τ_a and τ_R^*.

For $\xi > a$ and $t > \tau_a$, the component chains exhibit the entanglement relaxation. The entanglement segments of the size a, the coarse-grained motional units during this relaxation, exhibit the *interchain* motional cooperativity (or correlation) as considered in the CR model (cf. Figure 3.6), but this cooperativity is not as strong as that between the Rouse segments of PI and PtBS, the latter fully determining the equilibration rate for the former.

In contrast, for $\xi < b_R$ and $t < \tau_R^*$, the Rouse segments hardly relax, and the monomeric segments intimately related to the glass transition govern the blend dynamics. In general, the monomeric segments of different components have different $T_{g,eff}$ to relax separately at different relaxation times τ^* ($< \tau_R^*$), as explained in earlier sections. Recent quasi-elastic neutron scattering experiments by Doxastakis et al. (2002) suggested that the correlation of the local motion vanishes at $\xi < 0.8$ nm $\sim b_R$.

As noted from the above argument, the chemically different components in the entangled blends should exhibit a crossover from the almost independent, very local motion (for $\xi \leq b_R$) to the highly cooperative motion/equilibration (for $b_R < \xi \leq a$) and further to the moderately cooperative global motion (for $a < \xi$). This crossover is a very interesting subject of future research.

3.5.2.5 Thermorheological Behavior of Components in PI/PtBS Blends

For the entangled PI/PVE blends, the time-temperature superposition works for the viscoelastic and dielectric data of respective components in the global relaxation regime, as noted in Figures 3.11 and 3.15. (For the PVE290 chains examined in Figure 3.15, the superposition fails at high ω where the faster PI110 chains activates the CR relaxation for the PVE290 chains. However, the superposition still works for PVE290 at low ω where this CR relaxation completes.) Thus, for those blends exhibiting just a moderate dynamic asymmetry, the component relaxation is basically thermorheologically simple. This simplicity prevails because the component relaxation mechanism remains the same in the range of T examined for the blends.

The situation is different for the PI/PtBS blends exhibiting a significant dynamic asymmetry (reflecting the huge difference of $T_{g,eff}$ of PI and PtBS). Characteristic thermorheological complexities are noted for the PI and PtBS chains depending on their molecular weights. These complexities and the underlying mechanisms are summarized below.

3.5.2.5.1 High-M PI/PtBS Blends

As noted in Figure 3.17, the superposition fails in the terminal relaxation zone of the G^* data of the blend as a whole, which indicates the thermorheological complexity for the slow PtBS348 component governing this terminal behavior. The fast PI99 component also exhibits the complexity, as clearly noted from the high-ω entanglement plateau appearing at high T and the Rouse-like power-law behavior (lack of this plateau) at low T. This complexity reflects the change of the dominant relaxation mechanism of PI99 from the reptative relaxation much slower than the Rouse equilibration (at high T) to the Rouse equilibration that is hindered by the slow PtBS chains and masks the reptative relaxation immediately following this equilibration (at low T), as discussed earlier. Nevertheless, the dielectric mode distribution of PI is insensitive to this change, thereby satisfying the time-temperature superposition, as seen in Figure 3.17. This result reflects a fact that the dielectric mode distribution of PI, equivalent to the distribution of the end-to-end vector fluctuation modes, is similar for the reptation and Rouse dynamics (see, e.g., Watanabe, 1999, 2001). Thus, the change of the slow dynamics of PI still allowed the dielectric data to obey the superposition.

3.5.2.5.2 Low-M PI/PtBS Blends

For low-M PI20/PtBS16 blends ($M_{PI} = 2 \times 10^4$, $M_{PtBS} = 1.6 \times 10^4$) examined by Chen et al. (2008, 2011, 2012), the time-temperature superposability of the dielectric ε'' data is examined in Figure 3.21. For the blend with $w_{PI} = 50$ wt% (panel (a) of Figure 3.21), the superposition works satisfactorily for the ε'' data at $\omega > \omega_{peak}$ but fails at low ω to give branches in the ε'' plots. Clearly, the dielectric mode distribution narrows with increasing T and approaches that of bulk PI20 shown with the solid curve. This thermorheological complexity is attributed to the difference of the global relaxation times of PtBS and PI (longer for PtBS) and the rather scarce overlapping of the component chains having small M, as explained below (cf. Chen et al., 2008; Watanabe et al., 2007).

In the blend with $w_{PI} = 50$ wt%, the PtBS16 chains have the mass concentration $C_{PtBS} = 0.49$ g cm^{-3} and the overlapping concentration $C_{PtBS}{}^* = 0.21$ g cm^{-3} ($C_{PtBS}{}^* = \{M_{PtBS}/N_A\}/\{4\pi R_{g,PtBS}^3/3\}$ with N_A = Avogadro constant, and $R_{g,PtBS}$ = average radius of gyration = 3.1 nm for PtBS16). The $C_{PtBS}/C_{PtBS}{}^*$ ratio ($\cong 2.3$) is rather close to unity, indicating that the PtBS concentration C_{PtBS} cannot be perfectly uniform in space but exhibits dynamic undulation at a wavelength Λ_C comparable to $R_{g,PtBS}$: If we apply the blob picture (de Gennes, 1979) to the PtBS16 chains having the Gaussian conformation in the blend, we obtain an estimate of $\Lambda_C \sim \langle R_{PtBS}^2 \rangle_{eq}^{1/2} \{C_{PtBS}{}^*/C_{PtBS}\} = 3.3$ nm. Because PtBS16 is the slow component in the blends, the undulation of C_{PtBS} is quenched in the time scale of the global relaxation of PI20. In addition, the average end-to-end distance of the PI20 chain, $\langle R_{PI}^2 \rangle_{eq}^{1/2} = 12$ nm, is not significantly larger than Λ_C. Under these conditions, some PI20 chains (minority) are unavoidably localized in a PtBS-rich region to have a higher $T_{g,eff}$ compared to the remaining PI20 chains (majority). Then, the majority and minority PI chains have different friction coefficients of their whole backbone, ζ_{chain} (being larger for the minority). This distribution of ζ_{chain} broadens the dielectric mode distribution for the whole ensemble of the PI20 chains. Furthermore, the difference of the dielectrically detected global relaxation times τ_ε of the minority and majority PI decreases with increasing temperature (because of the WLF-type decrease of ζ_{chain} that is more significant for the minority having the higher $T_{g,eff}$). Thus, the dielectric mode distribution for the whole ensemble of the PI20 chains narrows with increasing T and approaches the distribution of bulk PI20, which results in the thermorheological complexity seen in panel (a) of Figure 3.21.

From the above argument, we expect that the complexity is reduced when the difference of the relaxation times of the PI and PtBS chains decreases. This expectation is confirmed for the PI20/PtBS16 blends with $w_{PI} = 70$ and 80 wt%, as shown in the panels (b) and (c) of Figure 3.21. The increase of w_{PI} enhances plasticization of PtBS16 due to PI20 to reduce the difference of the relaxation times of these components, thereby allowing the ensemble of PI20 chains to obey the superposition as a whole. In particular, in the blend with $w_{PI} = 80$ wt%, the PI20 chains relaxed slower than the PtBS chains (as noted from the coincidence of the dielectric and viscoelastic terminal relaxation times of the blend; cf. Chen et al., 2008; Takada et al., 2008) and the dielectric data exhibit excellent superposition.

We also expect that the complexity of the dielectric data is reduced when the PI chain dimension $\langle R_{PI}^2 \rangle_{eq}^{1/2}$ increases to smear the spatial frictional

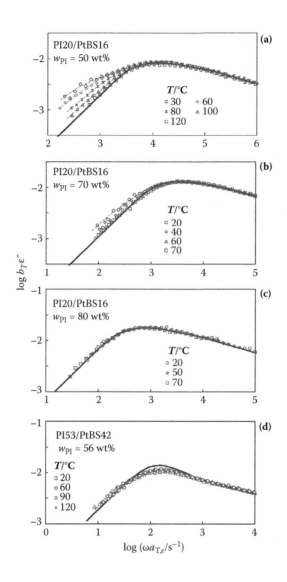

FIGURE 3.21
Test of time-temperature superposability for the dielectric ε″ data of low-*M* and middle-*M* poly-isoprene/poly(*p-tert* butyl styrene) (PI/PtBS) miscible blends as indicated. In panels (a)–(d), the sample code numbers of the blends denote $10^{-3}M$ of the components. The reference temperature is $T_r = 90°C$ for all blends. The solid curves indicate the ε″ data of bulk PI corrected for the PI volume fraction in the blends. These curves are shifted along the ω axis to match their peak frequency with that of the blends. (Data taken, with permission, from Chen, Q., Y. Matsumiya, Y. Masubuchi, H. Watanabe, and T. Inoue. 2008. Component dynamics in polyisoprene/poly(4-*tert*-butylstyrene) miscible blends. *Macromolecules* 41:8694–8711; Chen, Q., Y. Matsumiya, Y. Masubuchi, H. Watanabe, and T. Inoue. 2011. Dynamics of polyisoprene-poly(*p*-tert-butylstyrene) diblock copolymer in disordered state. *Macromolecules* 44:1585–1602; Chen, Q., Y. Matsumiya, K. Hiramoto, and H. Watanabe. 2012. Dynamics in miscible blends of polyisoprene and poly(*p*-tert-butyl styrene): Thermo-rheological behavior of components. *Polymer J.* 44:102–114.)

distribution within the chain. For this case, the spatial frictional distribution becomes equivalent to the distribution of the segmental friction ζ_s within the chain, not the distribution of ζ_{chain} among the chains, and all chains can exhibit the same dynamics. For a middle-M PI53/PtBS42 blends with M_{PI} = 5.3×10^4, M_{PtBS} = 4.2×10^4, and w_{PI} = 56 wt% examined by Chen et al. (2011) and the high-M PI99/PtBS348 blend examined in Figure 3.17, the spatial frictional distribution appears to be well smeared within the PI chain, as judged from the Λ_C and $\langle R_{PI}^2 \rangle^{1/2}$ values: $\Lambda_C \sim 3.7$ nm ($C_{PtBS}/C_{PtBS}{}^* = 3.3$) and $\langle R_{PI}^2 \rangle^{1/2}$ = 20 nm in the former blend, and $\Lambda_C \sim 3.3$ nm ($C_{PtBS}/C_{PtBS}{}^* = 10.8$) and $\langle R_{PI}^2 \rangle^{1/2}$ = 27 nm in the latter blend. As noted in Figure 3.17 and panel (d) of Figure 3.21, the dielectric data of these blends satisfactorily obey the superposition, confirming the above expectation. At the same time, it should be emphasized that the success of the superposition of the dielectric data does not necessarily indicate the thermorheological simplicity of the PI dynamics. In fact, the slow relaxation mechanism of PI in the high-M blend changes from the reptative mechanism much slower than the Rouse equilibration (at high T) to the retarded Rouse equilibration masking the reptative relaxation (at low T), but the dielectric mode distribution coincides for these two extreme mechanisms and thus remains insensitive to T.

Now, we turn our attention to the thermorheological behavior of PtBS chains in moderately entangled middle-M PI/PtBS blends. Utilizing the data for the entanglement length a_{PI} and complex modulus $G_{PI,bulk}{}^*(\omega)$ of bulk PI, we may estimate the modulus of PI (fast component) in the blends as $G_{PI}{}^*(\omega) = \phi_{PI}\{a_{PI}/a\}^2 G_{PI,bulk}{}^*(\omega\Lambda_{PI})$ with $\Lambda_{PI} = \tau_{G,PI}/\tau_{G,PI}^{[bulk]}$ (cf. Equation 3.63). The viscoelastic terminal relaxation time $\tau_{G,PI}^{[bulk]}$ of bulk PI included in the Λ_{PI} factor is experimentally determined. The viscoelastic $\tau_{G,PI}$ of PI in the blend, the other parameter included in Λ_{PI}, is estimated from the dielectric $\tau_{\varepsilon,PI}$ data after a correction for a change of the CR/DTD contribution to the PI relaxation on blending, as discussed by Chen et al. (2008, 2011). This correction, based on the data for PI/PI and PS/PS blends, is minor in a numerical sense to typically give $\tau_{\varepsilon,PI}/\tau_{G,PI}$ = 1 to 1.3 for the middle-M PI/PtBS blends (cf. Chen et al., 2011). Thus, $G_{PI}{}^*(\omega)$ of PI in the blends is experimentally estimated in a range of ω where the entanglement segment of PI and PtBS are internally Rouse-equilibrated and the above expression of $G_{PI}{}^*(\omega)$ based on the entanglement concept is valid. The modulus $G_{PtBS}{}^*(\omega)$ of PtBS (slow component) in the blends can be evaluated by subtraction of this $G_{PI}{}^*(\omega)$ from the $G^*(\omega)$ data of the blend.

Figure 3.22 shows $G_{PtBS}{}^*(\omega)$ thus estimated for the PI53/PtBS42 blend (M_{PI} = 5.3×10^4, M_{PtBS} = 4.2×10^4) and PI20/PtBS42 blend (M_{PI} = 2×10^4, M_{PtBS} = 4.2×10^4), both having w_{PI} = 56 wt% (cf. Chen et al., 2011, 2012). In fact, those $G_{PtBS}{}^*(\omega)$ were indistinguishable from the blend modulus data (because the PI relaxed much faster than PtBS and the PI contribution to the blend modulus was very small in the range of ω examined in Figure 3.22). PtBS42 is commonly included in these blends at the same concentration (w_{PtBS} = 44 wt%).

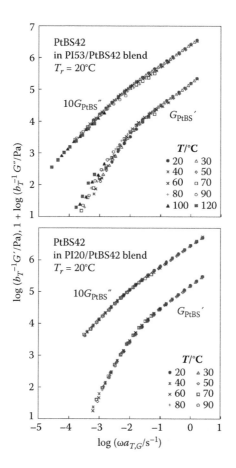

FIGURE 3.22

Test of time-temperature superposability for the modulus G_{PtBS}^* of poly(p-*tert* butyl styrene) (PtBS) components in the polyisoprene (PI)/PtBS miscible blends ($w_{PI} = 56$ wt%) as indicated. The sample code numbers of the blends denote $10^{-3}M$ of the components. G_{PtBS}^* were obtained by subtraction of the PI modulus from the blend modulus. The reference temperature is $T_r = 20°C$ for the two blends. (Redrawn, with permission, from Chen, Q., Y. Matsumiya, Y. Masubuchi, H. Watanabe, and T. Inoue. 2011. Dynamics of polyisoprene-poly(p-*tert*-butylstyrene) diblock copolymer in disordered state. *Macromolecules* 44:1585–1602.)

For clarity of the plots, G_{PtBS}'' is multiplied by a factor of 10. The G_{PtBS}' and $10\,G_{PtBS}''$ data are shifted along the ω axis by the same factor $a_{T,G}$.

The entanglement molecular weight in these blends, $M_{e,X} = \{a/a_X\}^2 M_{e,bulk\ X}$, is obtained from the a value (= 6.2 nm; Equation 3.61) as

$$M_{e,PI} = 5.7 \times 10^3 \text{ for PI}, M_{e,PtBS} = 1.1 \times 10^4 \text{ for PtBS} \qquad (3.74)$$

As judged from the molecular weights of the PI and PtBS chains, these chains are moderately entangled with each other in the blends. Clearly, $G_{PtBS}^*(\omega)$ of the PtBS42 chains satisfy the superposition in the PI20/PtBS42 blend (see the bottom panel of Figure 3.22) but not in the PI53/PtBS42 blend (top panel). The success of the superposition in the PI20/PtBS42 blend is indicative of lack of a change in the PtBS42 relaxation mechanism in the range of T examined. In this blend, the PI20 relaxation was much faster than the PtBS42 relaxation, thereby allowing the PtBS chains to relax through the same mechanism (pseudo-CR mechanism explained later) at all T. In contrast, in the PI53/PtBS42 blend, the PI53 relaxation time becomes comparable to the PtBS42 relaxation time at high T to force the PtBS42 relaxation mechanism to change at high T. This change of the mechanism with T, not observed for homopolymer systems, obviously leads to the failure of the superposition of the $G_{PtBS}^*(\omega)$ data. It should be emphasized that the difference of the PtBS behavior in the two blends results from a difference of the dynamic correlation of the PI and PtBS component chains due only to a difference of M_{PI}.

3.5.2.6 Relaxation Mechanism in Low-M PI/PtBS Blends

For the $G_{PtBS}^*(\omega)$ data of the PI20/PtBS42 blend satisfying the superposition (cf. bottom panel of Figure 3.22), the shift factor $a_{T,G}$ was well described by the WLF equation, Equation (3.28) with $C_1^{WLF} = 10.0$, $C_2^{WLF} = 116.5$ K, and the reference temperature $T_r = 25°C$, as reported by Chen et al. (2011, 2012). The equation with the same C_1^{WLF} and C_2^{WLF} values and $T_{r,bulk} = 180°C$ was valid for bulk PtBS42. Thus, the iso-τ_s state where the relaxation time τ_s ($\propto \zeta_s/T$) of the Rouse segment of PtBS has a given value is achieved in the blend and bulk PtBS42 at respective T_r.

In the top panel of Figure 3.23, the viscoelastic terminal relaxation times $\tau_{G,PtBS}$ of PtBS42 in the blend and bulk are plotted against a difference of the temperature from the iso-τ_s temperature T_{iso} (= T_r explained above). The T dependence of $\tau_{G,PtBS}$ is indistinguishable in the blend and iso-τ_s bulk, but the $\tau_{G,PtBS}$ value is larger in the blend by a factor of $\cong 10$. Similar results were found for the other middle-M PI/PtBS blends. Because the PtBS42 chains are barely entangled in their bulk state as noted from their molecular weight ($M_{PtBS42} = 4.2 \times 10^4$) and the $M_{e,bulk}$ value of PtBS (= 3.8×10^4), the difference of the $\tau_{G,PtBS}$ value in the blend and bulk is naturally attributed to the entanglement between the PtBS42 and PI20 chains in the blend (cf. Equation 3.74), as discussed by Chen et al. (2008, 2011). We note that the entanglement molecular weight in the PtBS42 solution equivalent to the PI20/PtBS42 blend after full relaxation of PI20, $M_{e,soln}^{[PtBS]} \cong M_{e,bulk}/\phi_{PtBS}^{1.3}$ (= 1.2×10^5), is much larger than M_{PtBS42}, and thus the PtBS42 chains in the blends are not entangled among themselves (which is quite different from the situation in the high-M blend examined in Figure 3.16). This lack of the PtBS42-PtBS42 entanglements results in the lack of the low-ω entanglement plateau noted in the bottom panel of Figure 3.22. Thus, the global motion of the PI20 chains entangled with the PtBS chains should activate the terminal relaxation of the PtBS42

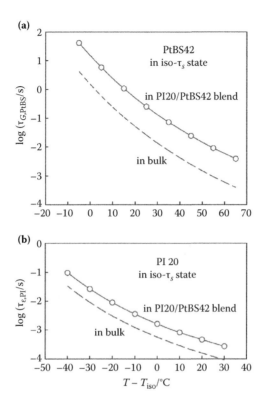

FIGURE 3.23

Comparison of the terminal relaxation time of the components PtBS42 (in panel (a)) and PI20 (in panel (b)) in the PI20/PtBS42 blend ($M_{PI} = 2 \times 10^4$, $M_{PtBS} = 4.2 \times 10^4$, $w_{PI} = 56$ wt%) and in respective bulk systems. The comparison is made in the iso-τ_s state where the Rouse segment of polyisoprene (PI) and poly(p-$tert$ butyl styrene) (PtBS) has the same relaxation time in the blend and in bulk. (Data taken, with permission, from Chen, Q., Y. Matsumiya, Y. Masubuchi, H. Watanabe, and T. Inoue. 2011. Dynamics of polyisoprene-poly(p-$tert$-butylstyrene) diblock copolymer in disordered state. *Macromolecules* 44:1585–1602; Chen, Q., Y. Matsumiya, K. Hiramoto, and H. Watanabe. 2012. Dynamics in miscible blends of polyisoprene and poly(p-$tert$-butyl styrene): Thermo-rheological behavior of components. *Polymer J.* 44:102–114.)

chains. This relaxation mechanism is somewhat similar to the CR mechanism explained earlier, but there is an essential difference. The terminal CR time τ_{CR} is essentially proportional to the global relaxation time of the fast component chains activating the CR relaxation. Nevertheless, the temperature dependence of the relaxation time is quite different for the PtBS42 and PI20 chains (as noted from the top and bottom panels of Figure 3.23), thereby severely violating this proportionality. Thus, the PtBS42 chains appear to have relaxed through a pseudo-CR mechanism for which the global motion of the fast PI20 chains stitching the neighboring PtBS42 chains triggers the PtBS42 relaxation, but the rate of this relaxation is determined by the hopping of the "PtBS42 entanglement segment in the blend" (with $M_e^{[PtBS]} = 1.1 \times 10^4$) slower than the global motion of PI20, as discussed by Chen et al. (2008). This hopping of the PtBS entanglement segment is governed by ζ_s of PtBS (obeying the WLF relationship for PtBS), which results in the pseudo-CR relaxation of PtBS42 being triggered by the PI20 motion but exhibiting the T dependence of $\tau_{G,PtBS}$ different from the dependence of the PI20 relaxation time $\tau_{\varepsilon,PI}$.

The time-temperature superposition (almost) holds for the dielectric data of the PI20 chains in the PI20/PtBS42 blend; cf. panel (d) of Figure 3.21. The corresponding shift factor $a_{T,\varepsilon}$ was well described by the WLF equation, Equation (3.28) with $C_1^{WLF} = 4.425$, $C_2^{WLF} = 140$ K, and $T_r = 60°C$, as reported by Chen et al. (2011, 2012). This WLF equation coincided with that for bulk PI20 with $T_{r,bulk} = 30°C$, so that the PI20 chains in the blend and bulk at respective T_r are in the iso-τ_s state. The bottom panel of Figure 3.23 compares the dielectric terminal relaxation times $\tau_{\varepsilon,PI}$ of PI20 in this state. The T dependence of $\tau_{\varepsilon,PI}$ is indistinguishable in the blend and iso-τ_s bulk, but the $\tau_{\varepsilon,PI}$ value is larger in the blend by a factor of $\cong 3$. This difference of $\tau_{\varepsilon,PI}$ can be attributed to the entanglement constraint from the slow PtBS42 chains that remains effective until the PI20 chains fully relax to trigger the pseudo-CR relaxation of PtBS42, as discussed by Chen et al. (2008).

3.5.2.7 Thermal Behavior Distinguishing Monomeric and Rouse Segments

For chemically uniform homopolymer systems, it is well known that the thermally activated motion of the monomeric segments is intimately related to the glass transition, whereas the motion of the Rouse segments is the basic process for the rubbery/terminal relaxation (see Section 3.3). The temperature dependence of the relaxation time is not identical for these two types of segments at low T, as noted from a difference of the shift factors in the glassy and rubbery relaxation regimes.

For miscible blends of chemically different chains, an additional interesting situation emerges. Because the two components have different $T_{g,eff}$, one component (A) can be in the glassy state while the other component (B) can be in the rubbery state in a range of T between $T_{g,eff}^{[A]}$ and $T_{g,eff}^{[B]}$ ($< T_{g,eff}^{[A]}$). This interesting situation can be noted for the thermal behavior of the blends. As an example, Figure 3.24 shows the differential scanning calorimetry (DSC) profile obtained for the PI53/PtBS42 blend ($w_{PI} = 56$ wt%) dielectrically and viscoelastically examined in panel (d) of Figure 3.21 and the top panel of Figure 3.22. $T_{g,eff}$ is higher for PtBS42 (slow component) than for PI20 (fast component). The DSC profiles for bulk components are also shown for comparison.

As seen in Figure 3.24, the DSC profile of the PI53/PtBS42 blend exhibits very broad glass transition. Such broad transition is well known for miscible blends having a large dynamic asymmetry between the components (see

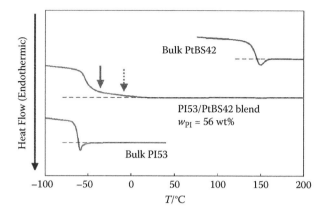

FIGURE 3.24
Differential scanning calorimetry (DSC) profile obtained for the PI53/PtBS42 blend (w_{PI} = 56 wt%) examined in Figure 3.21(d) and the top panel of Figure 3.22. The DSC profiles of bulk PI53 and PtBS42 are also shown for comparison. The solid and dotted arrows, respectively, indicate the effective $T_{g,eff}$ of the Rouse segments of polyisoprene (PI) and poly(*p-tert* butyl styrene) (PtBS) obtained from William–Landell–Ferry (WLF) analysis of their shift factors at high T ($>T_{g,eff}^{[PtBS]}$). Note that these $T_{g,eff}$ are the temperatures extrapolated from the high-T zone where the Rouse segments of both PI and PtBS exhibit active thermal motion. (Data taken, with permission, from Chen, Q., Y. Matsumiya, Y. Masubuchi, H. Watanabe, and T. Inoue. 2011. Dynamics of polyisoprene-poly(*p-tert*-butylstyrene) diblock copolymer in disordered state. *Macromolecules* 44:1585–1602.)

Section 3.3). More importantly, the shift factors of the viscoelastic and dielectric data of the PI53/PtBS42 blend measured at high $T \geq 20°C$ ($> T_{g,eff}^{[PtBS]}$) were of WLF type. The motion of the Rouse segments is the basic dynamic process underlying those high T data, and the WLF analysis of the shift factors specified the iso-τ_s temperature, T_{iso}, for the Rouse segments of PI and PtBS in the blend defined with respect to their bulk systems at a given T (cf. Chen et al., 2012). This T_{iso} is the temperature extrapolated from the high-T zone where the Rouse segments of both PI and PtBS exhibit active thermal motion. Thus, T_{iso} corresponding to $T_{g,bulk}$ of bulk PI53 and bulk PtBS42 ($T_{g,bulk}^{[PI]}$ $= -65°C$ and $T_{g,bulk}^{[PtBS]} = 147°C$) can be utilized as the effective $T_{g,eff}$ of the Rouse segments *extrapolated* from this high-T zone without quenching the motion of the PtBS segments. These $T_{g,eff}$ are shown with the arrows in Figure 3.24. The $T_{g,eff}$ for the Rouse segment of PtBS (dotted arrow) is located at the high-T end of the broad, thermal glass transition detected with DSC, the latter corresponding to the vitrification of the monomeric segments of PtBS occurring in a matrix of liquid PI. In this situation, the monomeric segments of PtBS have the maximum freedom of motion for vitrification (aimed by the rapid motion of the PI segments), which possibly allows the extrapolated $T_{g,eff}$ of the Rouse segment of PtBS to be close to the thermal $T_{g,eff}$ of the monomeric segments. The situation is different for PI. The thermal $T_{g,eff}$ of the monomeric segments of PI corresponds to the vitrification of those segments in a matrix of glassy PtBS, while the extrapolated $T_{g,eff}$ of the Rouse segment of PI does not include the effect of the glassy PtBS matrix. This difference probably resulted in the considerable deviation between the thermal $T_{g,eff}$ and the extrapolated $T_{g,eff}$ (solid arrow) seen in Figure 3.24. For similar PI/PtBS blends, Zhao et al. (2009) conducted detailed analysis of the thermal behavior to find a crossover of the dynamics of the *monomeric* PI segment from the WLF-type dynamics (at $T > T_{g,eff}^{[PtBS]}$) to the Arrhenius-type dynamics (at $T < T_{g,eff}^{[PtBS]}$). This crossover, corresponding to the effect of the glassy PtBS matrix explained above, deserves further study.

3.6 Concluding Remarks

This chapter summarizes characteristic features of the local and global dynamics in miscible blends of chemically different chains. For the local dynamics intimately related to the glass transition, the separate relaxation of the monomeric segments of different components is the most prominent feature. This separation is mainly attributed to the self-concentration due to the chain connectivity and partly to the intrachain torsional barrier. The self-concentration provides the segments of different components with different dynamic environments thereby allowing the separate relaxation of those segments. The concentration fluctuation further broadens the segmental relaxation. A model analysis based on the dielectric data suggests that the size of the monomeric segment is not very different from the size of the statistically well-defined Kuhn segment, although some uncertainty still remains.

For the global dynamics governing the rubbery/terminal relaxation, the thermorheological complexity of the components is one of the most prominent features. In PI/PVE blends associated with just a moderate dynamic asymmetry of the components, respective components exhibit very minor complexity and behave similarly to the components in chemically uniform blends such as PI/PI blends, as revealed from rheo-optical and dielectric studies. (PI chains have the type-A dipole so that their global motion is dielectrically detected.) The entanglement relaxation in the PI/PVE blends appears to occur through the mechanisms known for the chemically uniform blends, for example, through the reptation and constraint release (CR)/ dynamic tube dilation (DTD) mechanisms.

For PI/PtBS blends having a much larger dynamic asymmetry, respective components often exhibit significant thermorheological complexity according to their molecular weights and composition, as revealed from viscoelastic and dielectric studies. In particular, in well-entangled high-M PI/PtBS blends at low T, the fast component (PI) exhibits the cooperative Rouse equilibration retarded by the slow PtBS chains, and this Rouse relaxation process masks the high-ω entanglement plateau sustained by the PI and PtBS chains. At high T, the cooperative Rouse equilibration becomes much faster than the global motion of PI, thereby allowing the high-ω plateau and the successive CR/DTD/reptation relaxation process to be clearly observed. This change of the dominant relaxation mechanism of PI with T results in significant thermorheological complexity for the PI chains as well as for the PtBS chains entangled with those PI chains.

Furthermore, for low-M PI/PtBS blends, another type of complexity emerges for PI (fast component) because of the scarce overlapping of the low-M PtBS chains (slow component) that results in dynamic undulation of the PtBS concentration. This undulation is quenched to give a spatial heterogeneity of the frictional environment in the time scale of the global relaxation of the PI chains. For low-M PI chains, this frictional heterogeneity is not smeared/averaged within the chain backbone and thus gives a distribution of the chain friction coefficient for the ensemble of the PI chains. This distribution of the chain friction naturally results in the thermorheological complexity for the ensemble of the PI chains as a whole.

Thus, the miscible blends offer a very rich field for research of polymer dynamics. In particular, the data available by now suggest that the monomeric segments of different component chains relax in an essentially independent way in a very local scale, the Rouse segments of these components exhibit highly cooperative motion (equilibration) in the length sale of entanglement, and the component chain motion is moderately cooperative in a larger length scale for the global relaxation. However, the crossover of the magnitude of cooperativity has not been quantitatively understood. This crossover behavior and related subjects (such as the difference between the thermally determined $T_{g,eff}$ and viscoelastically/dielectrically extrapolated $T_{g,eff}$) deserve further studies.

References

Adachi, K., and T. Kotaka. 1993. Dielectric normal mode relaxation. *Prog. Polym. Sci.* 18:585–622.

Adam, G., and J. H. Gibbs. 1965. On temperature dependence of cooperative relaxation properties in glass-forming liquids. *J. Chem. Phys.* 43:139–146.

Alegria, A., J. Colmenero, K. L. Ngai, and C. M. Roland. 1994. Observation of the component dynamics in a miscible polymer blend by dielectric and mechanical spectroscopies. *Macromolecules* 27:4486–4492.

Arendt, B. H., R. Krishnamoorti, J. A. Kornfield, and S. D. Smith. 1997. Component dynamics in miscible blends: Equally and unequally entangled polyisoprene/polyvinylethylene. *Macromolecules* 30:1127–1137.

Ball, R. C., and T. C. B. McLeish. 1989. Dynamic dilution and the viscosity of star polymer melts. *Macromolecules* 22:1911–1913.

Cangialosi, D., A. Alegria, and J. Colmenero. 2006. Predicting the time scale of the component dynamics of miscible polymer blends: The polyisoprene/poly(vinylethylene) case. *Macromolecules* 39:7149–7156.

Chen, Q., Y. Matsumiya, Y. Masubuchi, H. Watanabe, and T. Inoue. 2008. Component dynamics in polyisoprene/poly(4-*tert*-butylstyrene) miscible blends. *Macromolecules* 41:8694–8711.

Chen, Q., A. Uno, Y. Matsumiya, and H. Watanabe. 2010. Viscoelastic mode distribution of moderately entangled linear polymers. *J. Soc. Rheol. Japan (Nihon Reoroji Gakkaishi)* 38:187–193.

Chen, Q., Y. Matsumiya, Y. Masubuchi, H. Watanabe, and T. Inoue. 2011. Dynamics of polyisoprene-poly(*p-tert*-butylstyrene) diblock copolymer in disordered state. *Macromolecules* 44:1585–1602.

Chen, Q., Y. Matsumiya, K. Hiramoto, and H. Watanabe. 2012. Dynamics in miscible blends of polyisoprene and poly(*p-tert*-butyl styrene): Thermo-rheological behavior of components. *Polymer J.* 44:102–114.

Chung, G. C., J. A. Kornfield, and S. D. Smith. 1994. Component dynamics in miscible polymer blends—A 2-dimensional deuteron NMR investigation. *Macromolecules* 27:964–973.

Colby, R. H., and J. E. G. Lipson. 2005. Modeling the segmental relaxation time distribution of miscible polymer blends: Polyisoprene/poly(vinylethylene). *Macromolecules* 38:4919–4928.

Cole, R. 1967. Correlation function theory of dielectric relaxation. *J. Phys. Chem.* 42:637–643.

Doi, M., and S. F. Edwards. 1986. *The Theory of Polymer Dynamics*. Oxford: Oxford University Press.

Doi, M., D. Pearson, J. Kornfield, and G. Fuller. 1989. Effect of nematic interaction in the orientational relaxation of polymer melts. *Macromolecules* 22:1488–1490.

Doi, M., and H. Watanabe. 1991. Effect of nematic interaction on the Rouse dynamics. *Macromolecules* 24:740–744.

Doxastakis, M., K. Chrissopoulou, A. Aouadi, B. Frick, T. P. Lodge, and G. Fytas. 2002, Segmental dynamics of disordered styrene-isoprene tetrablock copolymers. *J. Chem. Phys.* 116:4707–4714.

Ediger, M. D., T. R. Lutz, and Y. He. 2006. Dynamics in glass-forming mixtures: Comparison of behavior of polymeric and non-polymeric components. *J. Noncyrst. Solids* 352:4718–4723.

Ekanayake, P., H. Menge, M. E. Ries, and M. G. Brereton. 2002. Influence of polymer chain concentration and molecular weight on deformation-induced ^{2}H NMR line splitting. *Macromolecules* 35:4343–4346.

Endoh, M. K., M. Takenaka, T. Inoue, H. Watanabe, and T. Hashimoto. 2008. Shear small-angle light scattering studies of shear-induced concentration fluctuations and steady state viscoelastic properties. *J. Chem. Phys.* 128:Article # 164911.

Ferry, J. D. 1980. *Viscoelastic Properties of Polymers* (3rd ed.). New York: Wiley.

Fetters, L. J., D. J. Lohse, D. Richter, T. A. Witten, and A. Zirkel. 1994. Connection between polymer molecular-weight, density, chain dimensions, and melt viscoelastic properties. *Macromolecules* 27:4639–4647.

Fetters, L. J., D. J. Lohse, and W. W. Graessley. 1999. Chain dimensions and entanglement spacings in dense macromolecular systems. *J. Polym. Sci. Part B: Polym. Phys.* 37:1023–1033.

Fetters, L. J., D. J. Lohse, and R. H. Colby. 2007. Chain dimensions and entanglement spacings. In *Physical Properties of Polymer Handbook* (2nd ed.), Ed. J. E. Mark, 447–454. New York: Springer.

de Gennes, P. G. 1979. *Scaling Concept in Polymer Physics*. Ithaca: Cornell University Press.

Graessley, W. W. 1974. The entanglement concept in polymer rheology. *Adv. Polym. Sci.* 16:1–179.

Graessley, W. W. 1982. Entangled linear, branched and network polymer systems— Molecular theories. *Adv. Polym. Sci.* 47:67–117.

Graessley, W. W. 2008. *Polymeric Liquids and Networks: Dynamics and Rheology*. New York: Garland Science.

Haley, J. C., T. P. Lodge, Y. Y. He, M. D. Ediger, E. D. von Meerwall, and J. Mijovic. 2003. Composition and temperature dependence of terminal and segmental dynamics in polyisoprene/poly(vinylethylene) blends. *Macromolecules* 36:6142–6151.

Haley, J. C., and T. P. Lodge. 2005. Viscosity predictions for model miscible polymer blends: Including self-concentration, double reptation, and tube dilation. *J. Rheol.* 49:1277–1302.

Hirose, Y., O. Urakawa, and K. Adachi. 2003. Dielectric study on the heterogeneous dynamics of miscible polyisoprene/poly(vinyl ethylene) blends: Estimation of the relevant length scales for the segmental relaxation dynamics. *Macromolecules* 36:3699–3708.

Hirose, Y., O. Urakawa, and K. Adachi. 2004. Dynamics in disordered block copolymers and miscible blends composed of poly(vinyl ethylene) and polyisoprene. *J. Polym. Sci. Part B: Polym. Phys.* 42:4084–4094.

Inoue, T., H. Okamoto, and K. Osaki. 1991. Dynamic birefringence of amorphous polymers I: Measurement on polystyrene. *Macromolecules* 24:5670–5675.

Inoue, T., Y. Mizukami, H. Okamoto, H. Matsui, H. Watanabe, T. Kanaya, and K. Osaki. 1996. Dynamic birefringence of vinyl polymers. *Macromolecules* 29:6240–6245.

Inoue, T., H. Matsui, and K. Osaki. 1997. Molecular origin of viscoelasticity and chain orientation of glassy polymers. *Rheol. Acta* 36:239–244.

Inoue, T., T. Uemtasu, and K. Osaki. 2002. The significance of the Rouse segment: Its concentration dependence. *Macromolecules* 35:820–826.

Kamath, S. K., R. H. Colby, S. K. Kumar, K. Karatasos, G. Floudas, G. Fytas, and J. E. L. Roovers. 1999. Segmental dynamics of miscible polymer blends: Comparison of the predictions of a concentration fluctuation model to experiment. *J. Chem. Phys.* 111:6121–6128.

Kanaya, T., and K. Kaji. 2001. Dynamics in the glassy state and near the glass transition of amorphous polymers as studied by neutron scattering. *Adv. Polym. Sci.* 154:87–141.

Klein, J. 1986. Dynamics of entangled linear, branched, and cyclic polymers. *Macromolecules* 19:105–118.

Kornfield, J. A., Fuller, G. G., and Pearson, D. S. 1989. Infrared dichroism measurements of molecular relaxation in binary blend melt rheology. *Macromolecules* 22:1334–1345.

Kremer, F., and A. Schönhals. 2003. *Broadband Dielectric Spectroscopy*. Berlin: Springer.

Kubo, R. 1957. Statistical-mechanical theory of irreversible processes. I. General theory and simple applications to magnetic and conduction problems. *J. Phys. Soc. Japan* 12:570–586.

Kumar, S. K., R. H. Colby, S. H. Anastasiadis, and G. Fytas. 1996. Concentration fluctuation induced dynamic heterogeneities in polymer blends. *J. Chem. Phys.* 105:3777–3788.

Lodge, T. P., and T. C. B. McLeish. 2000. Self-concentrations and effective glass transition temperatures in polymer blends. *Macromolecules* 33:5278–5284.

Lutz, T. R., Y. He, M. D. Ediger, M. Pitsikalis, and N. Hadjichristidis. 2004. Dilute polymer blends: Are the segmental dynamics of isolated polyisoprene chains slaved to the dynamics of the host polymer? *Macromolecules* 37:6440–6448.

Marrucci, G. 1985. Relaxation by reptation and tube enlargement—A model for polydisperse polymers. *J. Polym. Sci. Polym. Phys. Ed.* 23:159–177.

Matsumiya, Y., A. Uno, H. Watanabe, T. Inoue, and O. Urakawa. 2011. Dielectric and viscoelastic investigation of segmental dynamics of polystyrene above glass transition temperature: Cooperative sequence length and relaxation mode distribution. *Macromolecules*, 44:4355–4363.

McLeish, T. C. B. 2002. Tube theory of entangled polymer dynamics. *Adv. Phys.* 51:1379–1527.

Miller, J. B., K. J. McGrath, C. M. Roland, C. A. Trask, and A. N. Garroway. 1990. Nuclear-magnetic-resonance study of polyisoprene poly(vinylethylene) miscible blends. *Macromolecules* 23:4543–4547.

Milner, S. T., and T. C. B. McLeish. 1998. Reptation and contour-length fluctuations in melts of linear polymers. *Phys. Rev. Lett.* 81:725–728.

Miura, N., W. J. MacKnight, S. Matsuoka, and F. E. Karasz. 2001. Comparison of polymer blends and copolymers by broadband dielectric analysis. *Polymer* 42:6129–6140.

Osaki, K., T. Inoue, T. Uematsu, and Y. Yamashita. 2001. Evaluation methods of the longest Rouse relaxation time of an entangled polymer in a semidilute solution. *J. Polym. Sci. Part B: Polymer Phys.* 39:1704–1712.

Pathak, J. A., R. H. Colby, S. Y. Kamath, S. K. Kumar, and R. Stadler. 1998. Rheology of miscible blends: SAN and PMMA. *Macromolecules* 31:8988–8997.

Pathak, J. A., R. H. Colby, G. Floudas, and R. Jerome. 1999. Dynamics in miscible blends of polystyrene and poly(vinyl methyl ether). *Macromolecules* 32:2553–2561.

Pathak, J. A., S. K. Kumar, and R. H. Colby. 2004. Miscible polymer blend dynamics: Double reptation predictions of linear viscoelasticity in model blends of polyisoprene and poly(vinyl ethylene). *Macromolecules* 37:6994–7000.

Rubinstein, M., and R. Colby. 2003. *Polymer Physics.* New York: Oxford University Press.

Sakaguchi, T., N. Taniguchi, O. Urakawa, and K. Adachi. 2005. Calorimetric study of dynamical heterogeneity in blends of polyisoprene and poly(vinylethylene). *Macromolecules* 38:422–428.

Sawada, T., X. Qiao, and H. Watanabe. 2007. Viscoelastic relaxation of linear polyisoprenes: Examination of constraint release mechanism. *J. Soc. Rheol. Japan (Nihon Reoroji Gakkaishi)* 35:11–20.

Schausberger, A., G. Schindlauer, and H. Janeschitz-Kriegl. 1985. Linear elastico-viscous properties of molten standard polystyrenes I. Presentation of complex moduli; role of short range structural parameters. *Rheol. Acta* 24:220–227.

Takada, J., H. Sasaki, Y. Matsushima, A. Kuriyama, Y. Matsumiya, H. Watanabe, K. H. Ahn, and W. Yu. 2008. Component chain dynamics in a miscible blend of low-M poly(p-t-butyl styrene) and polyisoprene. *Nihon Reoroji Gakkaishi (J. Soc. Rheol. Japan)* 36:35–42.

Tomlin, D. W., and C. M. Roland. 1992. Negative excess enthalpy in a Van der Waals polymer mixture. *Macromolecules* 25:2994–2996.

Uno, A. 2009. *Dielectric Determination of Segmental Size of Polystyrene*, MS Dissertation, Graduate School of Engineering, Kyoto University, Japan.

Urakawa, O., Y. Fuse, H. Hori, Q. Tran-Cong, and O. Yano. 2001. A dielectric study on the local dynamics of miscible polymer blends: Poly(2-chlorostyrene)/poly(vinyl methyl ether). *Polymer* 42:765–773.

Urakawa, O. 2004. Studies on dynamic heterogeneity in miscible polymer blends and dynamics of flexible polymer. *Nihon Reoroji Gakkaishi (J. Soc. Rheol. Japan)* 32:265–270.

Utracki, L. A. 1989. *Polymer Alloys and Blends*. Munich: Carl Hanser Verlag.

Watanabe, H., T. Kotaka, and M. Tirrell. 1991. Effect of orientational coupling due to nematic interaction on relaxation of Rouse chains. *Macromolecules* 24:201–208.

Watanabe, H. 1999. Viscoelasticity and dynamics of entangled polymers. *Prog. Polym. Sci.* 24:1253–1403.

Watanabe, H. 2001. Dielectric relaxation of type-A polymers in melts and solutions. *Macromol. Rapid Commun.* 22:127–175.

Watanabe, H., Y. Matsumiya, and T. Inoue. 2002. Dielectric and viscoelastic relaxation of highly entangled star polyisoprene: Quantitative test of tube dilation model. *Macromolecules* 35:2339–2357.

Watanabe, H., S. Ishida, Y. Matsumiya, and T. Inoue. 2004a. Viscoelastic and dielectric behavior of entangled blends of linear polyisoprenes having widely separated molecular weights: Test of tube dilation picture. *Macromolecules* 37:1937–1951.

Watanabe, H., S. Ishida, Y. Matsumiya, and T. Inoue. 2004b. Test of full and partial tube dilation pictures in entangled blends of linear polyisoprenes. *Macromolecules* 37:6619–6631.

Watanabe, H., T. Sawada, and Y. Matsumiya. 2006. Constraint release in star/star blends and partial tube dilation in monodisperse star systems. *Macromolecules* 39:2553–2561.

Watanabe, H., Y. Matsumiya, J. Takada, H. Sasaki, Y. Matsushima, A. Kuriyama, T. Inoue, K. H. Ahn, W. Yu, and R. Krishnamoorti. 2007. Viscoelastic and dielectric behavior of a polyisoprene/poly(4-tert-butyl styrene) miscible blend. *Macromolecules* 40:5389–5399.

Watanabe, H., Y. Matsumiya, E. van Ruymbeke, D. Vlassopoulos, and N. Hadjichristidis. 2008. Viscoelastic and dielectric relaxation of a Cayley-tree type polyisoprene: Test of molecular picture of tube dilation. *Macromolecules* 41:6110–6124.

Watanabe, H., and O. Urakawa. 2009. Component dynamics in miscible polymer blends: A review of recent findings. *Korean-Australian Rheol. J.* 21:235–244.

Watanabe, H. 2009. Slow dynamics in homopolymer liquids. *Polymer J.* 41:929–950.

Watanabe, H., Q. Chen, Y. Kawasaki, Y. Matsumiya, T. Inoue, and O. Urakawa. 2011. Entanglement dynamics in miscible polyisoprene/poly(*p-tert*-butylstyrene) blends. *Macromolecules* 44:1570–1584.

Ylitalo, C. M., J. A. Kornfield, G. G. Fuller, and D. S. Pearson. 1991a. Molecular-weight dependence of component dynamics in bidisperse melt rheology. *Macromolecules* 24:749–758.

Ylitalo, C. M., and G. G. Fuller. 1991b. Temperature effects on the magnitude of orientational coupling interactions in polymer melts. *Macromolecules* 24:5736–5737.

Yurekli, K., and R. Krishnamoorti. 2004. Thermodynamic interactions in blends of poly(4-*tert*-butyl styrene) and polyisoprene by small-angle neutron scattering. *J. Polym. Sci. Part B: Polym. Phys.* 42:3204–3217.

Zawada, J. A., G. G. Fuller, R. H. Colby, and L. J. Fetters. 1994a. Measuring component contributions to the dynamic modulus in miscible polymer blends. *Macromolecules* 27:6851–6860.

Zawada, J. A., G. G. Fuller, R. H. Colby, L. J. Fetters, and J. Roovers. 1994b. Component dynamics in miscible blends of 1,4-polyisoprene and 1,2-polybutadiene. *Macromolecules* 27:6861–6870.

Zetsche, A., and E. W. Fischer. 1994. Dielectric studies of the alpha-relaxation in miscible polymer blends and its relation to concentration fluctuations. *Acta Polym.* 45:168–175.

Zhang, R. Y., H. Cheng, C. G. Zhang, T. C. Sun, X. Dong, and C. C. Han. 2008. Phase separation mechanism of polybutadiene/polyisoprene blends under oscillatory shear flow. *Macromolecules* 41:6818–6829.

Zhao, J., L. Zhang, and M. D. Ediger. 2008. Poly(ethylene oxide) dynamics in blends with poly(vinyl acetate): Comparison of segmental and terminal dynamics. *Macromolecules* 41:8030–8037.

Zhao, J. S., M. D. Ediger, Y. Sun, and L. Yu. 2009. Two DSC glass transitions in miscible blends of polyisoprene/poly(4-tert-butylstyrene). *Macromolecules* 42:6777–6783.

4

Shape Memory Polymer Blends

Young-Wook Chang

Hanyang University
Ansan, Korea

CONTENTS

4.1 Introduction

A shape memory polymer (SMP) is a smart material that can memorize its original shape after being deformed into a temporary shape when it is heated or receives any other external stimuli such as light, electric field, magnetic field, chemical, moisture, and pH change [1–6]. Compared with shape memory alloys (SMAs), the SMP has many advantages of low cost, low density, substantially high elastic deformation, and facile tuning of switching temperature at which shape recovery occurs and elastic recovery stress can be tailored by the variation of structural parameters of the molecular architecture along with a good processibility. Moreover, the SMP can possess biofunctionality and biodegradability. These features make the SMP have diverse applications including smart textiles [7–9], self-deployable sun sails in spacecraft [10], biomedical devices [11,12], or implants for minimally invasive surgery [13,14].

Shape memory polymers basically have two structural features—that is, the cross-links that determine the permanent shape and the reversible segments acting as a switching phase. Molecular mechanisms describing thermally triggered shape memory process are shown in Figure 4.1. In the figure, T_{trans} is the thermal transition temperature of the reversible phase

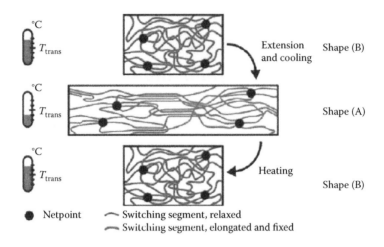

°C
T_{trans}

Extension
and cooling

Shape (B)

°C
T_{trans}

Shape (A)

°C
T_{trans}

Heating

Shape (B)

● Netpoint   ⌒ Switching segment, relaxed
              ⌒ Switching segment, elongated and fixed

FIGURE 4.1
Molecular mechanism of thermally triggered shape memory effect of polymers. T_{trans} = thermal transition temperature related to the switching phase. (Adapted from Lendlein, A., and Kelch, S. 2002. Shape-memory polymers. *Angewandte Chemie, International Edition* 41:2034–2057. Copyright Wiley-VCH Verlag GmbH & Co. KGaA. Reproduced with permission.)

(either glass transition temperature, T_g or crystalline melting temperature, T_m). If the increase in temperature is higher than T_{trans} of the switching segments, these segments are flexible and the polymer can be deformed elastically. The temporary shape is fixed by cooling down below T_{trans}. Thus, the work performed on the sample can be stored as latent strain energy if the recovery of the polymer chains is prohibited by vitrification or crystallization. The fixed state is stable for long times. Upon subsequent heating above T_{trans}, the stored strain energy can be released as the polymer chains are liberated. The strain or shape that the sample returns to is the original shape dictated during cross-linking, whether chemical (covalent bonds) or physical (associations). The rigidity of the polymer, represented by modulus, and the work that will be saved during deformation is determined by the cross-link density. The vitrification or crystallization of the polymer controls the fixing of the polymer chains and therefore allows setting of an arbitrary temporary shape. On the basis of the nature of the cross-links, SMPs are subdivided into physically cross-linked SMPs and chemically cross-linked SMPs.

Macroscopically, the shape memory effect can be described and evaluated using a cyclic tensile deformation experiment under a temperature programming, represented as a three-dimensional (3-D) plot of strain versus temperature and force as depicted in Figure 4.2. The sample is deformed to a certain level of strain at $T > T_{switch}$, and then can be fixed during cooling, as represented by the horizontal unloading curve at $T < T_{switch}$. Note that the shape fixing in this plot is achieved during cooling under fixed stress.

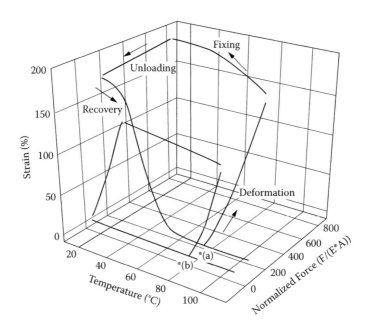

FIGURE 4.2
Three-dimensional (3-D) plot of the shape memory cycle for (a) a shape memory polymer (SMP) and (b) vulcanized natural rubber. The star indicates the start of the experiment (initial sample dimensions, temperature, and load). Both the SMP and the rubber were deformed under constant loading rate at constant temperature. The deformation step was then followed by a cooling step under constant load. At low temperature, the load was removed and shape fixing was observed for the SMP, but an instant recovery was seen for natural rubber. Shape recovery of the primary equilibrium shape was obtained by heating the SMP. (Adapted from Liu, C., Qin, H., and Mather, P. T. 2007. Review of progress in shape-memory polymers. *Journal of Materials Chemistry* 17:1543–1558. Copyright Royal Society of Chemistry. Reproduced with permission.)

In general, release of stress during the fixing stage will also lead to a slight strain decrease, depending on the extent of fixing. When the sample is heated at $T > T_{trans}$, the sample is recovered to its original dimension by releasing the stored strain energy as the polymer chains are liberated. Curve (b) in Figure 4.2b shows the behavior of natural rubber under the same thermomechanical cycle, which indicates that the rubber does not keep its fixing shape in the temperature range examined, instead, unloading returns the sample to its equilibrium strain instantly.

There are several excellent reviews on the SMP [15–20]. The present review focuses on the SMP synthesized from polymer blending technique. Polymer blending is an effective and economical way to develop the new polymeric materials with desirable properties. Numerous polymer blends exhibiting shape memory effects have been reported, which can be classified into miscible blend, immiscible blend, interpenetrating polymer network (IPN), and cross-linked blends.

4.2 Miscible Polymer Blends

A miscible polymer blend refers to a blend forming a single phase structure, which can be evidenced by the presence of one T_g between the component polymers.

SMP based on miscible blends of semicrystalline polymer/amorphous polymer was reported by the Mather research group, which included semicrystalline polymer/amorphous polymer such as polylactide (PLA)/polyvinylacetate (PVAc) blend [21,22], poly(vinylidene fluoride) (PVDF)/PVAc blend [23], and PVDF/polymethyl methacrylate (PMMA) blend [23]. These polymer blends are completely miscible at all compositions with a single, sharp glass transition temperature, while crystallization of PLA or PVDF is partially maintained and the degree of crystallinity, which controls the rubbery stiffness and the elasticity, can be tuned by the blend ratios. T_g of the blends are the critical temperatures for triggering shape recovery, while the crystalline phase of the semicrystalline PLA and PVDF serves well as a physical cross-linking site for elastic deformation above T_g, while still below T_m.

Another miscible semicrystalline polymer/amorphous polymer blend SMP is a polyethylene oxide (PEO)/novolac-type phenolic resin blend [24]. The blend was found to be completely miscible in the amorphous phase when the phenolic content is up to 30 wt%, and the crystalline melting temperature (T_m) of the PEO phase working as a transition temperature can be tuned.

A miscible amorphous/amorphous polymer blend exhibiting shape memory effects was reported by Choi and Chang [25]. The blend was composed of poly(epichlorohydrin) (PECH) rubber with a molecular weight of 700,000 g/mol and poly(styrene-co-acrylonitrile) (SAN) with an acrylonitrile (AN) content of 26% by weight. The blends are optically transparent and showed a single glass transition in the blends that varied from 6 to 99°C by increasing the SAN content in the blend from 20 to 80 wt%. Also, the modulus and strength of the blend can be tuned smoothly with composition of the blend. The PECH/SAN blends with a T_g well above room temperature, together with appreciable deformation and high strength, showed excellent shape retention and shape recovery when the sample deformed at $T > T_g$ was quenched at $T = T_g - 20°C$ in a stress-free state, and then heated above its T_g again. In the blend, chain entanglement in high molecular weight PECH was supposed to act as a cross-linking point.

Binary blends containing a multiblock copolymer as one component or both components in the blend, in which a certain block of the copolymer is miscible with the other polymer, were also reported to be an effective way to fabricate SMP with desired properties. This type of SMP blend include the segmented thermoplastic polyurethane (TPU)/phenoxy resin blend and TPU/polyvinyl chloride (PVC) blend [26,27]. In the blends, the soft segment of the TPU is miscible with the phenoxy resin or PVC, and T_g of this miscible amorphous domain can be tuned smoothly with composition of the blend. The blends had shape memory effect when the miscible domain was utilized as a reversible phase and phase separated hard segment domain in TPU was utilized as a fixed structure memorizing the original shape.

Ajili et al. reported on the shape memory effect of partially miscible blends composed of segmented polyurethane containing poly(ε-caprolactone) (PCL) diol as a soft segment as one component and PCL as another component [28]. It was shown that shape recovery temperature of the blends is around the melting temperature of PCL. The melting behavior of the PCL in the blends is strongly influenced by the blend composition and crystallization conditions; thus, the shape recovery temperature can be adjusted to the range of body temperature. Besides shape memory effects, the blend showed excellent biocompatibility, indicated by adhesion and proliferation of bone marrow mesenchymal stem cells, suggesting that the polyurethane (PU)/PCL SMP blend could be a potential candidate for stent implant applications.

Erden studied the shape memory behavior of a miscible blend of TPU-based SMP with polybenzoxazine (PB-a) [29]. The blends were prepared by in situ polymerization method; benzoxazine monomer, which is miscible with the PU prepolymer derived from 4,4'-methylenebis (phenyl isocyanate) (MDI) and poly(tetramethylene) glycol (PTMG) with average molecular weight of 650 g/mol, was polymerized into polybenzoxazine (PB-a) by thermal curing at 180°C to make TPU/PB-a blends. The blend showed a higher level of elastically stored energy than the TPU due to chemical and physical interactions of the PB-a with a polyurethane hard segment, thus TPU/PB-a blends produced better shape memory properties than the TPU.

Behl et al. reported on binary polymer blends composed of two different biodegradable multiblock copolymers, whereby the first one provides the segments forming hard domains and the second one provides the segments forming the switching domains [30]. As a structural concept for the two components, they selected multiblock copolymers each consisting of two different segment types. One segment forms either the hard or switching domains and therefore must be different in both components. Poly(p-dioxanone) (PPDO) is selected as the hard segment and PCL as the switching segment. Both segments are crystallizable, whereby the melting point of PPDO (T_m,PPDO) is higher than T_m,PCL. The amorphous aliphatic copolyester poly(alkylene adipate) from adipic acid and a mixture of diols (1,4-butanediol, ethylene glycol, and diethylene glycol) with a very low glass transition temperature (T_g « room temperature) was chosen as the soft segment. The second segment, which is the same in both components, is incorporated to mediate the miscibility of both components, contributing to the elasticity of the material. The permanent shape of the resulting polymer blends is determined by the PPDO-crystallites, which form physical cross-links associated with the highest thermal transition $T_{perm} = T_m$,PPDO. The fixation of the temporary shape of the polymer blends is obtained by the crystallization of the switching

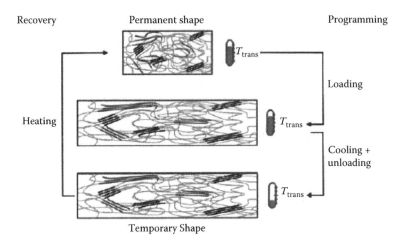

Recovery   **Permanent shape**   **Programming**

Heating

Temporary Shape

FIGURE 4.3
The shape memory effect for binary blends from two different multiblock copolymers; $T_{\text{trans}} = T_{\text{m}}$PCL determines the switching temperature. At a temperature $T > T_{\text{m}}$PCL the poly(ε-caprolactone (PCL) segments are viscoelastic. When the temperature is decreased to $T < T_{\text{m}}$PCL, the PCL domains become semicrystalline and fix the temporary shape of the blends, which is obtained after unloading the sample. At $T < T_{\text{m}}$PPDO the semicrystal-line poly(p-dioxanone) (PPDO) domains (black lines) act as permanent physical cross-links, determining the permanent shape. The amorphous soft segment PADOH (grey lines) con-tributes to the elasticity of the binary blends. The permanent shape is recovered as soon as T_{m}PPDO $> T > T_{\text{trans}}$s is reached by reheating. (Adapted from Behl, M., Ridder, U., Feng, Y., Kelch, S., and Lendlein, A. 2009. Shape memory capability of binary multiblock copolymer blends with hard and switching domains provided by different components. *Soft Matter* 5:676–684. Royal Society of Chemistry, Copyright Reproduced with permission.)

segment PCL. T_{m}PCL determines the T_{sw} of the polymer blends. The blends showed excellent shape memory properties, and the melting point associ-ated with the PCL switching domains (T_{m} of PCL) is almost independent of the weight ratio of the two blend components. At the same time the mechan-ical properties can be varied systematically. The mechanism of the shape memory effect of binary blends from multiblock copolymers is illustrated in Figure 4.3. In the programming step, a film of the polymer blend is deformed from its permanent shape at a temperature T_{high} of 50°C, which is above T_{m}PCL. The deformed sample is cooled below T_{m}PCL to a temperature T_{low} of 0°C either under constant strain control or under constant stress control. Upon removal of the external stress the temporary shape is obtained. In the recovery step, the polymer blend is reheated to $T_{\text{high}} = 50$°C, and its original permanent shape is recovered. Biodegradability, the variability of mechani-cal properties, and a response temperature around body temperature of this blend SMP are making this binary blend system an economically efficient, suitable candidate for diverse biomedical applications.

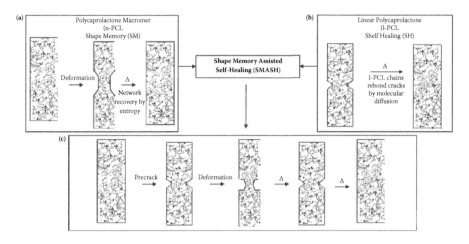

FIGURE 4.4

Shape memory assisted self-healing mechanism. (a) Virgin sample temporarily deformed. Shape recovery is attained by entropy elasticity of network. (b) Virgin sample is damaged and self-healing is triggered by molecular diffusion that rebonds crack surfaces. (c) Virgin sample is damaged and deformed. Shape memory and self-healing are simultaneously triggered when $T > T_m$ to attain full mechanical strength of blend. (Adapted from Rodriguez, E. D., Luo, X., and Mather, P. T. Shape memory miscible blends for thermal mending, *Behavior and Mechanics of Multifunctional Materials and Composites*, Eds. Z. Qunaies, and J. Li, Proceedings SPIE 7289z; 7289121–7289125. SPIE Copyright Reproduced with permission.)

Mather et al. reported an interesting miscible blend SMP exhibiting unique self-healing properties that was composed of linear poly(ε-caprolactone) (*l*-PCL) and chemically cross-linked network-PCL (*n*-PCL) [31]. They showed that photocured miscible blends of linear and cross-linked poly(ε-caprolactone) (*l*-PCL/*n*-PCL) demonstrate shape recovery for local crack closure and concomitant crack rebonding for recovery of full mechanical strength. The shape memory assisted self-healing mechanism is depicted in Figure 4.4. The cross-linked component, *n*-PCL, exhibits reversible plasticity, a form of shape memory where plastic deformation is fully recovered due to the entropy elasticity of the network that is liberated upon heating above the network melting temperature (T_m ~55°C) (Figure 4.4a). The linear component, *l*-PCL, interpenetrates the shape memory *n*-PCL component, yet freely diffuses above T_m to yield a tacky surface capable of rebonding any cracks formed during damage by molecular diffusion (Figure 4.4b). Importantly, both shape recovery and rebonding are initiated at the same temperature. Thus, a single heating event can simultaneously tackify crack surfaces while bringing them into molecular contact, resulting in shape memory assisted self-healing and recovery of mechanical strength (Figure 4.4c). Such shape memory assisted self-healing materials may be utilized for coatings and films where long-term use and facile repair is needed.

4.3 Immiscible Polymer Blends

It is not very often that two polymers are not mixed at a molecular level but are immiscible. Because the immiscible blends have a phase-separated structure, their physical properties including shape memory properties are influenced by phase morphology as well as the nature and relative amount of each phase. For the immiscible blend based SMP, in principle, one component acts as a reversible phase and the other component acts as a stationary phase.

Li et al. reported that immiscible high-density polyethylene (HDPE)/poly(ethylene terephthalate) (PET) blends, prepared by means of melt extrusion with ethylene–butyl acrylate–glycidyl methacrylate (EBAGMA) terpolymer as a reactive compatibilizer, can exhibit shape memory effects [32]. They observed that the compatibilized blends showed improved shape memory effects along with better mechanical properties as compared to the simple binary blends. In the blend, HDPE acts as a reversible phase, and the response temperature in the shape recovery process is determined by T_m of HDPE. The shape-recovery ratio of the 90/10/5 HDPE/PET/EBAGMA blend reached nearly 100%. Similar behavior was observed for immiscible HDPE/nylon 6 blends [33]. The addition of maleated polyethylene-octene copolymer (POE-g-MAH) increases compatibility and phase-interfacial adhesion between HDPE and nylon 6, and shape memory property was improved. The shape recovery rate of HDPE/nylon 6/POE-g-MAH (80/20/10) blend is 96.5% when the stretch ratio is 75%.

An interesting shape memory effect was observed in immiscible polylactide (PLA)/polyamide elastomer (PAE) blend [34]. In the blend, PAE forms a dispersed phase in the PLA matrix, and the PAE domains act as stress concentrators in the system with the stress release locally and lead to an energy-dissipation process. PAE acted as a plasticizer and decreased the T_g of the PLA, which promoted the orientation and reorganization of PLA molecules in the lower temperature. The blends showed unique shape memory effect that was different from the traditional SMPs. Figure 4.5 shows the shape recovery process of the PLA/PAE (90/10) blend. The sample was stretched to 100% at room temperature and the temporary shape was formed, which was different than traditional SMPs. When the sample is heated above the T_g of PLA, it shrinks and recovers to its original shape quickly. In the blend, the crystalline region of PLA acts as the stationary phase to keep the original structure. On the other hand, the amorphous region occurred to deformation involving molecular orientation during elongating upon the tensile load. When the tensile load is removed, the elongated shape can be kept and the stress was left in the system. This allows the molecules to have activity and to recover to their original shape with the stress releasing instantaneously when it is heated above its T_g.

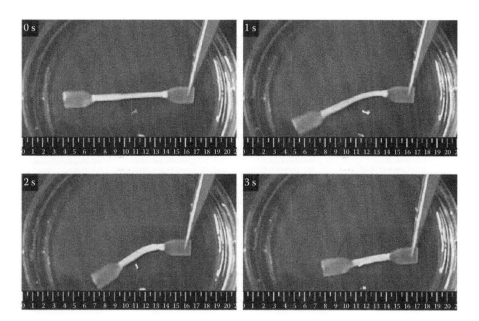

FIGURE 4.5
Shape memory effect of polylactide/polyamide elastomer (PLA/PAE) (90/10 wt/wt) blends. (The sample can recover to original shape in 8 s and 3 s at 80°C and 90°C, respectively.) (Adapted from Zhang, W., Chen, L., and Zhang, Y. 2009. Surprising shape-memory effect of polylactide resulted from toughening by polyamide elastomer. *Polymer* 50:1311–1315. Copyright Elsevier Ltd. Reproduced with permission.)

Zhang et al. examined the shape memory behavior of immiscible styrene-butadiene-styrene (SBS) triblock copolymer/poly(ε-caprolactone) (PCL) blends with a varying composition range and investigated the relationship between their phase morphology and shape memory properties [35]. In the blend, the two immiscible components separately contribute to shape memory performances, in which the SBS elastomer provides the stretching and recovery performances and the semicrystalline PCL provide the fixing and unfixing performances. Phase morphology greatly affects the shape recovery and shape fixing performance of this blend. Figure 4.6 describes the shape memory mechanism of this blend with different morphology schematically, and indicates that the best shape memory effect can be attained when the elastomer constitutes a continuous phase and the switch polymer constitutes a minor continuous phase. This study indicated that through careful design of the immiscible phase morphology, an ideal SMP system with both good stability and performances can be achieved.

In accordance with a concept derived from the SBS/PCL blend, Wang et al. developed a series of novel biodegradable SMP blends consisting of

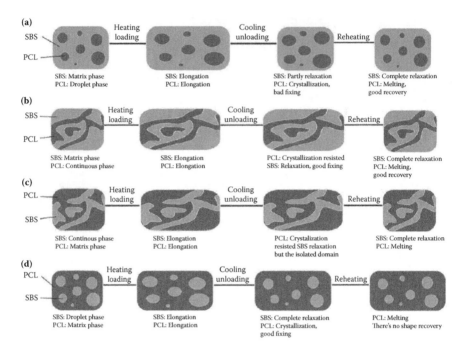

FIGURE 4.6
Schematic figures of the shape memory mechanism of the styrene-butadiene-styrene (SBS) triblock copolymer/poly(ε-caprolactone) (PCL) blend in (a)–(d). (Adapted from Zhang, H., Wang, H., Zhong, W., and Du, Q. 2009. A novel type of shape memory polymer blend and the shape memory mechanism. *Polymer* 50:1596–1601. Copyright Elsevier Ltd. Reproduced with permission.)

two immiscible components—poly(L-lactide-co-ε-caprolactone) (PLLCA) and poly (L-lactide-co-glycolide) (PLLGA) [36]. When the PLLGA content was less than 50 wt%, PLLCA and PLLGA behaved as the reversible phase and stationary phase, respectively, and the shape recovery rate increased with the PLLGA content. When the PLLGA content was more than 50 wt%, PLLGA was transformed from a droplet-like dispersion phase to a continuous phase, and the shape memory property of the blends was mainly derived from PLLGA. Increasing the concentration of the stationary component (PLLGA) to more than 50 wt% would transform the PLLCA–PLLGA shape memory system to a PLLGA-based system (i.e., when the PLLGA content was less than 50 wt%, the shape memory properties of the blend resulted from the associative effects of PLLCA and PLLGA, while the shape-memory properties of blends with more than 50 wt% PLLGA were primarily derived from PLLGA). In such a case, PLLCA primarily plays the role of adjusting the mechanical properties instead of acting as the reversible component.

The shape memory effect of immiscible chitosan/PLLA blend was studied [37]. In the blends, the shape memory effect arises from the viscoelastic properties of the PLLA composed of the amorphous structure and crystalline structure. It was found that chitosan does not significantly affect the glass and melting transition temperature of the PLLA, and the shape recovery ratio of the polymer decreases dramatically with increasing chitosan content due to the immiscibility between chitosan and PLLA. When the chitosan content is below 15 wt%, good shape memory effect of the composites was observed.

Luo and Mather developed a unique shape memory elastomeric composite by an interpenetrating combination of a crystallizable thermoplastic microfiber network (functioning as the "switch phase" for shape memory) with an elastomeric matrix [38]. The shape memory elastomer composite, composed of silicone rubber and PCL, was fabricated via a two-step process depicted in Figure 4.7. PCL was first electrospun from a 15 wt% chloroform/dimethyl formamide (DMF) solution. The resulting nonwoven fiber mat (thickness = 0.5 mm) was then immersed in a two-part mixture of silicone rubber (Sylgard 184) and vacuum was applied to ensure complete infiltration of the Sylgard 184 into the PCL fiber mat, followed by curing at room temperature for >48 h. This type of composite showed excellent shape memory performance, in which the PCL phase acts as a switching phase while the cured elastomer acts as a stationary phase and provides facile deformation. The manufacturing process of this type of SMP is simple, because it requires no specific chemistry or physical interactions. One can tune the properties of the individual components (the fibers and the matrix) to easily control the overall shape memory behavior and materials properties. For example, one can use thermoplastic fibers (either semicrystalline or amorphous) with different T_m's and T_g's to adjust the transition temperature or vary the cross-link density of the matrix to achieve different recovery stresses.

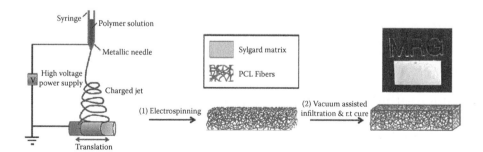

FIGURE 4.7
Two-step fabrication of Sylgard/poly(ε-caprolactone) (PCL) shape memory elastomeric composites. (Adapted from Luo, X., and Mather, P. T. 2009. Preparation and characterization of shape memory elastomeric composites. *Macromolecules* 42:7251–7253. Copyright ACS. Reproduced with permission.)

4.4 Interpenetrating Polymer Network

Full or semi-interpenetrating polymer networks (IPNs) can be designed to exhibit shape memory effects. In an IPN system, the network properties play an important role in controlling the shape memory performance. The hydrophilicity, transition temperatures, and mechanical properties of IPNs can be conveniently adjusted through variation of network compositions to match the desired applications.

Li et al. synthesized a PMMA-PEG semi-IPN by radical polymerization and cross-linking of PMMA in the presence of linear PEG, which exhibits two independent shape memory effects at two transition temperatures, the T_m of the PEG crystal and the T_g of the semi-IPN [39]. In the IPN, a single T_g appeared due to the miscibility of the amorphous phase of the two polymers. Based on a reversible order-disorder transition of the crystals below and above the T_m of PEG, and the large difference in storage modulus below and above the T_g of the semi-IPN, the polymer has a recovery ratio of 91 and 99%, respectively. For the shape-memory behavior at the T_m of PEG crystals, the fixing phase was the PMMA network and the reversible phase was PEG crystals. For the shape memory behavior at the T_g of the semi-IPNs, the fixing phase was the chemical cross-linked point, while the reversible phase was the PMMA-PEG complex phase.

They also synthesized a poly[(methyl methacrylate)-co-(N-vinyl-2-pyrrolidone)]/poly(ethylene glycol) semi-IPN based SMP by radical copolymerization of 37.5 to 59.5 wt% MMA and 26 wt% N-vinyl pyrrolidone in the presence of azobis isobutyronitrile (AIBN) as an initiator, ethylene glycol dimethacrylate (EGDMA) as a cross-linker, and 15 to 40 wt% linear PEG (MW: 400, 600, 800, and 1000) [40]. It was found that the network structure is stabilized by hydrogen-bonding interactions between the components. In this IPN, the fixing phase is the chemical cross-linked point, while the reversible phase is the PEG-PVP complex phase. The addition of PEG to P(MMA-co-VP) networks not only causes a depression in the T_g and stiffness of P(MMA-co-vinyl pyrrolidone (VP)) networks, but also increases the difference in storage modulus below and above T_g. The existence of a hydrogen bond causes the formation of a physical network between PVP and PEG, which partially restricted the side chain movements of PVP and slightly increased the stiffness of P(MMA-co-VP)/PEG semi-IPNs at room temperature. When the semi-IPNs were heated to T_{trans} (above T_g + 20°C), they can transform from the glass state to rubber-elastic state, but the hydrogen bonds between PVP and PEG were broken; moreover, the number of hydrogen bonds between the carbonyl groups of PVP and the terminal hydroxyl groups of PEG decreased with heating, which produces a diminution in the stiffness. All the above results make the semi-IPNs flexible; the polymer is easily deformed under external force. When the semi-IPNs were cooled to room temperature, they returned to the glass state and the broken hydrogen bonds recombined to favor the fixing of a temporary shape. It can be considered that the P(MMA-co-VP)/PEG semi-IPNs possessed shape memory effect due to a large difference in storage modulus below and above T_g.

The semi-IPN SMP composed of crystalline poly(ethylene oxide) (PEO) and cross-linked poly(methyl methacrylate) (x-PMMA) was synthesized by Ratna and Karger-Kocsis [41]. They prepared the semi-IPN with various PMMA/PEO ratios using different amounts of triethylene glycol dimethacrylate as a cross-linker. They observed that creep compliance of the semi-IPNs did not change significantly with the compositional change in the studied range but is significantly reduced with increasing degree of cross-linking of PMMA. Also, shape memory properties (shape recovery and shape fixity) of the semi-IPNs are observed to be able to be tailored upon the x-PMMA/PEO ratio and cross-linking degree of x-PMMA.

Zhang et al. synthesized a novel IPN composed of polyesterurethane and poly(ethylene glycol) dimethacrylate (PEGDMA) with good shape memory properties using a solvent casting method [42]. The star-shaped oligo[(rac-lactide)-co-glycolide] was coupled with isophorone diisocyanate to form a polyesterurethane (PULG) network, and PEGDMA was photopolymerized to form another polyetheracrylate network. PULG and PEGDMA networks were found to be miscible when PEGDMA content was below 50 wt% and only one T_g of the IPNs between T_g of PEGDMA and PULG was observed, which was proportional to PEGDMA content. The IPNs showed high strain recovery and strain fixity ability (above 93%).

More recently, Feng et al. synthesized thermally triggered and water triggered shape memory IPN through photopolymerization of hydrophobic poly[(D,L-lactide)-co-glycolide] tetraacrylate (PLGATA) and hydrophilic poly(ethylene glycol) dimethacrylate (PEGDMA) [43]. By means of adjusting the PEGDMA contents, the polymer networks could be changed from hard to rubbery material when the PEGDMA content reached 30 wt%. The materials can recover their original shape either quickly by heating above the T_m of PEGDMA crystal or slowly upon immersion in water. They synthesized similar biodegradable and biocompatible IPN SMP via photopolymerization of PEGDMA and thermal polymerization of star-shaped oligo[(D,L-lactide)-co-ε-caprolactone] (PCLA) with isophorone diisocyanate (IPDI) [44]. The IPNs have only one single T_g between the T_g of PEGDMA and polyesterurethane, which are adjusted in the range of −6 to 32°C. They are amorphous and are rubbery when PEGDMA content is above 10% at room temperature. These IPNs recover quickly their permanent form in 10 sec when the environment temperature is above its T_g. The strain recovery rate and the strain fixity rate are above 90%. The chemical cross-linking points act as a fixed phase to memorize the original shape, while the amorphous domains of PCLA and PEG act as a reversible phase. These biodegradable and biocompatible soft IPN SMPs offer a high potential for biomedical applications such as smart implants, site-specific controlled drug delivery systems, or stents in the treatment of cardiovascular disease. The hydrophilicity of PEGDMA plays a role in improved blood compatibility of polymer networks.

4.5 Cross-Linked Polymer Blends

Chemically cross-linked blends exhibiting shape memory effects were studied aiming at the effective cross-linking and improved shape memory behavior as well as development of a new SMP.

Zhu et al. prepared cross-linked blend SMP by the radiation cross-linking of the blend of PCL/polyfunctional poly(ester acrylate) (PEA) [45]. They observed that PEA with polyfunctional double bonds had a distinct promoting function for the radiation cross-linking of PCL. The greater the usage and functional group number were, the greater gel content and the more distinctive the radiation cross-linking effects were. This also indicated that the polyfunctional material directly participated in the cross-linking reaction. Dynamic mechanical analysis indicated that enhanced radiation cross-linking better raised the heat deformation temperature of PCL and presented a higher and wider rubbery-state plateau; it also produced greater strength at temperatures higher than the melting temperature and provided greater force for recovering the deformation than pure PCL. The shape memory results revealed that sensitizing cross-linked PCL presented 100% recoverable deformation and a quicker recovery rate.

A similar approach was applied for fabricating SMP with good biocompatibility. It was prepared by melt blending of 80 to 95% polycaprolactone (PCL) and 5 to 20% polymethylvinylsiloxane (PMVS) and a subsequent cross-linking of the blends by electron beam irradiation [46]. In the blend, PMVS promoted the radiation cross-linking. As the concentration of PMVS increased, the gelation dose and the ratio of degradation to cross-linking decreased and the efficiency of radiation cross-linking increased. The elastic modulus below the melting point of PCL of radiation cross-linked PCL/PMVS blends decreased with the increase of PMVS and increased above the melting point. The cross-linked PCL/PMVS blends exhibited excellent shape memory effects, and the ratios of deformation to recovery were more than 95%.

Kolesov and Radusch prepared peroxide cross-linked binary and ternary blend SMPs from high density polyethylene and two ethylene-1-octene copolymers with medium and high degrees of branching [47]. The blends were prepared by a melt mixing and subsequently are cross-linked with 2 wt% of liquid peroxide 2,5-dimethyl-2,5-di-(tertbutylperoxy)-hexane at 190°C. The blends showed multiple shape memory behavior that appeared only at consequent stepwise application of convenient programming strains and temperatures. Obviously, that is caused by multiple melting behavior of these blends with many poorly separated peaks.

Chang et al. prepared heat-triggered SMP blend networks from blends of end-carboxylated telechelic poly(ε-caprolactone) (XPCL) and epoxidized natural rubber (ENR), which can form cross-linked networked structure via interchain reaction between the reactive groups of each polymer during molding at high temperature [48]. Crystalline melting transition and degree of cross-links of the blend networks could be tuned by the blend composition and molecular weight of the XPCL. Such XPCL/ENR blend networks possessing crystalline domains together with appreciable degree of cross-linking showed good shape retention and shape recovery, and the recovery temperatures were well matched with T_m of each sample.

4.6 Miscellaneous

Besides conventional types of blends, some special blend SMPs have been synthesized. Cho et al. synthesized the electrically conductive SMP based on the blend of shape memory polyurethane with electrically conductive polypyrrole (PPy) through in situ polymerization of pyrrole in a surface layer of a polyurethane film using a chemical oxidative polymerization process [49]. They showed that this blend exhibited shape memory effects under electric fields. A similar electroactuated SMP based on polyethylene octene elastomer (POE)/polyaniline (PANi) blends was synthesized using the inverted emulsion method by Chang and Park [50]. As the electrically conductive PANi content is increased in the POE/PANi blends, tensile and dynamic storage moduli are increased with a decreased flexibility. The blends exhibited a melting transition at around 40°C, and it is varied marginally with the blend composition. As the PANi content is increased in the blend, the conductivity of the blends and the percolation threshold was found to be at the vicinity of 40 wt% of PANi, at which a high conductivity of the order of 1.310^{-4} S/cm was obtained. The POE/PANi blend could be heated above its melting transition temperature within 25 seconds when an electric field (80 V) was applied. The electroactuated shape recovery process of the POE/PANi blend sample with bending mode are shown in Figure 4.8.

A water-triggered SMP based on the blend of silicone rubber and slightly cross-linked polyacrylamide, a nonionic hydrogel, was introduced by Hron and Slechtova [51]. The blend was prepared by mixing the cross-linked hydrogel powder with silicone rubber in a Brabender mixer, followed by thermal curing. It was found that the composite material can be deformed at elevated temperature above T_g of cross-linked polyacrylamide, which is between 210°C and 220°C, and keeps its shape if cooled down in the deformed state. The composite material shaped in this way resumes its original shape if it is again heated above T_g of cross-linked polyacrylamide or allowed to swell in distilled water. In this blend SMP, the limiting concentration is 30

POE/PANi (40/60) at 80 V

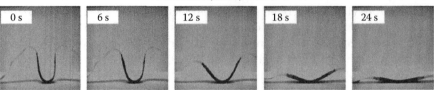

FIGURE 4.8
Electroactive shape recovery behavior of polyethylene octene elastomer/polyaniline (POE/PANi) blend. (From Chang, Y. W., and Park, K. 2011. Electroactuated shape memory polymer from polyethylene-octene elastomer/polyaniline composite synthesized by inverse emulsion polymerization, manuscript in preparation.)

phr of hydrogel phase, below which no shape memory exists, and above which it exists. Because both initial composite materials (silicone rubber and cross-linked polyacrylamide) are biocompatible, this composite material may be suitable for biomedical application, especially for medical devices that could be inserted into a body cavity in the thermally preshaped form, where it would swell by body fluids to the original shape.

Another special type of blend SMP is poly(acrylic acid-*co*-methyl methacrylate)/poly(ethylene glycol) (P(AA-*co*-MMA)/PEG) blends, which form hydrogen-bonded complexes [52]. It was found that both the concentration and molecular weight of PEG have a strong effect on the complexation with P(AA-*co*-MMA) gel. In such a system the minimum molecular weight of PEG required for the complex formation lowers to 1000.

A reactive blending technique can be employed to fabricate the SMP. Li et al. prepared shape memory graft copolymer HDPE-g-nylon by melt blending of maleated HDPE with nylon 6 at 230°C in a batched Haake mixer [53]. The graft copolymer was found to form in situ formation of polyethylene-*g*-nylon 6 graft copolymers bridged by maleic anhydride in the melt-blending process. The maleated polyethylene/nylon 6 blend specimens showed good shape memory effect under normal experimental conditions when the nylon contents of the blends are in the range from 5 to 20 wt%. The nylon domains, which serve as physical cross-links, play a predominant role in the formation of a stable network structure for the graft copolymers. The high crystallinity of the polyethylene segments at room temperature and the formation of a network structure in these specimens fulfill the structural requirement for this material to exhibit shape memory effects.

4.7 Summary

This review summarizes the numerous research on shape memory polymer blends conducted in recent decades, which demonstrates that polymer blending offers a convenient and economical way to fabricate a shape memory polymer possessing desired performance properties with adjusted T_{trans} in a wide range. Because concerns on the shape memory polymers are increasing in recent years, research and developments of the blend-based shape memory polymers are expected to continue rapid growth.

References

1. Hu, Z., Zhang, X., and Li, Y. 1995. Synthesis and application of modulated polymer gels. *Science* 269:525–527.
2. Osada, Y., and Matsuda, X. 1995. Shape-memory gel with order-disorder transition. *Nature* 376:219–223.
3. Jiang, H. Y., Kelch, S., and Lendlein, A. 2006. Polymers move in response to light. *Advanced Materials* 18:1471–1475.
4. Lendlein, A., Jiang, H., Junger, X., and Langer, R. 2005. Light-induced shape-memory polymers. *Nature* 434:879–882.
5. Buckley, P. R., McKinley, G. H., Wilson, T. S., Small, W., Benett, W. J., Bearinger, J. P., McElfresh, M. W., and Maitland, D. J. 2006. Inductively heated shape memory polymer for the magnetic actuation of medical devices. *IEEE Transactions Biomedical Engineering* 53:2075–2083.
6. Zhang, D., Liu, Y., and Leng, J. 2009. Magnetic field activation of SMP networks containing micro nickel (Ni) powder. *Second International Conference on Smart Materials and Nanotechnology in Engineering* 74931A.
7. Mondal, S., and Hu, J. L. 2007. Water vapor permeability of cotton fabrics coated with shape memory polyurethane. *Carbohydrate Polymers* 67:282–287.
8. Russel, A., Hayashi, S., and Yamada, T. 1999. Potential use of shape memory film in clothing. *Technical Textiles International* 8:17–19.
9. Ji, F. L., Zhu, Y., Hu, J. L., Liu, Y., Yeung, L. Y., and Ye, G. D. 2006. Smart polymer fibers with shape memory effect. *Smart Material Structures* 15:1547–1554.
10. Campbell, D., Lake, M., Scherbarth, M., Nelson, E., and Siv, R. 2005. Elastic memory composite material: An enabling technology for future furable space structures. In *46th AIAA/ASME/ASCE/AHS/ASC Structures, Structural Dynamics, and Materials Conference,* Austin, Texas.
11. Wache, H. M., Tartakowska, D. J., Hentrich, A., and Wagner, M. H. 2003. Development of a polymer stent with shape memory effect as a drug delivery system. *Journal of Materials Science: Materials in Medicine* 14:109–112.
12. Huang, W. M. 2010. Thermo-moisture responsive polyurethane shape memory polymer for biomedical devices. *The Open Medical Device Journal* 2:11–19.
13. Lendlein, A., and Langer, R. 2002. Biodegradable, elastic shape-memory polymers for potential biomedical applications. *Science* 296:1673–1676.
14. Metcalfe, A., Desfaits, A. C., Salazkin, I., Yahia, L., Sokolowski, W. M., and Raymond, J. 2003. Cold hibernated elastic memory foams for endovascular interventions. *Biomaterials* 24:491–497.
15. Lendlein, A., and Kelch, S. 2002. Shape-memory polymers. *Angewandte Chemie, International Edition* 41:2034–2057.

16. Liu, C., Qin, H., and Mather, P. T. 2007. Review of progress in shape-memory polymers. *Journal of Materials Chemistry* 17:1543–1558.
17. Behl, M., and Lendlein, A. 2007. Shape memory polymers. *Materials Today* 10:20–28.
18. Ratna, D., and Karger-Kocsis, J. 2008. Recent advances in shape memory polymers and composites: A review. *Journal of Materials Science* 43:254–269.
19. Behl, M., Zotzmann, J., and Lendlein, A. 2010. Shape-memory polymers and shape-changing polymers. *Advances in Polymer Science* 226:1–40.
20. Behl, M., Razzaq, M. Y., and Lendlein, A. 2010. Multifunctional shape-memory polymers. *Advanced Materials* 22:3388–3410.
21. Mather, P. T., Liu, C., and Campo, C. J. 2010. US Patent 7,795,350, 14 September, 2010.
22. Liu, C., and Mather, P. T. 2003. Thermomechanical characterization of blends of poly(vinyl acetate) with semicrystalline polymer for shape memory applications. *Proceedings of Annual Technical Conference SPE*, 61st, 2:1962–1966.
23. Campo, C. J., and Mather, P. T. 2005. PVDF/PMMA shape memory blends. *Polymeric Materials Science and Engineering* 93:933–934.
24. Ratna, D., Abraham, T. N., and Karger-Kocsis, J. 2008. Studies on polyethylene oxide and phenolic resin blends. *Journal of Applied Polymer Science* 108:2156–2162.
25. Chang, Y. W., and Choi, M. 2010. Thermomechanical properties and shape memory effect of PECH/SAN blend. *Polymeric Materials Science and Engineering* 162:826–827.
26. Jeong, H. M., Ahn, B. K., and Kim, B. K. 2001. Miscibility and shape memory effect of thermoplastic polyurethane blends with phenoxy resin. *European Polymer Journal* 37:2245–2252.
27. Jeong, H. M., Song, J. H., Lee, S. Y., and Kim, B. K. 2001. Miscibility and shape memory property of poly(vinyl chloride)/thermoplastic polyurethane blends. *Journal of Materials Science* 36:5457–5463.
28. Ajili, S. H., Ebrahimi, N. G., and Soleimani, M. 2009. Polyurethane/polycaprolactane blend with shape memory effect as a proposed material for cardiovascular implants. *Acta Biomaterialia* 5:1519–1530.
29. Erden, N. 2009. Polyurethane-polybenzoxazine based shape memory polymers, MS thesis, University of Akron, Ohio.
30. Behl, M., Ridder, U., Feng, Y., Kelch, S., and Lendlein, A. 2009. Shape memory capability of binary multiblock copolymer blends with hard and switching domains provided by different components. *Soft Matter* 5:676–684.
31. Rodriguez, E. D., Luo, X., and Mather, P. T. Shape memory miscible blends for thermal mending, *Behavior and Mechanics of Multifunctional Materials and Composites*, Eds. Z. Qunaies, and J. Li, Proceedings SPIE 7289z; 7289121–7289125.
32. Li, S. C., Lu, L. N., and Zeng, W. 2009. Thermostimulative shape-memory effect of reactive compatibilized high-density polyethylene/poly(ethylene terephthalate) blends by an ethylene-butyl acrylate-glycidyl methacrylate terpolymer. *Journal of Applied Polymer Science* 112:3341–3346.
33. Li, S. C., and Tao, L. 2010. Melt rheological and thermoresponsive shape memory properties of HDPE/PA6/POE-g-MAH blends. *Polymer—Plastics Technology and Engineering* 49:218–222.
34. Zhang, W., Chen, L., and Zhang, Y. 2009. Surprising shape-memory effect of polylactide resulted from toughening by polyamide elastomer. *Polymer* 50:1311–1315.

35. Zhang, H., Wang, H., Zhong, W., and Du, Q. 2009. A novel type of shape memory polymer blend and the shape memory mechanism. *Polymer* 50:1596–1601.
36. Wang, L. S., Chen, H. C., Xiong, Z. C., Pang, X. B., and Xiong, C. D. 2010. Novel degradable compound shape-memory polymer blend: Mechanical and shape-memory properties. *Materials Letters* 64:284–286.
37. Meng, Q., Hu, J., Ho, K., Ji, F., and Chen, S. 2009. The shape memory properties of biodegradable chitosan/poly(L-lactide) composites. *Journal of Polymers and the Environment* 17:212–224.
38. Luo, X., and Mather, P. T. 2009. Preparation and characterization of shape memory elastomeric composites. *Macromolecules* 42:7251–7253.
39. Liu, G., Ding, X., Cao, Y., Zheng, Z., and Peng, Y. 2005. Novel shape-memory polymer with two transition temperatures. *Macromolecular Rapid Communications* 26:649–652.
40. Liu, G., Guan, C., Xia, H., Guo, F., Ding, X., and Peng, Y. 2006. Novel shape-memory polymer based on hydrogen bonding. *Macromolecular Rapid Communications* 27:1100–1104.
41. Ratna, D., and Karger-Kocsis, J. 2011. Shape memory polymer system of semi-interpenetrating network structure composed of crosslinked poly (methyl methacrylate) and poly (ethylene oxide). *Polymer* 52:1063–1070.
42. Zhang, S., Feng, Y., Zhang, L., Sun, J., Xu, X., and Xu, Y. 2007. Novel interpenetrating networks with shape-memory properties. *Journal of Polymer Science, Part A: Polymer Chemistry* 45:768–775.
43. Feng, Y., Zhang, S., Zhang, L., Guo, J., and Xu, Y. 2011. Synthesis and characterization of hydrophilic polyester-PEO networks with shape-memory properties. *Polymers for Advanced Technologies*, in press, doi:10.1002/Pat1780.
44. Feng, Y., Zhao, H., Jiao, L., Lu, J., Wang, H., and Guo, J. 2011. Synthesis and characterization of biodegradable, amorphous, soft IPNs with shape-memory effect. *Polymers for Advanced Technologies*, in press, doi: 10.1002/pat.1885.
45. Zhu, G. M., Xu, Q. Y., Liang, G. Z., and Zhou, H. F. 2005. Shape memory behaviors of sensitizing radiation crosslinked polycaprolactone with polyfunctional poly(ester acrylate). *Journal of Applied Polymer Science* 95:634–639.
46. Zhu, G., Xu, S., Wang, J., and Zhang, L. 2006. Shape memory behavior of radiation-crosslinked PCL/PMVS blends. *Radiation Physics and Chemistry* 75:443–448.
47. Kolesov, I. S., and Radusch, H. J. 2008. Multiple shape-memory behavior and thermal-mechanical properties of peroxide cross-linked blends of linear and short-chain branched polyethylenes. *eXPRESS Polymer Letters* 2:461–473.
48. Chang, Y. W., Eom, J. P., Kim, J. G., Kim, H. T., and Kim, D. K. 2010. Preparation and characterization of shape memory polymer networks based on carboxylated telechelic poly(ε-caprolactone)/epoxidized natural rubber. *Journal of Industrial and Engineering Chemistry* 16:256–260.
49. Sahoo, N. G., Jung, Y. C., Goo, N. S., and Cho, J. W. (2005) Conducting shape memory polyurethane-polypyrrole composites for an electroactive actuator. *Macromolecular Materials and Engineering* 290:1049–1055.
50. Chang, Y. W., and Park, K. 2011. Electroactuated shape memory polymer from polyethylene-octene elastomer/polyaniline composite synthesized by inverse emulsion polymerization, manuscript in preparation.
51. Hron, P., and Slechtova, J. 1999. Shape memory of composites based on silicone rubber and polyacrylamide hydrogel. *Die Angewandte Makromolekulare Chemie* 268:29–35.

52. Liu, G., Ding, X., Cao, Y., Zheng, Z., and Peng, Y. 2004. Shape memory of hydrogen-bonded polymer network/poly(ethylene glycol) complexes. *Macromolecules* 37:2228–2232.
53. Li, F., Chen, Y., Zhu, W., Zhang, X., and Xu, M. 1998. Shape memory effect of polyethylene/nylon 6 graft copolymers. *Polymer* 39:6929–6934.

5

Synthesis and Properties of Ethylene Methacrylate (EMA) Copolymer Toughened Polymethyl Methacrylate (PMMA) Blends

Siddaramaiah

Department of Polymer Science and Technology
Sree Jayachamarajendra College of Engineering
Karnataka, India

P. Poomalai

Central Institute of Plastics Engineering and Technology
West Bengal, India

Johnsy George

Defence Food Research Laboratory
Karnataka, India

CONTENTS

5.1 Introduction

Polymer blends are mixtures of at least two macromolecular species, polymers, or copolymers. Blending of polymers together is a versatile route for producing newer materials with enhanced performance, which can be used for various applications in polymer engineering/technology. The blending of polymers will generally improve impact strength, tensile strength, modulus/rigidity, along with other properties [1–4]. Blending also benefits the manufacturer by offering: (1) Improved processability, product uniformity, and scrap reduction; (2) quick formulation changes; (3) plant flexibility and high productivity; (4) reduction of the number of grades that need to be manufactured and stored; (5) inherent recyclability, and so forth. Blending is also helpful in the manufacture of toughened plastics having the right combination of lightness and mechanical performance over a wide range of temperatures. The commercial development of polymer blends and alloys is driven by more favorable economics than in the more conventional chemical routes to new products. Blend systems, composed of existing materials, can be developed about twice as rapidly as new polymers, allowing manufacturers to respond more rapidly to new market requirements at reduced cost. Particularly, the polymer blends are designed and manufactured to modify certain properties in order to meet the requirements of newer end use applications. One of the properties most often to be improved is fracture toughness.

Polymethyl methacrylate (PMMA) is a hard, rigid, transparent, and good weather-resistant polymer. They are widely used as coatings and in glass and electronic materials due to their favorable properties like hardness, high T_g, and excellent transparency. But their use in engineering applications was limited due to brittleness and poor solvent resistance, which can be improved by toughening the polymer. The toughening can be achieved by copolymerization with a comonomer having low T_g, introducing glass fiber, or blending with rubbery materials. Enhanced properties were achievable by toughening PMMA with various rubbers or copolymers [5–10]. The impact strength of PMMA was improved by blending it with various acrylic elastomers.

For example, blends comprising 25 to 90 wt% of PMMA, with butadiene–butylacrylate–methylmethacrylate elastomeric copolymer showed excellent properties. Several other copolymers of methyl methacrylate with the ethyl-, butyl-, or octyl-methacrylates were also used for toughening PMMA [11].

Many researchers have investigated the toughening of PMMA by different routes, and it is observed that the brittle amorphous polymer (e.g., polystyrene, PS, and PMMA) can be toughened via the incorporation of extremely fine rubber particles [12]. Zachariah Oomen and Sabu Thomas studied the mechanical properties and fracture behavior of natural rubber/PMMA blends as thermoplastic elastomers (TPEs) [13]. Oomen and Thomas also reported compatibility studies on natural rubber (NR)/PMMA blends by viscometry and phase separation techniques [14]. Polybutadiene toughened PMMA and their characterizations have been studied by Raghavan et al. [15]. Archie et al. studied the use of cryogenic mechanical alloying of PMMA with up to 25% polyisoprene (PI) and poly (ethylene-alt-propylene) (PEP) and characterized by microscopic technique [16]. Micro hardness behavior of PMMA/NR blends prepared by solution method has also been reported [17]. Recently semi-interpenetrating polymer networks (SIPNs) based on nitrile rubber/PMMA has been reported [18]. Nakason et al. reported the studies of rheological and thermal properties of thermoplastic natural rubbers based on PMMA/epoxidized natural rubber blends [19]. Spadaro et al. studied the morphology and properties of PMMA/acrylonitrile-butadiene rubber (PMMA/NBR) blends, obtained through gamma-radiation-induced in situ polymerization of methyl methacrylate (MMA) in the presence of small amounts of a nitrile rubber [20]. Mina et al. studied the micromechanical behavior and T_g of PMMA-rubber blends [21]. Also they reported the interphase boundary of incompatible polymer blends such as PMMA/NR, PS/NR, and of compatible blends such as PMMA/NR/epoxidized NR and PS/NR/styrene-butadiene-styrene (SBS) block copolymer by means of micro indentation hardness and microscopy methods [22]. Suriyachi et al. reported the studies on mechanical properties and morphology of compatible NR/PMMA blends using NR-*g*-glycidyl methacrylate/styrene graft copolymer as compatibilizer [23]. The preparation of thermoplastic vulcanizates (TPUs) based on NR-*g*-PMMA/PMMA, PMMA/epoxidized natural rubber blends and characterization of their rheological, mechanical, morphological, and thermal properties has been reported by Nakason et al. [24,25]. Synthesis and characterization of polyisobutylene-PMMA interpenetrating polymer networks (IPNs) has been reported by Vancaeyzeele et al. [26]. Mina et al. investigated the morphology, micromechanical. and thermal properties of undeformed and mechanically deformed PMMA/natural rubber blends [27]. Dong et al. studied the morphology and mechanical properties of dihydroxy polydimethyl siloxane (PDMS)/PMMA blends [28]. Nakason et al. reported on preparation of reactive blending of maleated NR with PMMA and investigated their rheological, thermal, and morphological properties [29].

From the detailed literature survey, it is noticed that the toughening of PMMA was mostly carried out with NR/polybutadiene rubber (PBR)/ABS by solution blends, IPNs, reactive blending, and graft/copolymerization process. Some studies on blending through IPNs/solution process/ polymerization methods using a thermosetting-type of polyurethane (PU) and different acrylic materials have been reported [30,31]. Kosyanchuk et al. investigated the thermal, physical, and viscoelastic properties of PMMA and PU by in situ formation [32]. Lipatov et al. studied the blends of linear PU and PMMA obtained from in situ polymerization of their monomers and evaluated the two phase systems and their interfacial region [33]. Recently, Patricio et al. also reported the effect of blend compositions on microstructure, morphology, and permeability in PU/PMMA blends [34]. Siddaramaiah et al. investigated IPNs of PU with polyacrylates [35,36]. Study on the phase behavior, mechanical properties, and strain-rate effect of ethylene vinyl acetate (EVA) copolymer and PMMA blends by in situ polymerization was reported by Cheng and Chen [37]. Errico et al. studied the acrylate/EVA reactive blends and semi-IPN and characterized for chemical, physical, and thermo-optical properties [38]. Mayu Si et al. studied the blends of PS/PMMA, PC/styrene-acrylonitrile (SAN), and PMMA/EVA with and without modified organoclay Cloisite 2A and Cloisite 6A clays and compared their morphologies [39].

In addition to the above mentioned polymers, Ethylene methacrylate (EMA) copolymer is also an excellent choice for improving the performance of PMMA. EMA copolymer has excellent environmental stress crack resistance at elevated temperature with superior mechanical properties. Further, it has the advantage of a polar ester group and a reactive hydrogen atom, which can also provide some amount of interaction with PMMA. This chapter provides information about blending of EMA copolymer with PMMA for improving its performance, especially impact strength and chemical resistance while retaining its optical properties. The discussion focuses on the blend preparation and some important properties like physicomechanical and thermal properties of these blends. Attempts have been made to discuss the morphology–property correlations also.

5.2 Structure and Properties of Polymethyl Methacrylate (PMMA)

PMMA is the synthetic polymer of methyl methacrylate, which is widely known by a variety of trade names like Lucite, Oroglas, Perspex, Plexiglas, and so on. PMMA is a glassy polymer with an amorphous structure. It has a density of 1.19 g/cm^3 and has very low water absorption. The refractive index ranges from 1.49 to 1.51 depending on the type. PMMA shows high gloss, hardness, stiffness, rigidity, dimensional stability, excellent electrical insulating properties, and good abrasion resistance, but the surface scratches easily which can be polished out. It is also odorless, tasteless, and nontoxic and does not become brittle at low temperature. It has excellent resistance to dimensional changes, heat, oils, and so forth. It decomposes on heating strongly and burns slowly but does not flash-ignite. It can be easily moldable and machinable, unaffected by most household chemicals such as weak and strong alkalies, bleaching compounds and window cleaning solutions, salt, vinegar, animal and mineral oils, waxes, and common foodstuffs. The water absorption behavior is negligible. Dimensional stability and electrical properties remain good under humid conditions. Despite several advantages, PMMA has some disadvantages such as it is brittle under impact conditions and failure can occur by shattering. Thin-walled products are also difficult to mold using PMMA because of poor flow properties. Poor hot-melt strength limits processing methods, and it does not have significant elastic deformation before failure (i.e., goes straight to brittle fracture). Some of these limitations of PMMA can be overcome by blending with various polymers. The typical properties of PMMA are given in Table 5.1. Its structural formula is as shown in Scheme 5.1.

TABLE 5.1

Typical Properties of Polymethyl Methacrylate (PMMA)

Property	Value
Specific gravity	1.18
Refractive index	1.49
Total light transmittance (%)	92
Surface hardness (RHM)	90
Tensile strength at break (MPa)	65
Izod impact (Notched) (KJ/m)	0.02
Linear expansion (/°C $\times 10^{-5}$)	7
Elongation at break (%)	2.5
Water absorption (%)	0.3
Oxygen index (%)	19
Flammability UL 94	HB
Volume resistivity (ohm.cm)	$>10^{15}$
Dielectric strength (MV/m)	20
Dissipation factor @ 1 KHz	0.03
Dielectric constant @ 1 KHz	3.3
Glass transition temperature (°C)	105
Heat distortion temperature @ 0.45 MPa (°C)	103
Heat distortion temperature @ 1.8 MPa (°C)	95
Melting range (°C)	130–140

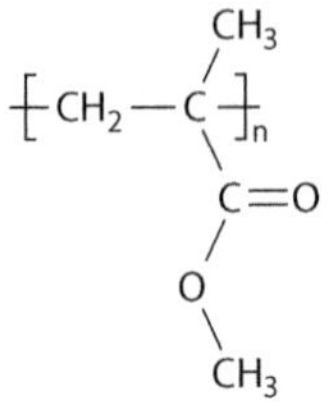

$$\left[CH_2-\underset{\underset{\displaystyle CH_3}{\overset{\displaystyle |}{\underset{\displaystyle O}{\overset{\displaystyle |}{C=O}}}}}{\overset{\displaystyle CH_3}{\overset{\displaystyle |}{C}}} \right]_n$$

SCHEME 5.1
Structure of polymethyl methacrylate (PMMA).

5.3 Structure and Properties of Ethylene Methacrylate (EMA)

Ethylene methacrylate (EMA) is a copolymer of ethylene and methacrylate. Copolymerizing ethylene with a small amount of methacrylate gives polarity to the polymer and reduces crystallinity leading to highly clear polymer as compared to polyethylene. Its structural formula is as shown in Scheme 5.2.

It can be used alone or compounded with other polymers as a modifying polymer for improved toughness. Incorporation of EMA into other polymers improves the properties such as flexibility, toughness, weatherability, and softness. The EMA polymers are characterized through the following features: high thermal stability against degradation (up to 400°C), high polarity, no reactive groups, low temperature flexibility that leads to benefits like superior processing stability and compatibility with a broad range of engineering polymers. Some of its typical properties are given in Table 5.2.

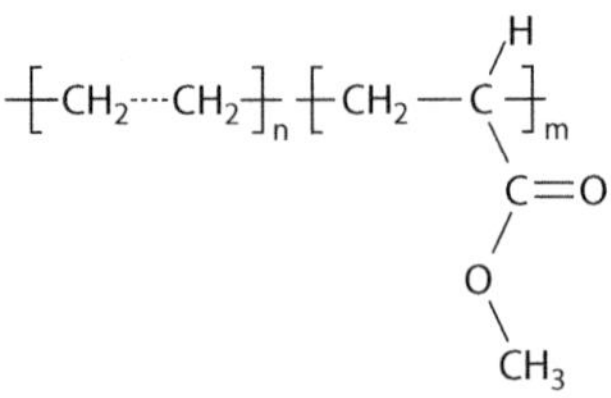

$$\left[CH_2\cdots CH_2 \right]_n \left[CH_2-\underset{\underset{\displaystyle CH_3}{\overset{\displaystyle |}{\underset{\displaystyle O}{\overset{\displaystyle |}{C=O}}}}}{\overset{\displaystyle H}{\overset{\displaystyle |}{C}}} \right]_m$$

SCHEME 5.2
Structure of ethylene methacrylate (EMA).

TABLE 5.2

Some of the Typical Properties
of Ethylene Methacrylate (EMA)

Property	Value
Density (g/cm^3)	0.938
Surface hardness (shore A/D)	87/28
Tensile strength (MPa)	11
Tensile modulus (MPa)	35
Izod impact (Notched) (J/m)	No break
Elongation at break (%)	780
Melting point (°C)	92
Melt flow index,	
2.16 kg/190 (°C) g/10 min	8
Vicat softening temperature (°C)	54

5.4 Toughening of PMMA Using EMA

Toughened blends are either immiscible or partially miscible blends having two separate T_g's. They have a heterogeneous microstructure with dispersed phase size in the order of micrometers. The incorporation of EMA can increase the fracture rate of PMMA by enabling increased plastic deformation of the matrix or by undergoing fracture and protecting PMMA from fracture. The EMA phase in the form of small particles has to be dispersed properly in the PMMA phase to achieve toughening. The particle size and size distribution of the dispersed EMA in the blend will depend on the extent of miscibility of the two phases and on the way in which they are mixed. If the rubber is having excellent miscibility with PMMA, then the dispersed rubber particles will be too small and even they will be distributed on a molecular scale throughout the PMMA phase. If the two phases are immiscible, then the rubber particles will be dispersed as macroscopic particles, and the toughening ability will be reduced [40]. Some of the factors influencing the toughening behavior are briefly discussed below.

5.4.1 Volume Fraction of Rubber Phase

The optimum volume fraction of the rubber phase in PMMA is the one that yields maximum mechanical properties. Increasing the concentration of rubber phase beyond this limit decreases the mechanical properties of blends irrespective of the properties of PMMA. In most of the rubber-thermoplastic blends, 20 to 30% rubber loading is acceptable. Higher rubber loading may also even lead to a reduction in impact strength.

5.4.2 Particle Size and Distribution of Rubber Phase in the Blend

If properly dispersed, the rubber phase can act as an effective stress concentrator and enhances the yielding characteristics of PMMA phase. The shear rate developed in the polymer blend is inversely proportional to the diameter of the rubber particle. Hence, the rubber particle size should be as low as possible to achieve effective toughening.

5.4.3 Interfacial Adhesion between Rubber and PMMA Phase

Interfacial adhesion is the adhesion in which interfaces between phases or components are maintained by intermolecular forces, chain entanglements, or both, across the interfaces. Interfacial adhesion between rubber and PMMA must be sufficient to permit the effective transfer of stress to the rubber particles and also to provide multiple sites for crazing and localized shear yielding for effective impact energy dissipation.

5.5 Mechanism of Toughening

The mechanism of toughening is usually explained by the phenomenon of shear yielding or crazing or voiding of the matrix. Shear yielding involves the macroscopic drawing of material without a change in volume and occurs at about 45° to the tensile axis. Depending on the polymer, the yielding may be localized or diffused throughout the stress region. This is initiated by a region of high stress concentration that occurs due to the incorporation of toughening agents like rubber particles. Rubber with low modulus acts as stress concentrators and enhances shear yielding. Crazing is a more localized form of yielding in the form of many minute cracks and occurs in the plane normal to the tensile stress. Crazes are initiated at points of maximum strain, which are usually near to the rubber particles and then propagate outward. The process is terminated when the stress concentration falls below the critical level for propagation or when a large particle or other obstacle is encountered. Thus, the rubber particle controls craze growth by initiating and terminating crazes. If craze formation extends over the entire cross section of the polymer blend, the material yields and flows on several planes and converts energy through irreversible deformation and becomes toughened. When many small crazes are formed at the same time, the crazes encounter other rubber particles so that strain at failure and also the energy absorption of the entire material remain lower. Voiding reduces the hydrostatic stress at the crack tip, and thus initiation of voids and their subsequent growth enhances the matrix flow.

5.6 Synthesis of Toughened PMMA/EMA Blends

Binary blends of PMMA and EMA with compositions viz., 100/0, 95/5, 90/10, 85/15, and 80/20 by wt% (PMMA/EVA) were prepared by melt mixing technique using Haake Rheocord 9000 (Germany), corotating twin-screw extruder (L/D ratio 18:1). PMMA (Gujpol–P 876G) with a density of 1.19 g/cc and melt flow index of 1.35 g/10 min was obtained from M/s Gujarat State Fertilizers Company Limited (India) and ethylene methacrylate (EMA) copolymer (Elvaloy 1820 AC) with a density of 0.94 g/cc, was obtained from M/s. DuPont (United States). Prior to mixing, the polymers were predried in a hot air oven at 80°C for 4 hr, and these predried granules of PMMA and EMA were mixed mechanically and fed into the extruder. The processing temperature range was 145 to 210°C, while a screw speed of 30 rpm was employed. The extruded strands were passed through a water bath for faster cooling and further cut into pellets and dried. These dried pellets were compression moulded into suitable shapes and used for further characterization.

5.7 Physicomechanical Properties of PMMA/EMA Blends

The prepared PMMA/EMA blends were characterized for physicomechanical properties such as density, tensile properties, impact strength, and so forth, and the results are briefly discussed below.

5.7.1 Density

PMMA and EMA have an average density of 1183 and 938 kg/m^3 respectively, while that of the blends lies in the range of 1132 to 1175 kg/m^3. Because EMA has a lower density as compared to PMMA, the density of PMMA/EMA blends decreased linearly with an increase in EMA concentration. The theoretical densities were calculated by the volume additive rule using the expressions $1/\rho = W_1/\rho_1 + W_2/\rho_2$, where ρ is the density of the blend, ρ_1 and ρ_2 are the densities of the virgin components, and W_1 and W_2 are the weight fractions of the respective parent polymers. The variation of experimental and theoretical density values as a function of blend compositions is shown in Figure 5.1. The experimental densities of PMMA/EMA blends are very close to the theoretical values, and the variation was found to be linear in nature. Furthermore, the actual density values of these blends are less than that of its theoretical values. Generally, reduction in density values of the blends and lower experimental density values as compared to theoretically

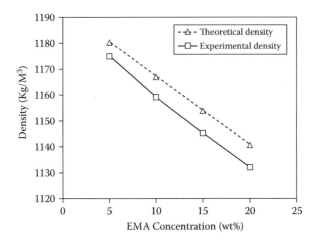

FIGURE 5.1
The plot of density as a function of ethylene methacrylate (EMA) content in poly methyl methacrylate (PMMA)/EMA blends.

obtained density are due to the incompatibility of the constituents, microvoid formation, and poor interfacial adhesion or phase separations. Similar observations were made for PMMA/EVA and PMMA/co poly ether ester (COPE) blends [40,41].

5.7.2 Tensile Properties

A tensile test, in a broad sense, is a measurement of the ability of a material to withstand forces that tend to pull it apart and to determine to what extent the material stretches before breaking. The basic understanding of stress-strain behavior of plastic material is of utmost importance to design engineers. The variation of tensile strength as a function of EMA weight percentage is as shown in Figure 5.2. Virgin PMMA has the highest tensile strength and Young's modulus in comparison with all the PMMA/EMA blends. It was observed that as EMA content increased, the tensile strength and Young's modulus gradually decreased. The decreasing trend of mechanical properties with increased EMA content in the blends is due to the incorporation of the soft elastomeric EMA phase, which gives a plasticization effect to rigid PMMA, The poor adhesion between the polar PMMA and less polar EMA phases also contributes to the poor stress transfer between the continuous phase and the dispersed phase. The extent of reduction in tensile strength and modulus significantly depends on the nature and amount of the elastomeric phase [42].

Various composite models such as parallel model, series model, Halpin–Tsai equation, and Kerner's model can be used to predict and compare the mechanical properties of polymer blends [43–45]. For the theoretical prediction of the tensile behavior of PMMA/EMA blends, some of these models

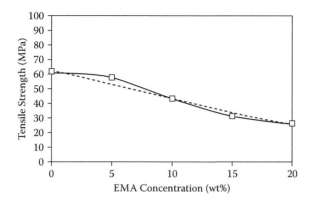

FIGURE 5.2
Effect of ethylene methacrylate (EMA) compositions on the tensile strength of polymethyl methacrylate (PMMA)/EMA blends.

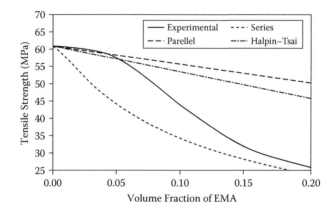

FIGURE 5.3
Experimental and theoretical values of tensile strength as a function of volume fraction of ethylene methacrylate (EMA) phase in polymethyl methacrylate (PMMA)/EMA blends.

were used. Figures 5.3 and 5.4 showed the experimental and theoretical curves of tensile strength and tensile modulus of these blends as a function of the volume fraction of EMA. The experimental value for tensile strength is very close to the series model as shown in Figure 5.3. In Figure 5.4, the tensile modulus as a function of volume fraction of EMA indicates that the experimental data are between a series and Kerner's model.

5.7.3 Impact Strength

The impact strength of the polymeric materials is directly related to the overall toughness of the material. Toughness is defined as the ability of the polymer to absorb applied energy. The incorporation of elastomeric phase, such

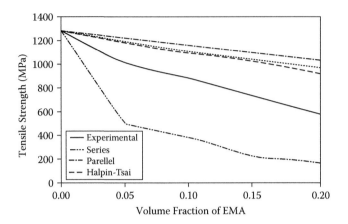

FIGURE 5.4
Experimental and theoretical tensile modulus as a function of volume fraction of ethylene methacrylate (EMA) phase in polymethyl methacrylate (PMMA)/EMA blends.

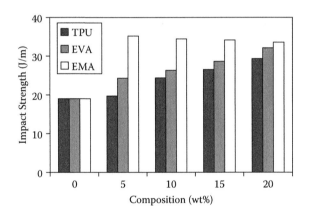

FIGURE 5.5
The effect of thermoplastic polyurethane (TPU), EVA, and ethylene methacrylate (EMA) on the impact strength of polymethyl methacrylate (PMMA) blends.

as thermoplastic polyurethane (TPU), EVA, and EMA into PMMA matrix enhanced the izod impact strength (Figure 5.5). The impact strength of all these blends increased considerably more than virgin PMMA. The addition of 5% EMA (95/5 PMMA/EMA) shows 82% improvement in impact strength. But at higher concentration of EMA (10 and 20 wt%) improvement in impact strength is insignificant. From these results it is very clear that a balanced formulation of rubbery phase in plastics is important for obtaining impact modifications [46,47]. The improvement in impact strength normally implies reduction in stiffness and increase in yield strain. In general, a balanced behavior between toughness and stiffness is always required for optimum performance of the toughened polymer. The order of impact strength with elastomeric phase is: EMA > EVA > TPU. These results clearly indicate that the impact strength of PMMA blends strongly depends on the nature of elastomeric phase and their volume fraction.

5.7.4 Melt Flow Index

The melt index test measures the rate of extrusion of a thermoplastic material through an orifice of specific length and diameter under prescribed conditions of temperature and pressure. This test is primarily used as a means of measuring the uniformity of the flow rate of the material. In this study, the melt flow index (MFI) value increased from 1.35 to 1.69 g/10 min with increase in EMA content from 0 to 10% in the PMMA/EMA blends (Table 5.3). This may be due to the reduction in cohesive strength as well as the plasticizing effect of EMA in PMMA/EMA blends. However, further addition of EMA content above 10% by weight shows no more increase in MFI.

5.7.5 Vicat Softening Temperature

The vicat softening temperature (VST) is one of the important parameters to assess the elevated temperature performance of plastics. The vicat softening temperature (VST) values of PMMA/EMA blends lies in the range of 95 to 96°C (Table 5.3). These data indicate that there is no significant influence of the blend compositions on VST. A slight reduction or retention in VST values of the blends was observed, which may be attributed to the presence of flexible EMA component in the PMMA matrix. From the results, it can be concluded that the addition of EMA does not significantly alter the service temperature of PMMA.

TABLE 5.3

Melt Flow Index and Vicat Softening Temperature of Polymethyl Methacrylate (PMMA), Ethylene Methacrylate (EMA), and Their Blends

Composition of PMMA/EMA Blends (wt/wt%)	Melt Flow Index at 230°C/1.2 kg (g/10 min) (±1%)	Vicat Softening Temperature at 10N Load (±1°C)
100/0	1.35	96.5
95/05	1.50	96.0
90/10	1.69	95.8
85/15	1.68	95.5
80/20	1.67	95.0
0/100	—	—

5.7.6 Effect of Chemical Aging

The resistance of plastics to various chemical reagents is measured by simple immersion of plastic specimens in chemical reagents/solvents as per the standard procedure throughout the plastic industries. This method can be used to compare the relative resistance of various plastics to typical chemical reagents. Resistance of plastics to chemical reagents is best understood through study of its basic polymer structure, the type of polymer bonds, the degree of crystallinity, branching, the distance between the bonds, and the energy required to break the bonds. PMMA is a polar material and its water absorption affects the mechanical properties, but it is not so in case of blends. The virgin polymers (PMMA and EMA) and their blends were exposed to various chemical environments, and the influence of chemical aging on mechanical performance has been evaluated. The tensile behavior of PMMA and all its blends before and after chemical aging are summarized in Tables 5.4 through 5.6. It is observed that after chemical aging, the tensile strength of PMMA is reduced by about 5 to 15%, whereas their blend shows marginal increase in tensile strength in the range of 0 to 8% (Table 5.4). The tensile elongation results of PMMA and all its blends before and after chemical aging are given in Table 5.5. The results indicated that elongation properties of PMMA were retained even after chemical aging, but its blends showed a reduction in percentage elongation at break after chemical aging.

TABLE 5.4

Influence of Chemical Aging on Tensile Strength of
Polymethyl Methacrylate (PMMA) and Its Blends with
Ethylene Methacrylate (EMA)

Chemical Resistance	Tensile Strength (MPa) of PMMA/EMA Blends (±3%)				
	100/0	95/5	90/10	85/15	80/20
Before Aging	60.9	53.0	43.2	31.5	25.8
After Aging in Alkali					
5% Na_2CO_3	51.8	55.2	42.5	30.5	38.8
5% NaOH	54.1	57.1	47.9	37.3	30.8
5% NH_4OH	53.2	55.4	45.7	36.4	24.2
After Aging in Acids					
5% CH_3COOH	55.8	57.1	49.7	33.9	31.9
5% HCl	56.2	55.9	39.2	29.8	24.8
5% H_2SO_4	57.8	55.3	40.3	27.7	25.2
After Aging in Neutral Solvents					
Water	52.3	53.2	41.3	29.2	24.9
n-Hexane	53.0	52.9	37.9	24.2	22.7

TABLE 5.5

Influence of Chemical Aging on Tensile Elongation of
Polymethyl Methacrylate (PMMA) and Its Blends with
Ethylene Methacrylate (EMA)

Chemical Resistance	Tensile Elongation (%) of PMMA/EMA Blends (±3%)				
	100/0	95/05	90/10	85/15	80/20
Before Aging	10.96	15.96	12.69	10.38	11.57
After Aging in Alkali					
5% Na_2CO_3	10.40	15.61	3.48	3.22	3.58
5% NaOH	10.70	11.84	3.73	3.44	3.69
5% NH_4OH	11.20	13.55	4.15	3.70	3.11
After Aging in Acids					
5% CH_3COOH	10.80	5.42	4.45	4.10	2.47
5% HCl	12.30	12.01	2.78	2.75	2.9
5% H_2SO_4	12.80	12.68	2.97	2.98	3.2
After Aging in Neutral Solvents					
Water	12.10	7.20	3.44	2.28	3.05
n-Hexane	12.60	12.02	2.55	1.72	2.32

Whereas in the case of tensile modulus, a marginal reduction was observed for PMMA after chemical aging, but in the case of PMMA/EMA blends, the tensile modulus has increased significantly after chemical aging (Table 5.6). As the EMA concentration increases, the modulus values increased as compared to the respective samples before aging. The tensile property results after chemical aging revealed that the incorporation of EMA into PMMA matrix enhanced the chemical resistance of amorphous PMMA material.

The reduction in mechanical properties of PMMA after chemical aging was expected due to variation in structural characteristics. But the improvement or retention in tensile strength and tensile modulus of PMMA/EMA blends after chemical aging may be due to the influence of chemical reagents with the constituents of the blends, which helps to improve the interaction among the materials that leads to improvement in properties [48]. However, due to the reduction in percentage of elongation at break after chemical aging obviously due to the effect of chemical reagents, the materials may undergo physical or chemical changes and hence, elongation is reduced.

5.7.7 Effect of Heat Aging

Plastic materials exposed to heat undergo many types of physical and chemical changes. The severity of the exposures in both time and temperature

TABLE 5.6

Influence of Chemical Aging on Tensile Modulus of
Polymethyl Methacrylate (PMMA) and Its Blends with
Ethylene Methacrylate (EMA)

Chemical Resistance	Tensile Modulus (MPa) of PMMA/EMA Blends (±6%)				
	100/0	95/05	90/10	85/15	80/20
Before Aging	1280	1077	960	896	755
After Aging in Alkali					
5% Na_2CO_3	1108	2506	2392	1952	1553
5% NaOH	1218	2571	2143	1815	1462
5% NH_4OH	1209	2398	2166	1872	1467
After Aging in Acids					
5% CH_3COOH	1180	2613	2187	1486	1777
5% HCl	1208	2391	2266	1866	1578
5% H_2SO_4	1231	2513	2378	1788	1550
After Aging in Neutral Solvents					
Water	1201	2571	2217	1981	1551
n-Hexane	1189	2457	2455	1916	1386

determines the extent and type of changes that take place. Extended periods of exposure of plastics to elevated temperature will generally cause some degradation with progressive changes in physical properties. Virgin PMMA and its blends with EMA were subjected to heat aging at 80°C for 168 hr, and their tensile properties such as tensile strength, tensile modulus, and elongation at break are as shown in Table 5.7. In the case of PMMA, tensile modulus reduced after heat aging, whereas tensile modulus of the blends showed significant improvement after heat aging.

TABLE 5.7

Influence of Heat Aging on Tensile Properties of Polymethyl Methacrylate
(PMMA) and Its Blends with Ethylene Methacrylate (EMA)

Composition of PMMA/EMA Blends (wt/wt%)	Tensile Strength (MPa) (±3%)		Tensile Elongation (%) (±4%)		Tensile Modulus (MPa) (±6%)	
	Before Aging	After Aging	Before Aging	After Aging	Before Aging	After Aging
100/0	60.9	54.8	10.96	8.23	1280	1216
95/05	51.9	44.0	15.96	3.17	1077	2231
90/10	47.9	30.4	12.69	2.74	960	1936
85/15	44.4	26.7	10.38	3.30	896	1533
80/20	37.8	27.7	11.57	3.15	755	1553

5.8 Thermal Properties

5.8.1 Thermogravimetric Analysis

Thermogravimetric analysis (TGA) of PMMA, EMA, and their blends was performed, and the thermograms of EMA, PMMA, and their blends 95/5 and 90/10 PMMA/EMA (Figure 5.6) showed one-step thermal degradation, whereas 85/15 and 80/20 PMMA/EMA blends showed two-step thermal degradation processes. The characteristic degradation temperatures obtained from TGA thermograms of the PMMA, EMA, and their blends are summarized in Tables 5.8 and 5.9. The initial decomposition temperature

TABLE 5.8

Thermogravimetric Analysis (TGA) Thermal Characteristics of Polymethyl Methacrylate (PMMA), Ethylene Methacrylate (EMA), and Their Blends

Composition of PMMA/EMA Blends (wt/wt%)	Temperature at Different Weight Loss (±2°C)					Ash Content (%)	Oxidation Index (OI)
	T_0	T_{10}	T_{20}	T_{50}	T_{max}		
100/0	378	383	388	410	438	0.5	0.035
95/05	382	385	388	410	430	0.5	0.035
90/10	381	382	390	400	430	0.5	0.035
85/15	377	380	390	400	480	0.5	0.035
80/20	376	380	385	400	480	0.5	0.035
0/100	461	462	465	485	515	1.5	0.105

Notes: T_0, temperature of onset of decomposition; T_{10}, temperature for 10% mass loss; T_{20}, temperature for 20% mass loss; T_{50}, temperature for 50% mass loss; and T_{max}, temperature for maximum weight loss.

TABLE 5.9

Thermal Data Obtained from Derivative Thermogravimetric Analysis (TGA) Curve of Polymethyl Methacrylate (PMMA), Ethylene Methacrylate (EMA), and Their Blends

Composition of PMMA/EMA (wt/wt%)	Transition Temperature Range (±2°C)		
	T_i	T_p	T_c
100/0	285	400	440
95/5	278	400	430
90/10	324	398	430
85/15	312	400	480
80/20	300	402	485
0/100	365	485	515

Notes: T_i, temperature at which decomposition starts; T_p, temperature at which decomposition rate is maximum; and T_c, temperature at which decomposition is completed.

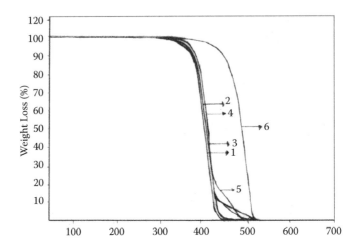

FIGURE 5.6
Thermogravimetric analysis (TGA) thermograms of polymethyl methacrylate (PMMA), ethylene methacrylate (EMA), and their blends.

(T_i) corresponding to 1% decomposition reached a lowest (278°C) for the PMMA/EMA (95/5) blend and highest (365°C) for EMA. This shows that a higher proportion of EMA in the blends makes them thermally more stable than the individual components. The temperature at which the maximum rate of decomposition is higher in the case of EMA (485°C), but decomposition temperature for maximum weight loss of blends increased as EMA content in the blends increased. This result indicated that the blends containing higher concentration of EMA content is less susceptible to thermal degradation (Figure 5.6).

5.8.2 Differential Scanning Calorimetry (DSC) Studies

DSC thermograms of virgin PMMA, EMA, and their blends are shown in Figure 5.7. It is very clear that the virgin PMMA and EMA shows single T_g, whereas their corresponding blends exhibit two T_g values for all the compositions. For PMMA, T_g values occurred at 100.6°C and for EMA shifted to a higher temperature at −32.1°C. In blends, the T_g values of PMMA shifted to a lower temperature for all the blends from 100.6°C to 76°C and for EMA from −32.1°C to −21.8°C. The variation in T_g values for the blends and the individual components are given in Table 5.10. A marked reduction in T_g was noticed after incorporation of 5% EMA. Further increase in EMA content in blends, increased T_g values gradually, which may be due to plasticization effect and the possibility of some kind of physical interaction among the components at higher concentrations of EMA content.

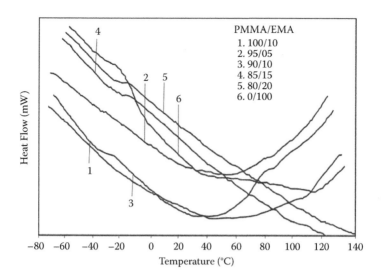

FIGURE 5.7
Differential scanning calorimetry (DSC) thermograms of polymethyl methacrylate (PMMA), ethylene methacrylate (EMA), and their blends.

TABLE 5.10

T_g Data Obtained from Differential Scanning Calorimetry (DSC) of Polymethyl Methacrylate (PMMA), Ethylene Methacrylate (EMA), and Their Blends

Composition of PMMA/EMA (wt/wt%)	T_g (±1°C)	
	T_{g1}	T_{g2}
100/0	—	100.6
95/05	−34.7	78.5
90/10	−34.2	78.0
85/15	−21.3	91.0
80/20	21.8	93.0
0/100	−32.1	—

5.8.3 Dynamic Mechanical Analysis (DMA)

DMA is a very useful thermoanalytical tool for studying the structure–property relationship versus performance of polymeric materials. The important information concerning relaxation transitions occurring on a molecular scale can be derived by subjecting polymers to a small amplitude cyclic deformation. Data obtained over a broad temperature range help to understand the influence of one polymer on the properties of another polymeric component

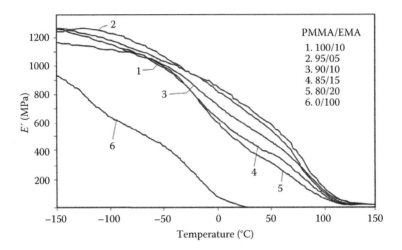

FIGURE 5.8

Effect of ethylene methacrylate (EMA) content on storage modulus of polymethyl methacrylate (PMMA) blends.

in the blend and ultimately aid in the design of a polymer blend with desirable properties [49,50]. The dynamic modulus indicates the inherent stiffness of material under dynamic loading conditions. The mechanical damping indicates the amount of energy dissipated as heat during the deformation of the material. The storage modulus and temperature curves of virgin PMMA, EMA, and their blends containing 5, 10, 15, and 20 wt% of the EMA polymer are shown in Figure 5.8. The two steps in the dynamic modulus–temperature curves that are very much visible with an increase in EMA content are the characteristics of an immiscible two-phase system [51]. The DMA analysis has been carried out to study the effect of EMA on storage modulus of PMMA/EMA blends. For all the blend compositions, the storage modulus decreased, indicating that the introduction of EMA reduced the storage modulus of PMMA proportionately. The thermograms of the blends also showed two-step reductions in storage modulus, one at the T_g of EMA and the other one at the T_g of PMMA. As EMA content increased, the reduction in storage modulus at the T_g region of EMA is more. The modulus of elasticity in the glassy state for the two blends is shown to decrease with increasing EMA content. This result is expected because the EMA is a low modulus flexible material.

Figure 5.9 shows the effect of EMA addition into PMMA on tan δ of blends. The EMA has maximum dampness at around –4.1°C, whereas PMMA has dampness at 98.6°C (Table 5.11). The PMMA/EMA blends have maximum dampness values at 106.8, 112.4, 104.9, and 100.7°C, for 5, 10, 15, and 20% EMA blends, respectively. The magnitude of the peak is more or less characteristic of the relative concentration of the components and whether or not the phases are dispersed or continuous [40,52–55]. For instance, at a given concentration, tan δ is greater if the polymer with high T_g value is in continuous phase [52].

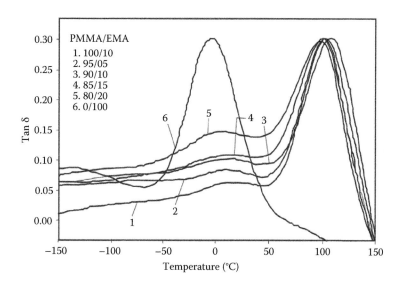

FIGURE 5.9
Effect of ethylene methacrylate (EMA) content on Tan δ of polymethyl methacrylate (PMMA) blends.

TABLE 5.11

T_g Obtained from Dynamic Mechanical Analysis (DMA) Thermograms for Polymethyl Methacrylate (PMMA), Ethylene Methacrylate (EMA), and Their Blends

Composition of PMMA/EMA (wt/wt%)	T_g from DMA (±1°C)	
	T_{g1}	T_{g2}
100/0	—	98.6
95/05	13.9	106.8
90/10	15.8	112.4
85/15	18.6	104.9
80/20	10.2	100.7
0/100	−4.1	—

5.9 Fourier Transform Infrared Spectroscopy

Information regarding the possible interaction between the components in a polymer blend can be obtained by comparison of the infrared spectra of the base material with that of the blends [20,21]. Figures 5.10a and 5.10b represent the typical Fourier transform infrared (FTIR) spectra of PMMA and PMMA/EMA (90/10) blends, respectively. The spectrum of PMMA shows a very sharp peak around 3398 cm⁻¹ due to overtones of stretching vibrations

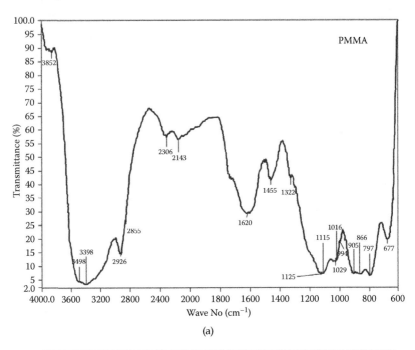

(a)

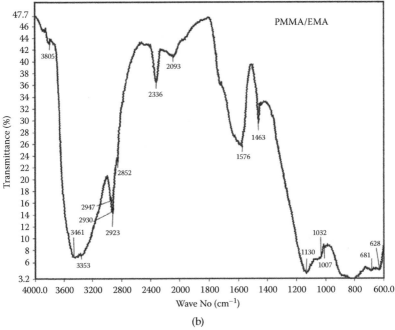

(b)

FIGURE 5.10
Fourier transform infrared (FTIR) spectra of (a) polymethyl methacrylate (PMMA) and (b) PMMA/ethylene methacrylate (EMA) (90/10) blend.

of ester groups. A stronger peak at 1600 to 1740 cm^{-1} is due to the carbonyl stretching of ester groups of PMMA. In PMMA, due to the presence of two methyl (electron donors) groups close to the ester group, the C=O stretching frequency has shifted to 1620 cm^{-1} rather than at 1730 cm^{-1}. The introduction of EMA onto the PMMA matrix shows further shifting of this peak from 1620 to 1576 cm^{-1}. Generally, good interactions between two compatible blends are expected to weaken the C=O stretching frequency by withdrawing electrons from it. So, this can be called incompatible blends. These results were supported by impact strength analysis, where impact strength increased with increased EMA content in the blends.

5.10 Optical Properties

The measured optical properties such as percentage of transmittance and clarity of virgin PMMA and its blends are given in Table 5.12. It is observed that the percentage transmittance of PMMA is 90.5 and for its blends, the transmittance values lies in the range of 84.2% to 86.8%, and the incorporation of EMA has not significantly affected the transmittance. The percent of clarity slightly reduces after the incorporation of EMA. The clarity values of the blends lie in the range of 70.8% to 77.3%. The EMA toughened PMMA blends retain optical properties.

5.11 Surface Morphology

The mechanical properties of the blends can be correlated with the morphology of the blends. The variation in morphology of heterogeneous polymer blends depends on composition of blend, melt viscosity of the individual polymers, and processing history [42]. When polymers with different viscosities are mixed to form blends, less viscous polymer will form the continuous phase [54]. The optical photomicrographs of PMMA/EMA blends are shown in Figures 5.11a through 5.11c, respectively. EMA is found to be

TABLE 5.12

Optical Properties of Polymethyl Methacrylate (PMMA) and Its Blends with Ethylene Methacrylate (EMA)

	Composition of PMMA/EMA Blends (wt/wt%) (±1%)				
Optical Property	**100/0**	**95/05**	**90/10**	**85/15**	**80/20**
Transmittance (%)	90.5	86.8	84.4	84.5	84.2
Clarity (%)	79.2	77.3	77.1	75.8	70.8

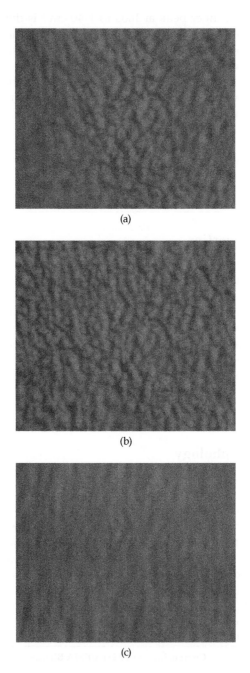

(a)

(b)

(c)

FIGURE 5.11
Optical micrographs of (a) 95/5, (b) 90/10, and (c) 80/20 polymethyl methacrylate (PMMA)/ethylene methacrylate (EMA).

dispersed as domains in the continuous PMMA matrix. This is due to the higher melt viscosity of EMA as well as the lower concentration of EMA in blend as compared to PMMA. As the EMA content in the blend increased from 5 to 20 wt%, the average domain size of the dispersed EMA phase also increased. The higher concentration of rubber content as well as the higher viscosity of the EMA and lower viscosity of the PMMA phase are attributed to the reagglomeration or coalescence of the dispersed rubber particles [42].

In Figure 5.11a, it is observed that the surface of the dispersed phase has stretched and became rough owing to better adhesion between EMA and PMMA. This indicates that at 5 wt% addition of EMA in PMMA matrix has very good adhesion among the constituents. This observation supports the increase in impact strength at 5 wt%. Figure 5.11b also reveals that the rubber phase dispersed in the rigid PMMA phase with less stretching/roughness in comparison with 95/5 PMMA/EMA might have contributed to the lower mechanical properties. Figure 5.11c exhibits more or less a cocontinuous phase, which also may be the reason for lower mechanical properties. This observation also supports the observation that higher EMA content in PMMA matrix decreases the impact strength of PMMA.

5.12 Conclusions

The binary blends of PMMA and EMA with different compositions were prepared by melt mixing technique using a corotating twin-screw extruder. The physicomechanical, thermal, chemical, optical, and morphological properties of the blends were evaluated. The addition of EMA increased the impact strength of PMMA blends without significantly affecting the optical properties of these blends. It was observed that as the EMA content increased, the tensile strength and modulus of the blends were reduced. Various composite models have been used to fit the experimental data for tensile strength and modulus. Theoretically calculated tensile properties showed agreement with the experimental results. The chemical and heat aging behavior of PMMA/EMA blends revealed that the EMA content in the blend enhanced the chemical and heat resistance properties of PMMA. The thermal stability of the blends had shown an increasing trend with an increase in EMA content in PMMA matrix. The DSC scans indicated that as the EMA content increased in the blends, T_g values increased gradually. The DMA analysis showed that for all the blend compositions, the storage modulus decreased, indicating that the introduction of EMA reduced the storage modulus of PMMA proportionately. The morphology of the blends indicates that the elastomeric phase is dispersed in the continuous PMMA matrix at lower (5%) composition of EMA. But at higher (20%) concentrations of EMA, it forms a continuous phase resulting more or less homogeneous morphology.

References

1. Paul, D. R., and Newman, S. 1978. *Polymer Blends*. New York: Academic Press.
2. Utracki, L. A. 1998. *Commercial Polymer Blends*. London, New York: Chapman and Hall.
3. Utracki, L. A. 2002. *Polymer Blends Handbook,* Vol. 1. Dordrecht: Kluwer Academic.
4. Poomalai, P., and Siddaramaiah. 2005. Studies on poly (methyl methacrylate) (PMMA) and thermoplastic polyurethane (TPU) blends. *Journal of Macromolecular Science, Part A—Pure and Applied Chemistry* 42:1399–1407.
5. Kumar, H., Anilkumar, A., and Siddaramaiah. 2006. Physico-mechanical, thermal and morphological behaviour of PU/PMMA semi interpenetrating polymer networks. *Polymer Degradation and Stability* 91:1097–1104.
6. Poomalai, P., Ramaraj, B., and Siddaramaiah. 2007. Thermal and mechanical properties of poly (methyl methacrylate) and ethylene-vinyl acetate co-polymer blends. *Journal of Applied Polymer Science* 106 (1):684–691.
7. Bucknall, C. B., Partridge, I. K., and Ward, M. V. 1984. Rubber toughening of plastics. *Journal of Materials Science* 19:2064–2072.
8. Smith, A. P., Ade, H., Balik, C. M., Kock, C. C., Smith, S. D., and Spontak, R. J. 2000. Cryogenic mechanical alloying of poly (methyl methacrylate) with polyisoprene and poly (ethylene-*alt*-propylene). *Macromolecules* 33:2595–2604.
9. Poomalai, P., and Siddaramaiah. 2006. Thermal and morphological studies on poly (methyl methacrylate)/thermoplastic polyurethane blends. *Journal of Macromolecular Science, Part A—Pure and Applied Chemistry* 43(4):695–702.
10. Poomalai, P., Varghese, T. O., and Siddaramaiah. 2009. Ethylene methacrylate (EMA) co-polymer toughened polymethyl methacrylate blends: Physicomechanical, optical, thermal and chemical properties. *Polymer-Plastics Technology and Engineering* 48(9):958–965.
11. Yemel Yanov, D. N., Myachev, V. A., and Dreval, V. Y. 1990. Glass transition, compatibility and viscosity of blends of poly methyl methacrylate with its copolymers. *Polymer Science USSR* 32:1532–1538.
12. Collyer, A. A. 1994. *Rubber Toughened Engineering Plastics*. London: Chapman and Hall.
13. Oommen, Z., and Thomas, S. 1997. Mechanical properties and failure mode of thermoplastic elastomers from natural rubber/poly (methyl methacrylate)/natural rubber-g-poly(methyl methacrylate) blends. *Journal of Applied Polymer Science* 65(7):1245–1255.
14. Oommen, Z., and Thomas, S.no access 1997. Compatibility studies of natural rubber/poly(methyl methacrylate) blends by viscometry and phase separation techniques. *Journal of Materials Science* 32(22):6085–6094.
15. Raghavan, D., Van Landingham, M., Gu, X., and Nguyen, T. 2000. Characterization of heterogeneous regions in polymer systems using tapping mode and force mode atomic force microscopy. *Langmuir* 16(24):9448–9459.
16. Archie, P. S., Harald, A., Carl, C. K., and Richard, J. S., 2000. Solid-state blending of polymers by cryogenic mechanical alloying. *Materials Research Society Symposium Proceedings* 629, FF 6.9.

17. Mina, M. F., Ania, F., Balta Calleja F. J., and Asano, T. 2004. Microhardness studies of PMMA/natural rubber blends. *Journal of Applied Polymer Science* 91(1): 205–210.

18. Samui, A. B., Dalvi, V. G., Parti, M., Chakraborty, B. C., and Deb, P. C. 2004. Studies on semi-interpenetrating polymer network based on nitrile rubber and poly (methyl methacrylate). *Journal of Applied Polymer Science* 91(1):354–360.

19. Nakason, C., Panklieng, Y., and Kamesamman, A. 2004. Rheological and thermal properties of thermoplastic natural rubbers based on poly (methyl methacrylate)/epoxidized-natural-rubber blends. *Journal of Applied Polymer Science* 92(6):3561–3572.

20. Spadaro, G., Dispenza, C., Mc Grail, P., and Valenza, A. 2004. Submicron structured polymethyl methacrylate/acrylonitrile-butadiene rubber blends obtained via gamma radiation induced in situ polymerization. *Advances in Polymer Technology* 23(3):211–221.

21. Mina, M. F., Ania, F., Huy, T. A., Michler, G. H., and Balta Calleja, F. J. 2004. Micromechanical behavior and glass transition temperature of poly(methyl methacrylate)-rubber blends. *Journal of Macromolecular Science B Physics* 43(5):947–961.

22. Mina, M. F., Haque, M. E., Balta Calleja, F. J., Asano, T., and Alam, M. M. 2004. Microhardness studies of the interphase boundary in rubber-softened glassy polymer blends prepared with/without compatibilizer. *Journal of Macromolecular Science B Physics* 43(5):1005–1014.

23. Suriyachi, P., Kiatkamjornwong, S., and Prasassarakich, P. 2004. Natural rubber-*g*-glycidyl methacrylate/styrene as a compatibilizer in natural rubber/PMMA blends. *Rubber Chemistry and Technology* 77(5): 914–930.

24. Nakason, C., Pechurai, W., Sahakaro, K., and Kaesaman, A. 2005. Rheological, mechanical and morphological properties of thermoplastic vulcanizates based on NR-g-PMMA/PMMA blends. *Polymers for Advanced Technologies* 16(8):592–599.

25. Nakason, C., Tobprakhon, A., and Kaesaman, A. 2005. Thermoplastic vulcanizates based on poly(methyl methacrylate)/epoxidized natural rubber blends: Mechanical, thermal, and morphological properties. *Journal of Applied Polymer Science* 98(3):1251–1261.

26. Vancaeyzeele, C., Fichet, O., Boileau, S., and Teyssie. D. 2005. Polyisobutene–poly(methylmethacrylate) interpenetrating polymer networks: Synthesis and characterization. *Polymer* 46(18):6888–6896.

27. Mina, M. F., Alam, A. K. M. M., Chowdhury, M. N. K., Bhattacharia, S. K., and Balta Calleja, F. J. 2005. Morphology, micromechanical, and thermal properties of undeformed and mechanically deformed poly (methyl methacrylate)/rubber blend. *Polymer Plastics Technology and Engineering* 44(4):523–537.

28. Dong, J., Liu, Z., and Zheng, C. 2006. Preparation, morphology, and mechanical properties of elastomers based on ω-dihydroxy-polydimethylsiloxane/poly(methyl methacrylate) blends. *Journal of Applied Polymer Science* 100(2): 1547–1553.

29. Nakason, C., Saiwaree, S., Tatun, S., and Kaesaman, A. 2006. Rheological, thermal and morphological properties of maleated natural rubber and its reactive blending with poly(methyl methacrylate). *Polymer Testing* 25(5):656–667.

30. Wu, L., Yu, H., Yan, J., and You, B. 2001. Structure and composition of the surface of urethane/acrylic composite latex films. *Polymer International* 50(12):1288–1293.
31. Kukanja, D., Golob, J., Zupancic-Valant, A., and Krajnc, M. 2000. The structure and properties of acrylic-polyurethane hybrid emulsions and comparison with physical blends. *Journal of Applied Polymer Science* 78(1):67–80.
32. Kosyanchuk, L. F., Babkina, N. V., Yarovaya, N. V., Kozak, N. V., and Lipatov, Y. S., 2008. Phase separation in semi-interpenetrating polymer networks based on crosslinked poly(urethane) and linear poly(methyl methacrylate) containing iron, copper, and chromium chelates. *Polymer Science Series A* 50(4):434–433.
33. Lipatov, Y. S., and Kosyanchuk, L. F. 2004. Interfacial region in blends of linear polymers formed in situ. *Composite Interfaces* 11(5–6):393–402.
34. Patricio, P. S. O., De Sales, J. A., Silva, G. G., Windmoller, D., and Machado, J. C. 2006. Effect of blend composition on microstructure, morphology, and gas permeability in PU/PMMA blends. *Journal of Membrane Science* 271(1–2):177–185.
35. Muneera, B., and Siddaramaiah. 2004. Synthesis and characterization of polyurethane/polybutyl methacrylate interpenetrating polymer networks. *Journal of Materials Science* 39(14):4615–4623.
36. Jeevananda, T., Muneera, B., and Siddaramaiah. 2001. Studies on polyaniline filled PU/PMA interpenetrating polymer networks. *European Polymer Journal* 37(6):1213–1218.
37. Cheng, S. K., and Chen, C. Y. 2004. Mechanical properties and strain-rate effect of EVA/PMMA in situ polymerization blends. *European Polymer Journal* 40(6):1239–1248.
38. Errico, M. E., Greco, R., Laurienzo, P., Malinconico, M., and Viscardo, D. 2006. Acrylate/EVA reactive blends and semi-IPN: Chemical, chemical–physical, and thermo-optical characterization. *Journal of Applied Polymer Science* 99(6):2926–2935.
39. Mayu, Si., Tohru, A., Harald, A., Kilcoyne, A. L. D., Robert, F., Jonathan, C. S., and Miriam, H. R. 2006. Compatibilizing bulk polymer blends by using organoclays. *Macromolecules* 39:4793–4801.
40. Poomalai, P., Ramaraj, B., and Siddaramaiah. 2007. Poly (methyl methacrylate) toughened by ethylene-vinyl acetate co-polymer: Physico-mechanical, thermal and chemical properties. *Journal of Applied Polymer Science* 104:3145–3150.
41. Poomalai, P., Varghese, T. O., and Siddaramaiah. 2008. Investigation on thermoplastic co poly (ether-ester) elastomer toughened poly (methyl methacrylate) blends. *Journal of Applied Polymer Science* 109(6):3511–3518.
42. George, S., Joseph, R., Thomas, S., and Varghese, K. T. 1995. Blends of isotactic polypropylene and nitrile rubber: Morphology, mechanical properties and compatibilization. *Polymer* 36:4405–4416.
43. Coran, A. Y. 1998. *Handbook of Elastomers—New Development and Technology*, (Eds.) Bhowmick, A. K., and Stephens, H. L. New York: Marcel Decker.
44. Dickie, R. A. 1978. *Polymer Blends*, (Eds.) Paul, D. R., and Newman, S. New York: Academic Press.
45. Kerner, E. H. 1956. The elastic and thermo-elastic properties of composite media. *Proceedings of Physics Society (B)* 69:808–813.
46. Bucknall, C. B. 1977. *Toughened Plastics*. London: Applied Science.
47. Singh, Y. P., and Singh, R. P. 1989. Investigation on multiphase polymeric systems of poly (vinyl chloride). I. PVC–chlororubber-20-*gp*-styrene–acrylonitrile (2:1) blends. *Journal of Applied Polymer Science* 37:2491–2515.

48. Pendyala, V. N. S., Xavier, S. F., and Utracki, L. A. Eds. 2002. *Polymer Blends Handbook.* Dordrecht: Kluwer Academic.
49. Kalkar, K., and Roy, N. K. 1992. Dynamic mechanical properties of bisphenol-A polycarbonate/poly (p-t-butylphenol formaldehyde) blends. *Polymer Science Contemporary Themes*, II, Tata New Delhi: McGraw-Hill..
50. Varghese, K. T., Nando, G. B., De. P. P., and De, S. K. 1988. Miscible blends from rigid poly (vinyl chloride) and epoxidized natural rubber. Part 1. Phase morphology. *Journal of Materials Science* 23:3894–3902.
51. Ward, I. M. 1971. *Mechanical Properties of Solid Polymers.* New York: John Wiley and Sons.
52. Deutsch, K., Hoff, E. A. W., and Reddish, W. 1954. Relation between the structure of polymers and their dynamic mechanical and electrical properties. Part I. Some alpha-substituted acrylic ester polymers. *Journal of Polymer Science* 13:565–582.
53. Powles, J. G., Hunt, B. I., and Sandiford, D. J. H. 1964. Proton spins lattice relaxation and mechanical loss in a series of acrylic polymers. *Polymer* 5:505–515.
54. Heijboer, J. 1965. *Physics of Non-crystalline Solids.* Amsterdam: North-Holland, 231.
55. Miyamoto, T., Kodama, K., and Shibayama, K. 1970. Structure and properties of a styrene-butadiene-styrene block copolymer. *Journal of Polymer Science A-2 Polymer Physics* 8:2095–2103.

6

Molecular Dynamics Simulation Studies of Binary Blend Miscibility

Hua Yang

Tianjin Key Laboratory of Structure and Performance for Functional Molecules
Tianjin Normal University
Tianjin, People's Republic of China

CONTENTS

6.1 Introduction

Polymer blends are expected to create many useful novel materials with specific properties that cannot be achieved by individual polymers in the future. And blending is an easy and inexpensive method of modifying various properties of polymer. A polymer blend or polymer mixture is a kind of material analogous to metal alloys, in which at least two polymers are blended together to create a new material with tailored physical properties, which can be broadly divided into three categories: immiscible, compatible, and miscible polymer blends. The immiscible blends are made of two polymers, and two glass transition temperatures will be observed. Compatible polymer blends are immiscible polymer blends that exhibit macroscopically uniform physical properties. The macroscopically uniform properties are usually caused by sufficiently strong interactions between the component polymers. Miscible polymer blends are single-phase structures. In this case, one glass transition temperature will be observed [1].

It is generally very difficult or even impossible to predict the properties of a blend; however, this is only the second step. The first problem is an understanding of the mixing properties (i.e., knowledge of under which conditions two polymers will form either a homogeneous phase or a two-phase structure) [2]. The blend miscibility is an important effect to determine the structure and properties of the new material. Polymer blend miscibility is therefore a central subject in the science and technology of polymers.

Besides great industrial significance, polymer blend miscibility entails many peculiar problems of academic interest. Unlike small molecule mixtures, most polymers are immiscible when blending. This is attributable to the much smaller favorable entropy of mixing in a polymer than small molecule mixtures arising from the chain connectivity of the macromolecules. In most miscible blends, the heat of mixing is negative due to specific long-range interactions such as hydrogen bonding or van der Waals forces [3]. The dynamical behavior of each component in a polymer blend could also be modified by blending. This behavior to a large extent is a consequence of increased local concentration of a given segment around itself due to chain connectivity [4–7].

Usually, blend miscibility is characterized by the Flory–Huggins parameter, χ [8], which gives a theoretical description of the phase separation curves for polymers of different chain length and composition. Parameter χ can be obtained through different methods such as experiments, theoretical predictions, and computer simulations. In addition, solubility parameter (δ), cohesive energy density (CED) of the pure components in blends and energy of mixing (ΔE_{mix}), and glass transition temperatures (T_g) may also be used to show the compatibility of different polymers. The determination of heat of mixing, T_g and morphology by electron microscopy, scanning electron microscopy, dynamic mechanical response, and differential scanning calorimetry are some of the methods extensively used in the literature [9–12]. Other methods also exist to study blend compatibility, which include viscometry refractometry, sonic velocity, fluorescence spectra, and so forth, which revealed various aspects of polymer blend miscibility of polymer blends in a viscous state [13,14]. The experimental protocols, though accurate, are time consuming and expensive; sometimes the results are contradictory.

Besides extensive experiments, many computer simulations have been carried out on polymer blends, primarily, including Monte Carlo (MC) [15–18], molecular dynamics (MD) [12,19–35], mesoscopic dynamics (MesoDyn) [12,24,25], and dissipative particle dynamics (DPD) [33,36,37]. In the area of theoretical polymer physics, MesoDyn and DPD have been used to treat polymeric chains in a coarse-grained (mesoscopic) level by grouping atoms together up to the persistence length of polymers. Recent trends in the use of MD simulations on bulk polymers have led to the calculations of important

properties of polymeric blends with greater accuracy that are otherwise difficult to study experimentally or for which experimental data are not available or often when such results are conflicting. In this chapter, we briefly
survey several MD simulations on binary blend miscibility and seek future
prospects of modeling studies.

6.2 Molecular Dynamics Simulation Strategies

MD simulations are performed to obtain information about properties
and processes on the atomic or molecular level [38] in which force field is
an important effect to determine the accuracy of the computational results.
Before simulation, one should choose a proper force field first. COMPASS
[39] (condensed-phase optimized molecular potentials for atomistic simulation studies) force field based on PCFF (polymer consistent force field) may
be a nice choice for modeling interatomic interactions. In the COMPASS force
field, total potential energy (E_T), which includes bonding (E_{Bond}), nonbonding
$(E_{nonbond})$, and cross-coupling (E_{cross}) energies, can be represented as following:

$$E_T = E_{bond} + E_{nonbond} + E_{cross} = E_b + E_\theta + E_\varphi + E_{oop} + E_{vdw} + E_q + E_{cross} \qquad (6.1)$$

Here, the first four terms of the right hand of the equation represent the
bonded interactions, which correspond to energies associated with the bond
stretching, E_b; bond angle bending, E_θ; torsion angle rotations, E_φ; out of
loop, E_{oop}; respectively. The last two terms represent nonbonded interactions,
which consist of van der Waals energy, E_{vdw}, and electrostatic energy, E_q. The
Lennard-Jones 6-12 potential is used to describe E_{vdw}, while E_q is calculated
from the partial charges of atoms in the system as estimated by the charge-
equilibration method [40]. Electrostatic interaction can also be calculated by
the Ewald summation method [41], because it calculates long-range interactions more accurately.

The simulation procedure is usually as follows:

1. Model constructions (pure polymer chain or polymer blends under
   periodical boundary condition and in an amorphous conformation).
2. Annealing and energy minimization (to eliminate the unfavorable
   contacts and to ensure sufficient contacts).
3. Cool down the system stepwise from a high temperature to low
   temperature at a constant interval. At each temperature, MD simulations perform in the canonical (NVT) ensemble and isothermal-
   isobaric (NPT) .
4. Save the trajectories for analysis at last.

6.2.1 Polymer Chain Length in Simulations

Polymer systems may contain thousands of atoms or repeated units. Even if MD simulation on such a system were possible, in many cases much of the information generated would be discarded. Due to computer limitations, simulations could not be performed using the actual size of the polymer. Hence, the size of the polymer used in simulations is important when computing thermodynamic quantities. The minimum level molecular size chosen should be sufficient to represent the real polymer system. Many reports usually used short oligomers with term or slightly more repeated units to simulate the miscibility of polymers [30,42]. Such short chains might lead to end effects and cannot represent the real systems accurately. On the other hand, one also does not want to use very long chains because of demanding much more simulation time. To determine the appropriate polymer chain length in the simulation, the density and solubility parameter may be examined as a function of polymer chain length, which usually increases or decreases with increasing chain length, respectively. When the two physical parameters are almost constant ($\geq N_{eq}$), we usually believe that the corresponding chain length is sufficient to present a polymer chain.

The solubility parameter, as defined by Hildebrand and Scott [43], is the square root of the cohesive energy density. The equation is

$$\delta = \sqrt{CED} = \sqrt{\frac{E_{coh}}{V}} = \sqrt{\frac{E_{vac} - E_{bulk}}{V}} \tag{6.2}$$

where V is the mole volume of the polymer system, and E_{coh} is the cohesive energy per mole obtained from the energy difference between the molecule in a vacuum (E_{vac}) and in an amorphous bulk state (E_{bulk}). In order to obtain a density and solubility parameter, polymer models of different chain lengths should be built first. After MD simulations, we can calculate their density and solubility parameter from the MD trajectories. Luo and Jiang [33] calculated the density and parameter δ of polyvinyl chloride (PVC) and poly(ethylene oxide (PEO) with different chain lengths, as shown in Figure 6.1. They found that for PEO, when the number of monomers reaches 40, the density and δ are almost constant; thus 40 repeat units are sufficient to present a PEO chain. For PVC, the δ does not change significantly for 50-mers and longer chains. Therefore, PVC with a length of 50 is reasonable to present a polymer.

6.2.2 Blend Miscibility

The solubility parameter can be used to measure the miscibility between different polymers. If the difference of δ between two polymers is large, it indicates that these two polymers are immiscible, whereas it has the probability to be miscible when the difference is small. Empirically, if $(\delta_A - \delta_B)^2$ between

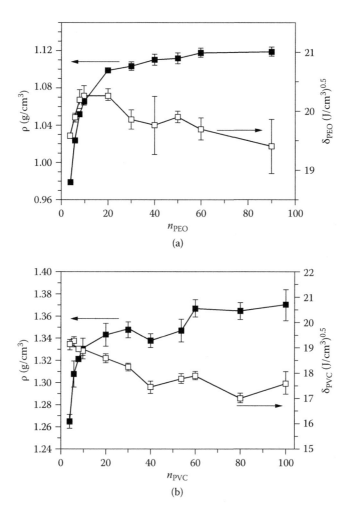

FIGURE 6.1
Density and solubility parameter as a function of chain length for (a) polyethylene oxide (PEO) and (b) polyvinyl chloride (PVC). (From Luo, Z. L., and Jiang, J. W. 2010. Molecular dynamics and dissipative particle dynamics simulations for the miscibility of poly(ethylene oxide)/poly(vinyl chloride) blends. *Polymer* 51:291–299.)

two polymers A and B is less than 4 J/cm³, one may believe that the two polymers are miscible [44]. This approach is only suitable to judging the miscibility of simple systems in which the specific interactions (such as H-bonds) or noncombinatorial entropy effects (such as a volume change induced by blending) do not play a dominant role. Moreover, the Flory–Huggins parameter, χ, enthalpy, entropy, and Gibbs free energy of mixing can also be calculated to understand the energetics in mixing of polymers, which indicated favorable interactions. The Flory–Huggins parameter χ could be simply estimated from

$$\chi = \frac{V_{mono}}{RT}(\delta_A - \delta_B)^2 \qquad (6.3)$$

where V_{mono} is monomer unit volume per mole; R is gas constant; and T is temperature in Kelvin. χ estimated from above equation is always positive. This is not exactly correct.

If a binary blend is sufficiently equilibrated, the energy of mixing ΔE_{mix} could be calculated [45]:

$$\Delta E_{mix} = \varphi_A\left(\frac{E_{coh}}{V}\right)_A + \varphi_B\left(\frac{E_{coh}}{V}\right)_B - \left(\frac{E_{coh}}{V}\right)_{mix} \qquad (6.4)$$

Here, the subscripts A, B, and *mix* represent two kinds of polymers and the blend of them, respectively. The symbols φ_A and φ_B represent the volume fractions of the two polymers A and B, $\varphi_A + \varphi_B = 1$. The Flory–Huggins parameter χ could also be calculated by [45,46]

$$\chi = \frac{z\Delta E_{mix}}{RT} \qquad (6.5)$$

or

$$\chi = \left(\frac{\Delta E_{mix}}{RT}\right)V_{mono} \qquad (6.6)$$

where z is the coordination number (its value for the cubic lattice model is taken as 6). Unlike Equation (6.3), Equations (6.5) and (6.6) give either positive or negative χ. A positive χ does not necessarily mean that the blend is immiscible. Base on the Flory–Huggins theory, the critical value of χ could be calculated using the following equation:

$$\chi_{critical} = \frac{1}{2}\left(\frac{1}{\sqrt{n_A}} + \frac{1}{\sqrt{n_B}}\right)^2 \qquad (6.7)$$

where n_A and n_B are degrees of polymerization (actual number of repeated units) of A and B. If $\chi < \chi_{critical}$, the blend is considered to be miscible and vice versa. If χ is slightly greater than the critical value, the blends could be partially miscible. Comparing χ with critical values will provide an indication about the degree of miscibility of the polymer blend.

Figure 6.2 displays the results of χ versus the weight fraction of poly(vinyl alcohol) (PVA) in PVA/chitosan (CS) blends calculated from Equations (6.5) and (6.7) [25]. $\chi_{critical}$ exhibits a linearity (around 0.18) over their studied composition range of the blends. For the blends from 10% to 40% weight fraction

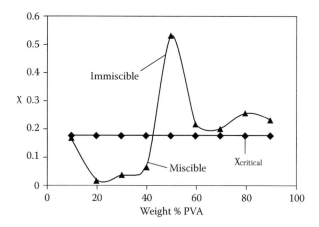

FIGURE 6.2
Flory–Huggins interaction parameter versus weight fraction of PVA in PVA/chitosan blends. (From Jawalkar, S. S., Raju, K. V. S. N., Halligudi, S. B., Sairam, M., and Aminabhavi, T. M. 2007. Molecular modeling simulations to predict compatibility of poly(vinyl alcohol) and chitosan blends: A comparison with experiments. *Journal of Physical Chemistry B* 111:2431–2439.)

of PVA, the value of χ lies below the $\chi_{critical}$ line, so they are miscible blends. For other compositions—that is, 1:1 and above (with increasing composition of PVA)—χ lie above the critical value indicating blend immiscibility.

The Gibbs free energy of mixing at constant pressure ΔG_m governs the process of dissolution between two components and is related to the changes in the enthalpy ΔH_m and the entropy ΔS_m^{comb} of the blend system. In general, for a miscible blend system, the Gibbs free energy is negative [35]. To compute ΔG_m we need the enthalpy of mixing ΔH_m and entropy of mixing ΔS_m^{comb}. According to Hildebrand and Scott [43], ΔH_m is expressed as

$$\Delta H_m = V_m \left[\sqrt{CED_A} - \sqrt{CED_B} \right]^2 \varphi_A \varphi_B \qquad (6.8)$$

where V_m is the total volume of the blend. CED values used in computing ΔH_m could be taken from the MD simulations like δ. And ΔS_m^{comb} is obtained from the combinatorial entropy term of the binary blends using the Flory–Huggins theory [46,47]:

$$\Delta S_m^{comb} = -R(x_A \ln x_A + x_B \ln x_B) \qquad (6.9)$$

Here, x_A and x_B are the mole fractions of A and B. Then, by using the identity $\Delta G_m = \Delta H_m - T\Delta S_m^{comb}$, Gibbs free energy of mixing is calculated. The ΔS_m^{comb} is usually positive; hence, ΔH_m dominates the value of ΔG_m. Jawalkar et al. calculated ΔG_m and found that both ΔG_m and ΔH_m exhibit almost similar trends over the entire compositions of poly(L-latide) (PLL) in the PLL/PVA blend as

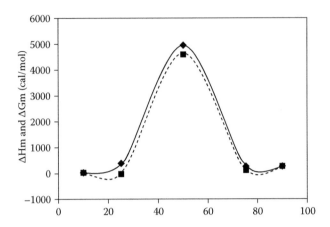

FIGURE 6.3

Molar enthalpy (solid curve) and Gibbs free energy (dashed curve) versus weight fraction of PLL in PLL/PVA blend. (From Jawalkar, S. S., and Aminabhavi, T. M. 2006. Molecular modeling simulations and thermodynamic approaches to investigate compatibility/incompatibility of poly(L-latide) and poly(vinyl alcohol) blends. *Polymer* 47:8061–8071.)

shown in Figure 6.3 [24]. For only a narrow blend composition, the calculated ΔG_m is slightly below zero, thus showing a slight miscibility of the blend. This is in conformity with the reported data of experiments [48].

Another commonly used method for establishing miscibility of polymer blends is through determination of the glass transition in the blend versus those of the unblended constituents. A miscible polymer blend will exhibit a single glass transition. In case of borderline miscibility, broadening of transition will occur. With cases of limited miscibility, two separate transitions may occur [49]. At glass transition temperature T_g, there is a characteristic change in the motion of polymer chains. The glass transition is mainly caused by the freezing of the motion of chain segments (local chain movement). The occurrence of T_g can be used as a rule of thumb to determine the compatibility of a blend. If a binary blend is miscible, only a single T_g appears. Otherwise, two T_g could be detected, and each T_g indicates the frozen temperature of one component.

NPT MD simulations are often performed to calculate the volume-temperature properties of the studied system. Figures 6.4 and 6.5 show the specific volume versus temperature plots of miscible poly(hydroxybutyrate) (PHB)/PEO blend (1:2 blends in terms of repeated units) and PHB/PE blend (1:2 blends in terms of repeated units) from the literature [34,36]. The intersect point of the two straight lines segments marks the T_g value. For the miscible PHB/PEO blend, only one T_g (about 258) could be found, and it is a little higher than the experimental values, 230 to 252 K. For the immiscible PHB/PE blend, two T_g (249 K and 187 K) were found. Comparing to the T_g of pure PHB (268 to 278 K) and PE (140 to 170 K), the two transition temperatures correspond to PHB-rich (249 K) and PE-rich domains' (187 K) T_g, respectively.

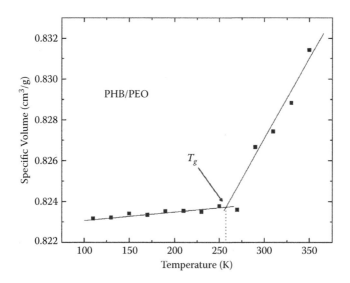

FIGURE 6.4
Specific volumes versus temperature using molecular dynamics simulation for blend system PHB/PEO (1:2 blends in terms of repeated units).

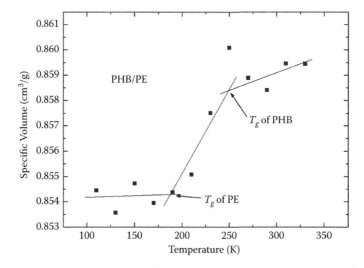

FIGURE 6.5
Specific volumes versus temperature using molecular dynamics simulation for blend system PHB/polyethylene (PE) (1:2 blends in terms of repeated units).

6.2.3 Packing Modes of Polymer Chain in a Blend

Why are two polymers compatible with each other? Their packing details may show much important information. The radial distribution function $g(r)$ is commonly used to characterize molecular structure. This function gives a measure of the probability that, given the presence of an atom at the origin of an arbitrary reference frame, there will be an atom with its center located in a spherical shell of infinitesimal thickness at a distance, r, from the reference atom. It is defined as [50]

$$g_{AB}(r) = \frac{1}{Nx_Ax_B\rho} \left\langle \sum_{i=1}^{N_A} \sum_{j=1}^{N_B} \delta(r - r_i - r_j) \right\rangle_{i \neq j} \tag{6.10}$$

where x_A and x_B are the mole fraction of polymer A and B, respectively. N_A and N_B are the number of atoms of polymer A and B, respectively. N is the total number of atoms, and ρ is the overall number density. Note that A and B could be the same type of atoms. It has been observed that if a binary system is compatible, the intermolecular $g(r)$ of the AB pair between two different polymers is larger than those of AA and BB pairs [51].

Jawalkar et al. found that the miscibility between PVA and CS polymers is attributed to hydrogen bond formation between groups of CS ($-CH_2OH$ or $-NH_2$) and hydrogen atom of PVA by MD simulation of radical distribution functions for atoms involved in interaction. Figure 6.6 shows the radial distribution function as a function of r for these atoms in the miscible blend of PVA/CS (1:1) [25].

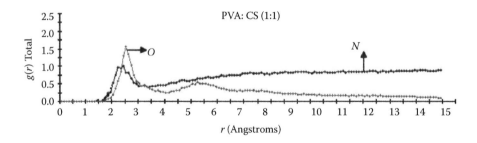

FIGURE 6.6
Radial distribution function calculation for PVA/chitosan (CS) blend representing oxygen atom of hydroxyl methyl group of CS and nitrogen atom of CS relative to the distance of the hydrogen atom of the hydroxyl group of PVA. (From Jawalkar, S. S., Raju, K. V. S. N., Halligudi, S. B., Sairam, M., and Aminabhavi, T. M. 2007. Molecular modeling simulations to predict compatibility of poly(vinyl alcohol) and chitosan blends: A comparison with experiments. *Journal of Physical Chemistry B* 111:2431–2439.)

6.3 Miscibility of Several Binary Blends

6.3.1 PHB/PEO and PHB/PE [34,36]

In 2004, our group calculated the solubility parameters for two polymers (PHB and PEO). The solubility parameters of the two polymers are similar and are consistent with the literature values. This means that PHB may be compatible with PEO. Then the volume–temperature curve of PHB/PEO blend system (1:2 blends in terms of repeated units) is simulated. From the curve, we got only one T_g for the blend system. The calculated T_g is consistent well with the experimental results. Based on the above two points, we concluded that PHB/PEO is a miscible blend. To confirm this conclusion further, the MD simulation was also carried out for an immiscible PHB/PE blend. Two T_g are observed for the immiscible blend. This is qualitatively in agreement with the experiment and supports our conclusion.

6.3.2 PVA-Based and Poly(Vinyl Pyrrolidone) (PVP)-Based Blends [12,24,25,27,28]

An MD simulation approach has been used to understand the blend compatibility/incompatibility of PVA/PLL, PVA/polymethylmethacrylate (PMMA), PVA/CS, PVP/poly(ether sulfones) (PES), and PVP/CS blends by Jawalkar et al. They calculated solubility parameters and Flory–Huggins interaction parameters to assess the blend miscibility at different compositions. It was found that at 1:9 blend composition of PLL/PVA, miscibility was observed, but increasing immiscibility was prevalent at higher compositions of PLL component. For PVA/PMMA blends, MD simulations confirmed blend miscibility at >60 wt% of PVA. The blends of PVP and PES are almost incompatible over all the compositions. Miscibility was observed for PVA/CS blends below 50% of PVA, while immiscibility was prevalent at composition between 50% and 90% PVA. The computed results confirmed the experimental findings. MD simulations were performed to clarify the most favorable site to form a hydrogen bond. When simulations were carried out on the PVP/chitosan blends, the computational results are consistent with the experimental findings that intermolecular interactions take place via the C=O of PVP. The simulations also verify that the hydroxyl groups of chitosan are the more favorable sites for hydrogen bond formations than the NH-C_2 group. Additionally, the hydrogen atom of OH-C_6 is more favorable than that of OH-C_3 as a proton donor group. The nitrogen atom does not serve as a proton acceptor in these blends.

6.3.3 PEO/PVC [33]

The miscibility of PEO/PVC blends has been investigated by Luo and Jiang by MD simulations. It was found that 40 repeated units are sufficient to represent a PEO chain and 50 for a PVC chain. Based on the temperature dependence of a specific volume, a single glass transition occurs in three blends at compositions of 70/30, 50/50, and 30/70, which indicates PEO and PVC are miscible. The glass transition temperature increases with increasing PVC composition. The Flory–Huggins χ parameter for PEO/PVC 50/50 blend is positive in the whole temperature range and increases with decreasing temperature. Nevertheless, χ for PEO/PVC 30/70 and 70/30 blends are negative and vary slightly with temperature. The simulations reveal that PEO/PVC 50/50 blend is less miscible than 70/30 and 30/70 blends. The radial distribution functions of the intermolecular carbon atoms also indicate that 50/50 blend is less miscible. The specific electrostatic attractions between O and Cl atoms, as well as the H-bonds between O atoms of PEO and H atoms of PVC, contribute to the miscibility of PEO and PVC.

6.4 Conclusion

Since the mid-1990s, MD simulation studies of polymer blends have rapidly grown in number and in variety of problems treated. We have given a brief survey of the present status of MD simulation studies of predicting the miscibility of a binary blend. The main interest was in understanding basic elementary effects in binary blend miscibility. Many important parameters could be calculated to assess the compatibility of two polymers. The targeted systems of the simulations have changed greatly from oligermer (a few repeated units) to rather large systems that can match the behavior of real polymer systems. Our objective is to develop a technique to understand the miscibility of blends. Still remaining are problems to extend the simulation to polymers with more complex chemical structures. Though our journey to making it a sufficiently useful and reliable tool may be long, it is undoubtedly very challenging.

Molecular dynamics simulation of binary polymer blends is still a young and uncultivated area of research. In addition to contributions to the design of conventional polymer blend materials, the MD simulation method for prediction miscibility of a binary polymer blend must have great potential in various molecular-level designs of functional materials, such as high drug loading [35] or nanocomposites [29]. MD simulation in polymer blends is only just beginning.

Acknowledgment

This work is supported by the National Natural Science Foundation of China (20803052), the Foundation of Tianjin Educational Committee

(20070605), and the Foundation for the Teacher by the Tianjin Normal University (5RL065).

References

1. Paul, D. R., and Bucknall, C. B. 1999. *Polymer Blends*. New York: Wiley.
2. Strobl, G. R. 2007. *The Physics of Polymers Concepts for Understanding Their Structure and Behavior*. Berlin Heidelberg: Springer-Verlag.
3. Coleman, M. M., Graf, J. F., and Painter, P. C. 1991. *Specific Interactions and the Miscibility of Polymer Blends*. Technomic: Lancaster, PA.
4. Colmenero, J., and Arbe, A. 2007. Segmental dynamics in miscible polymer blends: Recent results and open questions. *Soft Matter* 3:1474–1485.
5. Maranas, J. K. 2007. The effect of environment on local dynamics of macromolecules. *Current Opinion in Colloid and Interface Science* 12:29–42.
6. Lodge, T. P., and McLeish, T. C. B. 2000. Self-concentrations and effective glass transition temperatures in polymer blends. *Macromolecules* 33:5278–5284.
7. Leroy, E., Alegría, A., and Colmenero, J. 2003. Segmental dynamics in miscible polymer blends: Modeling the combined effects of chain connectivity and concentration fluctuations. *Macromolecules* 36:7280–7288.
8. de Gennes, P.-G. 1979. *Scalling Concepts in Polymer Physics*. Ithaca, NY: Cornell University Press.
9. Naidu, B. V. K., Sairam, M., Raju, K. V. S. N., and Aminabhavi, T. M. 2005. Thermal, viscoelastic, solution and membrane properties of sodium alginate/hydroxyethylcellulose blends. *Carbohydrate Polymers* 61:52–60.
10. Papke, N., and Karger-Kocsis, J. 2001. Thermoplastic elastomers based on compatibilized poly(ethylene terephthalate) blends: Effect of rubber type and dynamic curing. *Polymer* 42:1109–1120.
11. George, J., Joseph, R., Thomas, S., and Varughese, K. T. 1995. High density polyethylene/acrylonitrile butadiene rubber blends: Morphology, mechanical properties, and compatibilization. *Journal of Applied Polymer Science* 57:449–465.
12. Jawalkar, S. S., Adoor, S. G., Sairam, M., Nadagouda, M. N., and Aminabhavi, T. M. 2005. Molecular modeling on the binary blend compatibility of poly(vinyl alcohol) and poly(methyl methacrylate): An atomistic simulation and thermodynamic approach. *Journal of Physical Chemistry B* 109:15611–15620.
13. Naidu, B. V. K., Mallikarjuna, N. N., and Aminabhavi, T. M. 2004. Blend compatibility studies of polystyrene/poly(methyl methacrylate) and polystyrene-acrylonitrile by densitometry, viscometry, refractometry, ultraviolet absorbance, and fluorescence techniques at 30°C. *Journal of Applied Polymer Science* 94:2548–2550.
14. Wali, A. C., Naidu, B. V. K., Malikarjuna, N. N., Sainkar, S. R., Halligudi, S. B., and Aminabhavi, T. M. 2005. Miscibility of chitosan-hydroxyethylcellulose blends in aqueous acetic acid solutions at 35°C. *Journal of Applied Polymer Science* 96:1996–1998.
15. Clancy, T. C., Putz, M., Weinhold, J. D., Curro, J. G., and Mattice, W. L. 2000. Mixing of isotactic and syndiotactic polypropylenes in the melt. *Macromolecules* 33:9452–9463.

16. Xu, G., Clancy, T. C., Mattice, W. L., and Kumar S. K. 2002. Increase in the chemical potential of syndiotactic polypropylene upon mixing with atactic or isotactic polypropylene in the melt. *Macromolecules* 35:3309–3311.

17. Choi, P., and Mattice, W. L. 2004. Molecular origin of demixing, prior to crystallization, of atactic polypropylene/isotactic polypropylene blends upon cooling from the melt. *The Journal of Chemical Physics* 121:8647–8651.

18. Choi, P., Rane, S. S., and Mattice, W. L. 2006. Effect of pressure on the miscibility of polyethylene/poly(ethylene-*alt*-propylene) blends. *Macromolecular Theory and Simulations* 15:563–572.

19. Jaramillo, E., Wu, D. T., Grest, G. S., and Curro, J. G. 2004. Anomalous mixing behavior of polyisobutylene/polypropylene blends: Molecular dynamics simulation study. *The Journal of Chemical Physics* 120:8883–8886.

20. Genix, A. C., Arbe, A., Alvarez, F., Colmenero, J., Willner, L., and Richter, D. 2005. Dynamics of poly(ethylene oxide) in a blend with poly(methyl methacrylate): A quasielastic neutron scattering and molecular dynamics simulations study. *Physical Review E* 72:031808-20.

21. Ao, Z. M., and Jiang, Q. 2006. Size effects on miscibility and glass transition temperature of binary polymer blend films. *Langmuir* 22:1241–1246.

22. May, A. F., and Maranas, J. K. 2006. The single chain limit of structural relaxation in a polyolefin blend. *The Journal of Chemical Physics* 125:024906-13.

23. Bedrov, D., and Smith, G. D. 2006. A molecular dynamics simulation study of segmental relaxation processes in miscible polymer blends. *Macromolecules* 39:8526–8535.

24. Jawalkar, S. S., and Aminabhavi, T. M. 2006. Molecular modeling simulations and thermodynamic approaches to investigate compatibility/incompatibility of poly(L-latide) and poly(vinyl alcohol) blends. *Polymer* 47:8061–8071.

25. Jawalkar, S. S., Raju, K. V. S. N., Halligudi, S. B., Sairam, M., and Aminabhavi, T. M. 2007. Molecular modeling simulations to predict compatibility of poly(vinyl alcohol) and chitosan blends: A comparison with experiments. *Journal of Physical Chemistry B* 111:2431–2439.

26. Wu, C. F., and Xu, W. J. 2007. Atomistic molecular simulations of structure and dynamics of crosslinked epoxy resin. *Polymer* 48:5802–5812.

27. Suknuntha, K., Tantishaiyakul, V., Vao-Soongnern, V., Espidel, Y., and Cosgrove, T. 2008. Molecular modeling simulation and experimental measurements to characterize chitosan and poly(vinyl pyrrolidone) blend interactions. *Journal of Polymer Science Part B: Polymer Physics* 46:1258–1264.

28. Jawalkar, S. S., Nataraj, S. K., Raghu, A. V., and Aminabhavi, T. M. 2008. Molecular dynamics simulations on the blends of poly(vinyl pyrrolidone) and poly(bisphenol-A-ether sulfone). *Journal of Applied Polymer Science* 108:3572–3576.

29. Yani, Y., and Lamm, M. H. 2009. Molecular dynamics simulation of mixed matrix nanocomposites containing polyimide and polyhedral oligometric silsesquioxane (POSS). *Polymer* 50:1324–1332.

30. Ahmadi, A., and Freire, J. J. 2009. Molecular dynamics simulation of miscibility in several polymer blends. *Polymer* 50:4973–4978.

31. Liu, W. J., Bedro, D., Kumar, S. K., Veytsman, B., and Colby, R. H. 2009. Role of distributions of intramolecular concentrations on the dynamics of miscible polymer blends probed by molecular dynamics simulation. *Physical Review Letters* 103:037801-4.

32. Brodeck, M., Alvarez, F., Moreno, A. J., Colmenero, J., and Richter, D. 2010. Chain motion in nonentangled dynamically asymmetric polymer blends: Comparison between atomistic simulations of PEO/PMMA and a generic bead-spring model. *Macromolecules* 43:3036–3051.

33. Luo, Z. L., and Jiang, J. W. 2010. Molecular dynamics and dissipative particle dynamics simulations for the miscibility of poly(ethylene oxide)/poly(vinyl chloride) blends. *Polymer* 51:291–299.

34. Yang, H., Li, Z. S., Qian, H. J., Yang, Y. B., Zhang, X. B., and Sun, C. C. 2004. Molecular dynamics simulation studies of binary blend miscibility of poly(3-hydroxybutyrate) and poly(ethylene oxide). *Polymer* 45:453–457.

35. Gupta, J., Nunes, C., Vyas, S., and Jonnalagadda, S. 2011. Prediction of solubility parameters and miscibility of pharmaceutical compounds by molecular dynamics simulations. *Journal of Physical Chemistry B* 115:2014–2023.

36. Yang, H., Li, Z. S., Lu, Z. Y., and Sun, C. C. 2005. Computer simulation studies of miscibility of poly(3-hydroxybutyrate)-based blends. *European Polymer Journal* 41:2956–2962.

37. Wang, Y. C., Lee, W. J., and Ju, S. P. 2009. Modeling of the polyethylene and poly(L-lactide) triblock copolymer: A dissipative particle dynamics study. *The Journal of Chemical Physics* 131:124901-10.

38. Allen, M. P., and Tildesley, D. J. 1987. *Computer Simulation of Liquids*. Oxford: Clarendon Press.

39. Sun, H. 1998. COMPASS: An ab initio force-field optimized for condensed-phase applications—Overview with details on alkane and benzene compounds. *Journal of Physical Chemistry B* 102:7338–7364.

40. Rappe, A. K., and Goddard, W. A. 1991. Charge equilibration for molecular dynamics simulations. *Journal of Physical Chemistry* 95:3358–3363.

41. Ewald, P. P. 1921. Die berechnung optischer und elektrostatischer gitterproten-tiale. *Annalen der Physik* 369:253–287.

42. Spyriouni, T., and Vergelati, C. 2001. A molecular modeling study of binary blend compatibility of polyamide 6 and poly(vinyl acetate) with different degrees of hydrolysis: An atomistic and mesoscopic approach. *Macromolecules* 34:5306–5316.

43. Hildebrand, J. H., and Scott, R. L. 1950. *The Solubility of Nonelectrolytes*, 3rd ed. New York: Reinhold.

44. Mason, J. A., and Sperling, L. H. 1976. *Polymer Blends and Composites*. New York: Plenum Press.

45. Case, F. H., and Honeycutt, J. D. 1994. Will my polymers mix? Applications of modeling to study miscibility, compatibility, and formulation. *Trends in Polymer Science* 2:259.

46. Flory, P. J. 1953. *Principles of Polymer Chemistry*. Ithaca, NY: Cornell University Press.

47. Flory, P. J. 1989. *Statistical Mechanics of Chain Molecules*. Munich, Germany: Hanser.

48. Shuai, X., He, Y., Asakawa, N., and Inoue, Y. 2001. Miscibility and phase structure of binary blends of poly(L-lactide) and poly(vinyl alcohol). *Journal of Applied Polymer Science* 81:762–772.

49. Olabisi, O., Robeson, L. M., and Shaw, M. T. 1979. *Polymer-Polymer Miscibility*. New York: Plenum Press.

50. Hansen, J.-P., and McDonald, I. R. 1990. *Theory of Simple Liquids,* 2nd ed. London: Academic Press.
51. Akten, E. D., and Mattice, W. L. 2001. Monte Carlo simulation of head-to-head, tail-to-tail polypropylene and its mixing with polyethylene in the melt. *Macromolecules* 34:3389–3395.

7

Conformation and Topology of Cyclic-Linear Polymer Blends

Gopinath Subramanian

Los Alamos National Laboratory
Los Alamos, New Mexico

Sachin Shanbhag

Florida State University
Tallahassee, Florida

CONTENTS

7.1 Introduction

Theoretical treatment of the statistical properties of linear polymers (LPs), shown schematically in Figure 7.1a, have been in existence since the 1950s (Rouse, 1953), and have undergone continuous refinement. In particular, tube theories (viz. reptation) (de Gennes, 1971; Doi and Edwards, 1986), and subsequent refinements, like constraint release (Viovy et al., 1991) and contour length fluctuation (Frischknecht and Milner, 2000) are some of the greatest

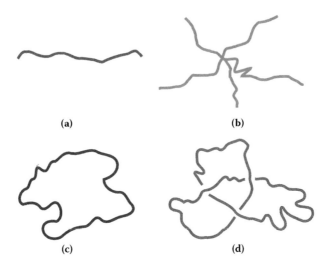

(a)   (b)

(c)   (d)

FIGURE 7.1
Different polymer architectures and two possible topological states of a cyclic polymer (CP).
(a) linear, (b) star, (c) unknotted cyclic, (d) trefoil knotted cyclic.

successes of polymer theory. In these theories, the underlying picture of a test chain immersed in a sea of obstacles is established by solving the problem of curvilinear diffusion along the contour of a tube. Branched polymers (Figure 7.1b) have also been described successfully by tube theories, where the arms can retract into their confining tubes and poke out in potentially different directions, thereby renewing their configurations (Ball and McLeish, 1989; Doi and Kuzuu, 1980; Milner and McLeish, 1997). Conceptually, the presence of polymer ends is a key requirement in analyzing the dynamics of both linear and branched polymers within the framework of tube theories.

Cyclic polymers (CPs) are simply linear polymers (LPs), with one extra constraint—the ends are forced to be attached to each other. This single extra constraint introduces a new degree of complexity in the analysis. The absence of chain ends forces us to reexamine our conceptual understanding of the mechanisms that lead to CP dynamics. Topological problems also present themselves—knots on CPs force the polymers to adopt more compact and less flexible conformations (Micheletti et al., 2006). The excluded volume constraint ensures that these knots are permanent, unless the CP contour is broken and then reattached. This kind of topological isomerism has been observed in vivo in knotted plasmid DNA (Dean et al., 1985; Trigueros et al., 2004; Yan et al., 1999). Figure 7.1c shows a schematic of the simplest type of CP, the unknot, or trivial knot, and Figure 7.1d shows the simplest type of knotted CP, the trefoil knot. The field of knots is an extremely rich subject that has been studied since the times of Carl Friedrich Gauss (Gauss, 1833) and is beyond the scope of this manuscript. We refer the reader elsewhere for the analysis of knots (Adams, 2004; Menasco and Thistlethwaite, 2005; Whitehead, 1963), and limit our attention on the simplest type of CP, the unknot.

Theoretical studies (Cates and Deutsch, 1986; Iyer et al., 2006; Obukhov et al., 1994; Rubinstein, 1986) have considered the motion of CPs in a gel, and it has been suggested that CPs thrust and pull on unentangled loops, executing amoeba-like motion. There is also a substantial body of viscoelastic experimental data on polystyrene and polybutadiene CPs (McKenna et al., 1987, 1989; Roovers, 1985; Roovers and Toporowski, 1988) comparing and contrasting the properties of LPs and CPs. Numerous computational studies of CPs (Brown et al., 2001; Brown and Szamel, 1998a, 1998b; Hur et al., 2006; Müller et al., 1996, 2000a, 2000b; Ozisik et al., 2002) have explored the conformational properties of CPs quite extensively.

Blends of CPs and LPs (referred to as CLBs), however, have not been studied extensively, and thus, only a relatively small body of literature is available (Geyler and Pakula, 1988; Iyer et al., 2007; Subramanian and Shanbhag, 2008a, 2008c, 2009). Blend systems are important, because most experimental data on "pure CPs" are actually data on CLBs due to either contamination, or limitations of purification methods. The dynamics of CPs are extremely sensitive to the presence of LPs (McKenna and Plazek, 1986). It was shown experimentally by Kapnistos et al. (2008) that the shear modulus of a melt of CPs can increase by an *order of magnitude*, when only 0.7% LPs are added to the melt. Furthermore, Robertson and Smith (2007b) used fluorescence microscopy to track entangled solutions of linear and cyclic DNA, and showed that the diffusivity of a tracer CP drops by two orders of magnitude when the surrounding matrix of CPs is replaced by LPs. This extreme sensitivity was hypothesized to be a result of two distant LPs being bridged by CPs (Kapnistos et al., 2008), and the percolation threshold was examined more closely by numerical studies (Vasquez and Shanbhag, 2011).

In the following sections, we describe the bond fluctuation model, and results obtained from this model on the statics and dynamics of CLBs. We also present results on the conformational free energy of CPs, and conclude with some recent results that discuss deviations from the theory at high molecular weights.

7.2 Model and Methods

7.2.1 Model

The bond fluctuation model proposed by Shaffer (1994), henceforth referred to as S-BFM, was used to study CLBs. Other models, such as the ones proposed by Carmesin and Kremer (1988) can be used, and indeed, properties measured in one model can be mapped into the other (Subramanian and

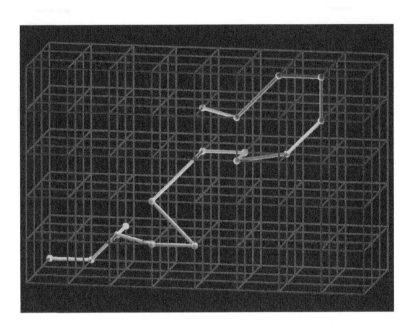

FIGURE 7.2
Shaffer's bond fluctuation model (Shaffer, J. S., 1994. Effects of chain topology on polymer dynamics—Bulk melts, *J. Chem. Phys.*, 101:4205–4213). Polymers are grown as random walks on a simple cubic lattice, subjected to the excluded volume, chain connectivity, and chain uncrossability constraints described in the text.

Shanbhag, 2008b). We chose to use the S-BFM because representation of a polymer chain on the lattice is simpler and is slightly more coarse grained, allowing us to explore polymers with higher molecular weight. In the S-BFM, polymers are grown on a simple cubic lattice, shown in Figure 7.2. The side of each individual cube is taken to be unity and sets the length scale of the simulations. The simulation box is cubic with side L and is subjected to periodic boundary conditions. The molecular weight of CPs and LPs is chosen to be identical and is denoted by N. The number of CPs and the number of LPs in the system are denoted by n_C and n_L, respectively. (Henceforth, the subscripts C and L are used to refer to quantities pertaining to CPs and LPs, respectively.) Monomers (beads) are placed on cube vertices, and the system is subjected to the following three constraints:

1. Excluded volume: enforced by forbidding multiple occupancy of lattice sites
2. Chain connectivity: enforced by restricting the length of the bond connecting two consecutive beads to the set of values $\{1, \sqrt{2}, \sqrt{3}\}$
3. Chain uncrossability: enforced by forbidding bond midpoints to intersect

The fractional occupancy of lattice sites by CPs and by LPs is given by $\phi_C = n_C N L^{-3}$ and $\phi_L = n_L N L^{-3}$, respectively, and the total occupancy is maintained at $\phi = \phi_C + \phi_L = 0.5$. This total monomer density of $\phi = 0.5$ is adequate to describe polymer melts (Shaffer, 1994; Shanbhag and Larson, 2005, 2006).

Thermal motion of the system is simulated by picking a monomer at random and attempting a move to one of its six nearest neighbors, also chosen at random. The attempted move is accepted if it does not violate any of the three constraints described above. One unit of time is said to have elapsed when a number of moves equal to the number of monomers in the system have been attempted. This sets the timescale of the simulations.

7.2.2 Equilibration

Systems are initialized by growing n_C noncatenated, unknotted CPs in the simulation box. These CPs are shuffled for $25000N\,n_C$ trial moves to ensure that they are evenly distributed in space. n_L LPs are then grown as random walks that avoid all occupied lattice sites and do not violate any of the three constraints. The LPs thus grown are then shuffled for approximately $25000n_L$ N trial moves while the CPs are held frozen, with the three constraints enforced. In this initial configuration, the sampling of configuration space by the polymers is not necessarily the same as that of an equilibrium statistical mechanical ensemble. In other words, some entropic stress is present in the initial configuration, which is relaxed away by equilibration. Subjecting the system to thermal motion guarantees that the system will evolve to its minimum energy, stress-free configuration, when polymers in the systems have lost all memory of their initial position. Thus, during the equilibration phase, the autocorrelation functions of the diametrical vector (for CPs) and the end-to-end vector (for LPs) were monitored. These autocorrelation functions are defined by

$$p_i(t) = \frac{\langle \mathbf{r}_{ee}(t) \cdot \mathbf{r}_{ee}(0) \rangle}{\langle \mathbf{r}_{ee}(0) \cdot \mathbf{r}_{ee}(0) \rangle} \tag{7.1}$$

where for CPs ($i = C$), p_C is obtained by setting $\mathbf{r}_{ee}(t) = \mathbf{r}_{N/2}(t) - \mathbf{r}_1(t)$, the "diametrical" vector connecting bead 1 to bead $N/2$ of a CP. For LPs ($i = L$), p_L is obtained by setting $\mathbf{r}_{ee}(t) = \mathbf{r}_N(t) - \mathbf{r}_1(t)$, the vector connecting bead 1 to bead N of a polymer chain. The system is considered equilibrated when both p_C and p_L drop below 0.1. All measurements of size, entanglement statistics, and diffusivity are performed on this equilibrated configuration.

The timescales associated with this brute-force equilibration are on the order of the time it takes a polymer to diffuse a distance approximately equal to its size, which can be prohibitively large for ultra high molecular weight polymers. Alternate, more rapid methods for polymers with ends, like the connectivity altering algorithms, have been in existence for almost two decades (Alexiadis et al., 2008; Baig et al., 2010; Daoulas et al., 2005; de Pablo et al., 1992; Dodd et al., 1993; Karayiannis et al., 2002; Leontidis et al., 1994; Mavrantzas et al., 1999; Pant and Theodorou, 1995; Ramos et al., 2007; Sides et al., 2004; Siepmann and Frenkel, 1992; Uhlherr et al., 2002) but have not been extended to CPs. At the time of writing, the successive molecular weight doubling algorithm (Subramanian, 2010) is the only rapid equilibration algorithm that has been demonstrated to work for CPs. However, it is designed for off-lattice molecular dynamics models and cannot be easily extended to the S-BFM. Nevertheless, the integer operations that are characteristic of the S-BFM, in conjunction with modern computers, allow us to examine moderately entangled polymers using this brute-force method of equilibration.

7.2.3 Primitive Path Analysis

The annealing algorithm originally developed for off-lattice molecular dynamics simulations (Everaers et al., 2004) has been adapted and further developed to examine the primitive path (PP) network for the S-BFM (Shanbhag and Larson, 2005; Subramanian and Shanbhag, 2008a) for both CPs and LPs. In the lattice version of the annealing algorithm, the end beads of all the equilibrated LPs are held fixed while the interior beads are free to move. For CPs, all beads are free to move. The intramolecular excluded volume constraint, *between consecutive beads only*, is turned off, while the intermolecular excluded volume constraint is maintained. This is done to facilitate shrinkage of the chain contours while maintaining their noncrossability with other chains. The probability with which moves that increase the PP length are accepted is decreased according to the expression

$$p_{acc}(t) = \min\left\{1, \exp\left[-A\Delta L\left(\frac{t}{\tau_{anneal}}\right)^2\right]\right\} \tag{7.2}$$

where ΔL is the change in contour length (positive or negative) due to a trial move, and τ_{anneal} is the duration of the annealing process. Values of $A = 16$ and $\tau_{anneal} = 5 \times 10^4$ have been shown to be adequate (Shanbhag and Larson, 2005, 2006; Shanbhag et al., 2007; Subramanian and Shanbhag, 2008a).

After annealing is complete, the geometrical "identification of local deviations" (ILD) algorithm (Shanbhag and Kroger, 2007; Shanbhag and Larson, 2006) is used to identify individual entanglements. Here, if the length of any small segment of the PP deviates from the shortest possible path connecting its ends, the presence of a topological constraint in that neighborhood that prevents the PP segment from decreasing its length may be indirectly inferred. The smallest such element that can be examined is one containing three consecutive PP monomers, or two consecutive bonds. Thus, if the trajectory of the PP between a bead and its second-nearest neighbor does not follow the shortest possible path along the cubic lattice, it is due to an obstacle or entanglement. It was shown that this natural choice led to agreement with the number of entanglements Z calculated from ensemble averages of monodisperse LP chains (Shanbhag and Larson, 2006).

7.3 Size

The size of CPs in a melt of pure CPs and in CLBs has been observed to be affected by the topological constraint of noncatenation that affects only CPs. This is a manifestation of the excluded volume constraint that does not allow the transformation shown in Figure 7.3, whereby two CPs can link with each other. Using this concept of noncatenation, the size of both CPs and LPs in a CLB was examined by means of a scaling argument, originally presented by Iyer et al. (2007). This argument is based on the blob model, and CLBs are modeled as a semidilute solution of CPs in a Θ-solvent consisting of LPs. Because a CP does not experience excluded volume interactions in a melt of LPs, its conformation is identical to that in a Θ-solvent. At low concentrations, individual CPs are far apart and do not interact with each other. At the overlap or threshold concentration ϕ_C^*, the CPs begin to interact with each other. The noncatenation constraint takes effect, and CP concentration in solution is the same as the concentration of monomers (Kuhn segments) within an individual CP. This concentration is given by the ratio of CP molecular weight, N, to the pervaded volume of the CP in dilute solution, $(R_C^{dil})^3 \sim N^{3/2}$, where R_C is some characteristic size. The threshold concentration therefore scales as

$$\phi_C^* = \frac{N}{R_C^{dil}} \sim \frac{N}{N^{3/2}} \sim N^{-1/2} \tag{7.3}$$

At the threshold concentration, ϕ_C^*, the correlation length, ξ (i.e., the mean distance between Kuhn segments on neighboring CPs) is on the order of the

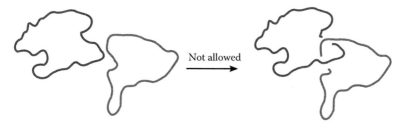

Not allowed

FIGURE 7.3
Catenation of two cyclic polymers (CPs) is a topological change forbidden by the excluded volume constraint in real polymers.

characteristic length scale of the CP, R_C^{dil}. Beyond the threshold concentration ϕ_C^*, the CLB enters the semidilute regime, and the correlation length ξ decreases from R_C^{dil} with increasing ϕ_C. Further, we expect the correlation length to be independent of N, beyond the overlap concentration. Thus, the concentration dependence of the correlation length is given by

$$\xi = R_C^{dil}\left(\frac{\phi_C^*}{\phi_C}\right)^m \sim N^0 \tag{7.4}$$

Because ϕ_C is independent of N, we obtain $m = 1$ by using $R_C^{dil} \sim N^{1/2}$ and the scaling relationship in Equation (7.3). Thus, the correlation length scales inversely as the concentration in the semidilute regime and is given by

$$\xi(\phi_C) = \left(\frac{\phi_C}{\phi_C^*}\right)^{-1} \tag{7.5}$$

Considering the CP to be composed of blobs of size ξ consisting of, say, N_K Kuhn segments, the CP can be examined at length scales less than ξ and on the order of ξ. Below the correlation length scale, ξ, we argue that the CP encounters on average only LP neighbors, and a chain section in a blob of size ξ behaves as if it was under Θ-conditions. Thus, from Equation (7.5),

$$\xi \sim N_K^{1/2} \tag{7.6}$$

$$N_K = \left(\frac{\phi_C}{\phi_C^*}\right)^{-2} \tag{7.7}$$

For a coarse-grained picture on length scales on the order of ξ, we argue that a CP, on average, only sees CP neighbors. The noncatenation constraint becomes operational at this length scale, and hence the CP bears a lattice-tree structure. At the length scale ξ, we consider the CP to be composed of N/N_K segments of length ξ with only CP neighbors. Based on a conjecture proposed by Cates and Deutsch (Cates and Deutsch, 1986), the characteristic size of such a structure would scale as

$$R_C \sim \left(\frac{N}{N_K}\right)^{2/5} \xi \sim \left(\frac{\phi_C^0}{(\phi_C/\phi_C^*)^{-2}}\right)^{2/5}(\phi_C/\phi_C^*)^{-1} \sim (\phi_C/\phi_C^*)^{-1/5} \tag{7.8}$$

Note that the molecular weight N is independent of concentration. The above scaling argument indicates that in the semidilute regime the characteristic size of a CP scales with CP concentration is $(\phi_C/\phi_C^*)^{-1/5}$.

We can follow a similar line of reasoning for the case of LPs in a CLB starting from a small fraction of LPs up to the semidilute regime. In the semidilute regime the LPs can also be considered to be composed of blobs of size ξ. Because of the absence of noncatenation for LPs, the scaling at length scales less than ξ persists up to and beyond length scales of the order of ξ. This yields a characteristic size of the LP in the semidilute regime as

$$R_L \sim \left(\frac{N}{N_K}\right)^{2/5} \xi \sim \left(\frac{\phi_L^0}{\phi)L^{-2}}\right)^{1/2} \phi_L^{-1} \sim \phi_L^0 \tag{7.9}$$

This demonstrates that in the case of LPs, the characteristic size is independent of concentration. This is expected for LPs as the constraints imposed on it do not change when a neighboring LP is replaced with a CP.

CP size has also been modeled using the concept of free energy by Cates and Deutsch (1986). Here, the conformational free energy of a CP in a melt of pure CPs is the sum of two terms and is given as

$$F_C(R_C) = \frac{R_C^3}{N} + \frac{N}{R_C^2} \tag{7.10}$$

where, as before, R_C is some characteristic linear dimension of a CP. The first term is proportional to the number of neighbors of a CP and denotes the increase in free energy due to the noncatenation constraint. The second term denotes the Gaussian free energy increase that favors expansion of the CP. Minimizing the free energy $(dF_C/dR_C = 0)$ leads to $R_C \sim N^{2/5}$. (Direct computation of the free energy is discussed in more detail in Section 7.6.)

For CLBs, the noncatenation constraint on a CP is imposed only by its neighbors that are themselves CPs, and we can easily modify Equation (7.10) for CLBs as

$$F_C(R_C) = \frac{\phi_C}{\phi} \frac{R_C^3}{N} + \frac{N}{R_C^2} \tag{7.11}$$

which at the minimum free energy configuration yields a scaling for CP size as

$$R_C \sim \left(\frac{\phi_C}{\phi}\right)^{-1/5} N^{2/5} \tag{7.12}$$

in the semidilute regime.

Predictions of the foregoing theory were validated using the S-BFM to probe the size of two series of CLBs, with degree of polymerization $N = 150$ and $N = 300$. The composition of the CLB was varied from $\phi_C = 0.5$ (pure CPs)

TABLE 7.1

Description of the Systems Simulated
(Simulation Box Size $L_{box} = 60$, and
Total Density $\phi_C + \phi_L = 0.5$)

ϕ_L	ϕ_C	n_L	n_C	R_L	R_C
N = 150					
0.500	0.000	720	0	7.95 ± 0.10	—
0.479	0.021	690	30	8.21 ± 0.07	5.85 ± 0.16
0.458	0.042	660	60	8.10 ± 0.07	5.82 ± 0.11
0.438	0.063	630	90	8.19 ± 0.08	5.76 ± 0.10
0.375	0.125	540	180	7.99 ± 0.08	5.80 ± 0.07
0.313	0.188	450	270	8.02 ± 0.09	5.50 ± 0.05
0.250	0.250	360	360	8.21 ± 0.10	5.54 ± 0.05
0.188	0.313	270	450	8.16 ± 0.12	5.46 ± 0.04
0.125	0.375	180	540	8.15 ± 0.15	5.30 ± 0.03
0.063	0.438	90	630	8.01 ± 0.18	5.25 ± 0.03
0.042	0.458	60	660	8.04 ± 0.25	5.25 ± 0.03
0.021	0.479	30	690	8.33 ± 0.34	5.16 ± 0.03
0.000	0.500	0	720	—	5.09 ± 0.14
N = 300					
0.500	0.000	360	0	11.20 ± 0.17	—
0.450	0.050	324	36	11.32 ± 0.15	8.60 ± 0.20
0.375	0.125	270	90	11.51 ± 0.18	8.71 ± 0.20
0.250	0.250	180	180	12.27 ± 0.25	8.10 ± 0.11
0.167	0.333	120	240	12.02 ± 0.29	7.50 ± 0.11
0.100	0.400	72	288	12.37 ± 0.33	7.24 ± 0.08
0.050	0.450	36	324	12.35 ± 0.55	7.10 ± 0.07
0.025	0.475	18	342	11.67 ± 0.56	6.96 ± 0.05
0.000	0.500	0	360	—	7.02 ± 0.06

to $\phi_C = 0.0$ (pure LPs), and Table 7.1 summarizes the details of the systems studied. The size of a polymer is characterized by its radius of gyration R, given in terms of the position vectors of the beads r_i:

$$R^2 = \frac{1}{2N^2} \sum_{i=1}^{N} \sum_{j=1}^{N} (r_i - r_j)^2 \tag{7.13}$$

As predicted by the scaling model, size of the LPs, R_L, was independent of the concentration of CPs ϕ_C, and was approximately equal to the size of a coil in a pure LP melt, where $R_L(\phi_C = 0) = 7.95 \pm 0.10$ and 11.20 ± 0.17 for $N = 150$ and 300, respectively.

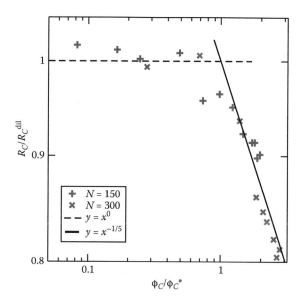

FIGURE 7.4
Radii of gyration of cyclic polymers (CPs) in the simulations are seen to follow the theory. In the dilute regime, CP size is unaffected. As CPs begin to interact with each other, the non-catenation constraint manifests itself as a "pressure" that leads to compression of the CPs, in accordance with the theory.

The size of CPs, however, increased with the fraction of LPs in the melt. For $N = 150$, R_C increased from 5.18 ± 0.03 (pure CPs) to 5.85 ± 0.17 when $\phi_L = 0.48$. For $N = 300$, R_C increased from 7.02 ± 0.06 (pure CPs) to 8.60 ± 0.20 when $\phi_L = 0.45$. Thus, it may be concluded that as ϕ_L increases, LPs infiltrate the volume occupied by CPs and force them to swell. The average volume of a CP, R_C^3, increased by approximately 45% for $N = 150$ and almost doubled (increased by 84%) for $N = 300$.

These data are also collapsed onto a master curve, as shown in Figure 7.4. The data for all sets of simulations follow the predicted scaling. In particular, in the semidilute regime, CP size remains unchanged. As the melts enter the semidilute regime, the data for all sets of simulations follow the scaling of $R_C \sim \phi_C^{-1/5}$.

7.4 Topology

In addition to size, the topological characteristics of CLBs play a significant role in determining the macroscopic properties of a melt. In particular, the interaction of CPs with LPs, and the effect of entanglements are important

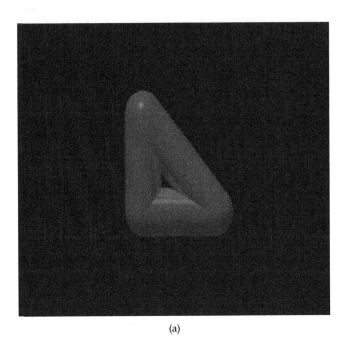

(a)

FIGURE 7.5
Snapshots of representative primitive paths of cyclic polymers (CPs) at different linear polymer (LP) fractions for $N = 300$ show increased infiltration of CP volume by entanglements. For clarity, only a single CP (solid), interacting matrix segments (translucent), and location of entanglement points as identified by the identification of local deviations (ILD) algorithm are depicted. When (a) $\phi_L = 0.0$, the individual CPs are completely disentangled. As ϕ_L increases through (b) 0.025, (c) 0.100, and (d) 0.375, the number of entanglements increases.

parts of the physics that govern the rheology of CLBs. Topological information is obtained by computing the primitive path (PP) network using the annealing algorithm, and identifying individual entanglements, as described in Section 7.2.3. When a melt of pure CPs is annealed, the PPs of all CPs collapse to a triangle, shown in Figure 7.5a. The structure does not get any simpler and collapse to a straight line or a point, because in order to prevent chains from passing through each other, midpoints of bonds are not allowed to overlap in S-BFM. This is an artifact of the particular lattice model chosen and is absent in off-lattice simulations. It cannot be avoided unless the underlying lattice description is altered. However, this small, nonzero PP length is inconsequential, because the resulting PP length is much smaller (on the order of 3) than the PP length of CPs as ϕ_L increases.

Figures 7.5b through 7.5d show representative snapshots of PPs of CPs when ϕ_L is increased from zero, for $N = 300$. The agreement of the location of entanglement points identified by the ILD algorithm and the associated PPs of matrix chains is excellent. As LPs are introduced into a melt of CPs ($\phi_L = 0.025$), the CPs become more entangled, and their PP becomes longer

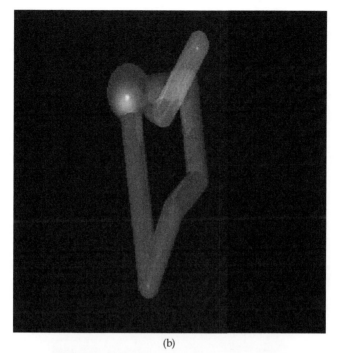

(b)

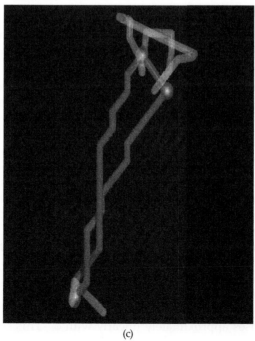

(c)

FIGURE 7.5 (continued)

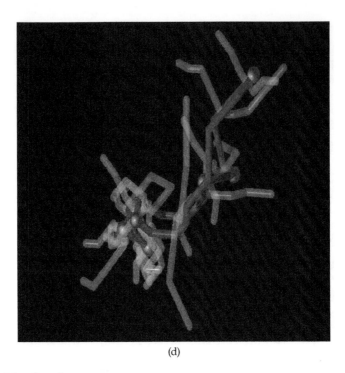

(d)

FIGURE 7.5 (continued)

and more tortuous. For some CPs, *in the presence of LPs*, the PPs of two CPs sometimes interact with each other topologically which, while not entirely unexpected, is an interesting departure from the case of pure CPs, where the PPs of noncatenated CPs do not interfere with each other. One such example is shown in Figure 7.5c, where the two entanglements at the top of the figure arise due to a neighboring CP, which is itself unable to shrink further. Thus, the effect of LPs is not merely confined to simple CP-LP interactions as shown in Figure 7.5b, but additionally manifested through CP-CP interactions that are otherwise absent when $\phi_L = 0$.

Figure 7.6 shows only the engaged polymers (i.e., polymers with entanglements) for both molecular weights, at the lowest nonzero concentration of LPs that were simulated. The entangled polymers form a percolating network that spans the entire simulation box, especially for $N = 300$. Therefore, it may be concluded that even in the presence of a small fraction of LPs, entanglement effects permeate the entire system and are in all probability the origin of the unexpected sensitivity of CLB dynamics.

Figure 7.7a shows, for $N = 300$, the average PP lengths, L_{pp} of the CP and LP species. L_{ppL} is essentially independent of ϕ_L. The effect on L_{ppC} is more dramatic, and increases approximately linearly with ϕ_L. This suggests that the

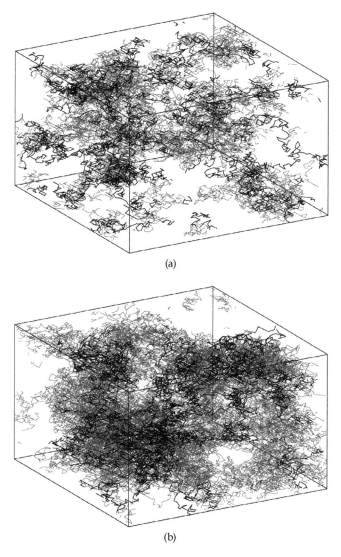

(a)

(b)

FIGURE 7.6
Simulation box showing only entangled polymers. Linear polymers (LPs) (solid black) thread through multiple cyclic polymers (CPs) inducing CP-CP interactions, and form a percolating network that spans the entire simulation box. (a) $N = 150$ with $\phi_L = 0.042$ and (b) $N = 300$ with $\phi_L = 0.05$. Note: The periodic boundary condition causes some CPs to appear as LPs with small molecular weight.

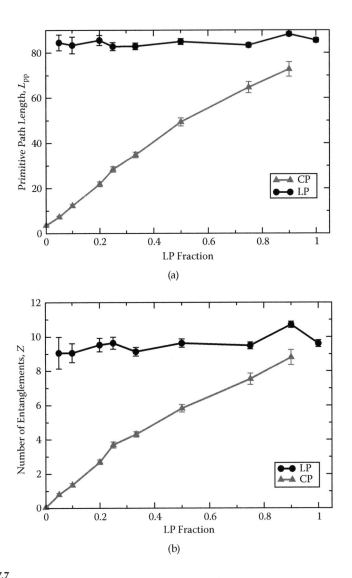

FIGURE 7.7

(a) Average primitive path length, L_{pp}, and (b) number of entanglements, Z for systems with $N = 300$. Both these quantities increase linearly with linear polymer (LP) fraction. Systems with $N = 150$ display the same trend. At the highest LP fraction, the number of entanglements on a cyclic polymer (CP) is comparable with the number of entanglements on an LP.

PP length of a single CP molecule embedded in a sea of LPs is comparable with L_{ppL}, especially as N increases—that is, for large N,

$$L_{ppC}(\phi_L) \approx \frac{\phi_L}{\phi} L_{ppL} \tag{7.14}$$

The average number of entanglements computed using the ILD algorithm reinforces this description. As shown in Figure 7.7b, for $N = 300$, the average number of entanglements on a tracer CP molecule in an LP matrix approaches the average number of entanglements on the LPs:

$$Z_C(\phi_L) \approx \frac{\phi_L}{\phi} Z_L \tag{7.15}$$

7.5 Dynamics

The equilibrated systems simulated above were subjected to additional thermal fluctuations, beginning at time $t = 0$ for a period τ_{Sim}, during which each polymer was tracked. The mean-square displacement of CPs and LPs was measured separately as in Equation (7.16), and the diffusion constant was determined from the slope of the $g_3(t)$ curve according to the relation in Equation (7.17). Figure 7.8 shows plots of the raw measured diffusivities of each species at various blend compositions:

$$g_3(t) = \left\langle (\mathbf{r}_{cm}(t) - \mathbf{r}_{cm}(0))^2 \right\rangle \tag{7.16}$$

$$D = \frac{1}{6} \frac{dg_3}{dt} \tag{7.17}$$

This raw diffusivity data can be explained using the following minimal model, which was originally proposed by the authors (Subramanian and Shanbhag, 2008c). Ignoring prefactors and other numerical details, the threaded LPs (see Section 7.4) restrain the mobility of a CP. As some of the LPs venture out, others arrive and form entanglements at the same rate, and the equilibrium structure of the melt is not disturbed (see Figure 7.9). Consequently, the primitive path of the CP undergoes local rearrangement as it relaxes by constraint release (CR) Rouse motion (Graessley, 1982). For a bidisperse or polydisperse blend of LPs, the processes of reptation and CR occur simultaneously, as widely discussed in the literature (Graessley, 1982;

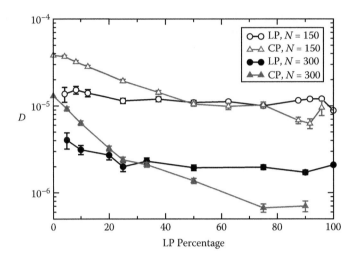

FIGURE 7.8
Self-diffusivity of cyclic polymers (CPs) (triangles) and linear polymers (LPs) (circles) for $N = 150$ (open symbols) and $N = 300$ (filled symbols) at different cyclic linear polymer blend (CLB) compositions.

Graessley and Struglinski, 1986; Green and Kramer, 1986; Green et al., 1984; Liu et al., 2006; Rubinstein and Colby, 1988; Viovy et al., 1991), and the faster process controls properties such as diffusivity and viscosity. A model based on a similar principle has also been proposed for the diffusion of CPs in a CLB (Klein, 1986).

In sharp contrast, based on the physical picture described above, the present model postulates that the relaxation of a CP in a CLB is determined by the *slower* process. Under this premise, one can argue that

$$\frac{1}{D_C(\phi_L)} = \frac{1}{D_C(\phi_L = 0)} + \frac{1}{D_{CR}(\phi_L)} \tag{7.18}$$

where $D_C(\phi_L)$ and $D_C(\phi_L = 0)$ are the diffusivities of a CP in a CLB, and in a pure melt (no LPs), respectively, and $D_{CR}(\phi_L)$ is the CR Rouse diffusivity, which will be specified shortly. When $\phi_L \approx 0$, it follows that $1/D_{CR} \approx 0$, and hence $D_C(\phi_L) \approx D_C(0)$.

Similarly, when $Z_C \gg 1$, we expect $D_{CR}(\phi_L) \ll D_C(\phi_L = 0)$, and $D_C(\phi_L) \approx D_{CR}(\approx)$. Using standard CR arguments, the local hopping time for the Z_C "effective Rouse" beads (entanglements on the CP primitive path) is set by the LP relaxation time τ_L. The self-diffusivity of an ordinary N-bead Rouse chain is given by $D = k_B T/\zeta N$, where k_B is Boltzmann's constant, T is the absolute temperature, and ζ is the frictional drag per bead. If we assume that the effective drag on a CP in a CLB is dominated by entanglements with LPs and is proportional to $\tau_L = R_L^2(\phi_L)/D_L(\phi_L)$, we obtain $D_{CR}(\phi_L) \sim 1/\tau_L Z_C$. From Section 7.3, $R_L(\phi_L)$ is independent of ϕ_L. Thus, ignoring prefactors and constants we get

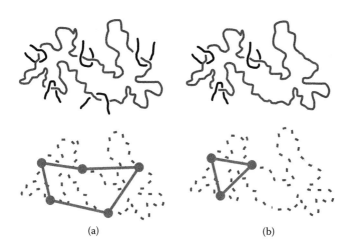

FIGURE 7.9
The number of entanglements on a cyclic polymer (CP), Z_C, depends on the concentration of linear polymers (LPs), ϕ_L. (a) and (b) correspond to large and small values of ϕ_L, respectively. Both Z_C (filled circles) and the length of the primitive path (straight lines) of a CP of fixed molecular weight are proportional to ϕ_L, as depicted schematically by the figures at the bottom. Each of these entanglements has a mean lifetime corresponding to the reptation time τ_L of the LP. In modeling the constraint release (CR) relaxation of the CP, we visualize the entanglements as Rouse beads with a frictional drag proportional to τ_L.

$$\frac{1}{D_C(\phi_L)} = \frac{1}{D_C(\phi_L = 0)} + \frac{Z_C}{D_L(\phi_L)}. \tag{7.19}$$

which can be rearranged to

$$Z_C \frac{D_C(\phi_L = 0)}{D_L(\phi_L)} = c_1 \frac{D_C(\phi_L = 0)}{D_C(\phi_L)} + c_2 \tag{7.20}$$

where c_1 and c_2 are constants that account for terms neglected in this minimal model.

From our simulations (Figure 7.8), and from prior primitive path analysis, presented in Section 7.4, all the parameters in Equation (7.20) can be determined, and the viability of the minimal model can be ascertained. As shown in Figure 7.10, all the available simulation data, independent of composition and molecular weight, collapse onto a linear master curve.

Experimental data have also been shown (Subramanian and Shanbhag, 2008c) to abide by the model. In particular, self-diffusivity data of CPs and LPs at varying concentrations of linear and cyclic DNA (Robertson and Smith, 2007a) were collapsed into the form of Equation (7.20), and the results are shown in Figure 7.10. While two of the 16 data points diverge from the trend line, the rest of the data are in agreement with the predicted linear dependence.

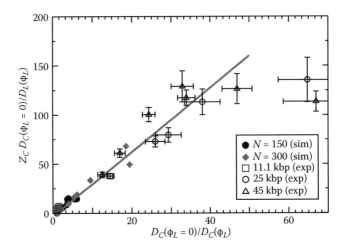

FIGURE 7.10
To validate Equation 7.20, the self-diffusivity simulation data (solid symbols) on $N = 150$ (circles) and $N = 300$ (diamonds) and experimental data (open symbols) on the self-diffusivity of CPs in an entangled linear DNA matrix (Robertson, R. M., and Smith, D. E., 2007a. Self-diffusion of entangled linear and circular DNA molecules: Dependence on length and concentration, *Macromolecules*, 40(9):3373–3377) at different concentrations, and three different molecular weights are replotted. The straight line confirms that there is a linear relationship between the quantities on the horizontal and vertical axes.

7.6 Free Energy

7.6.1 Preliminaries

The modified Cates–Deutsch expression for the conformational free energy of a CP in a CLB may be rewritten as

$$F(R_C) = F_{cat}(R_C) + F_{gauss}(R_C) \tag{7.21}$$

$$F_{cat}(R_C) = \phi_C \frac{R_C^3}{N} \tag{7.22}$$

$$F_{gauss}(R_C) = \frac{N}{R_C^2} \tag{7.23}$$

From a microscopic viewpoint, the total free energy can also be expressed as

$$\frac{F(R_C)}{k_B T} = -\ln \Omega(R_C) \tag{7.24}$$

where $\Omega(R_C)dR_C$ is the number of conformations with size between R_C and $R_C + dR_C$. Given the molecular weight and the composition of a CLB, the value of $\Omega(R_C|N; \phi_C)$ is related to the probability density function of R_C as

$$P(R_C|N;\phi_C) = \frac{\Omega(R_C|N;\phi_C)}{\int \Omega(R_C|N;\phi_C)dR_C} = \frac{\Omega(R_C|N;\phi_C)}{\lambda(N,\phi_C)\mathbf{Z}_L(N)} \qquad (7.25)$$

where $P(R_C|N;\phi_{R_C})dR_C$ represents the fraction of CPs with a size between R_C and $R_C + dR_C$.

Dropping the denominator $k_B T$, the free energy of a CP may thus be obtained as

$$F(R_C|N;\phi_C) = -\ln[\lambda(N,\phi_C)P(R_C|N;\phi_C)] - \ln \mathbf{Z}_L(N) \qquad (7.26)$$

$\mathbf{Z}_L(N)$ (not to be confused with Z, the number of entanglements on a polymer) is the partition function of a LP with N beads and is indicative of the total number of conformations that it may adopt. As shown in Sections 7.3 and 7.4, static properties of LPs are unaffected by the composition of the CLB; hence $\mathbf{Z}_L$ is assumed to be independent of ϕ_C. The partition function $\mathbf{Z}_L(N)$ does not enter into calculations explicitly, beyond an additive constant, which cancels out when we calculate F_{cat}.[*] $\lambda(N, \phi_C)$ is the probability with which an LP adopts a conformation in which its ends overlap. The second equality in Equation (7.25) offers a more tractable computational route to calculate $P(R_C)$. Computation of $F(R_C)$ thus requires knowledge of (a) $\lambda(N, \phi_C)$, the probability with which an LP adopts a conformation indistinguishable from that of a CP, which occurs when its ends essentially overlap, and (b) $P(R_C)$, the probability density function of CPs characterized using the size R_C as the macrostate variable. These quantities are expected to change with the composition of the CLB.

The free energy contribution of the catenation constraint, F_{cat} may be computed as

$$F_{cat}(R_C|N;\phi_C) = F(R_C|N;\phi_C) - F_{gauss}(R_C|N) \qquad (7.27)$$

where the expression for $F_{gauss}(R_C)$ is the free energy of Gaussian CPs and may be computed either analytically or via simulation (described in more detail in Sections 7.6.2 and 7.6.3), and is independent of ϕ_C.

---

[*] However, knowledge of $\mathbf{Z}_L$ is required to directly compare the free energies of two different molecular weights.

7.6.2 Implementation

To obtain the quantities $\lambda(N, \phi_C)$, $P(R_C)$, and $F_{gauss}(R_C)$, simulations were executed using the S-BFM. Fully equilibrated systems of CLBs with different concentrations of CPs were subjected to thermal motion. The CPs in each system are referred to as "true CPs." At the end of *each time step*, the LPs in the simulation box were examined one by one. If it were possible to connect the end beads of an LP by a valid lattice bond vector (cube edge, face diagonal, or solid diagonal) that did not lead to crossing of preexisting bonds (to preserve enforcement of the uncrossability constraint), the LP was considered to be indistinguishable from a CP. The closing bond was added to convert the LP into a CP, and such a converted CP is referred to as a "false CP."

False CPs were examined for self-knotting using the annealing algorithm, and for catenation with the true CPs using the linking number algorithm, which we will describe in some detail. Given a simulation snapshot, the annealing algorithm (Everaers et al., 2004; Shanbhag and Larson, 2005) prescribes a method to identify the primitive paths—the shortest contour that does not violate topological constraints—of all the polymers in the system. This information may be used to probe whether a CP is knotted or not. Alternatively, it is possible to use knot polynomials such as the Jones polynomial (Jones, 1997) to study the existence and complexity of knots in CPs. However, an analysis of the primitive path network that the annealing algorithm generates can diagnose some pathological knots that knot polynomials could in principle miss, because the question of whether there exists a nontrivial knot having the same Jones polynomial of an unknot is an open one (Adams, 2004).

Upon completion of annealing, any CP that is a trivial knot (i.e., not knotted with itself) should collapse into a single point. However, the lattice structure adopted for these simulations allows such a CP to collapse into a triangle, and no further, as discussed in Section 7.4. This is because of the bond uncrossability constraint that is enforced. Thus, unknotted CPs would collapse into a triangle with a maximum contour length of $3\sqrt{2}$ (three bonds along face diagonals of the lattice). These cases are illustrated in Figure 7.11a. If, on the other hand, a CP were "self-knotted," the annealing algorithm would be incapable of shrinking its contour length beyond a certain point without violating the three constraints and would yield conformations like the one shown in Figure 7.11b. This conformation is the trefoil knot. Such knotted conformations have contour lengths greater than $3\sqrt{2}$. This property may be exploited to determine whether a false CP is a knot other than the trivial knot. The annealing algorithm can also be used to test for catenation, but all the true CPs in the system have to be included in the annealing operation, which can be computationally expensive. However, an examination of the primitive path of false CPs, *inspected one at a time,* is sufficient to reveal self-knotting. Computationally, the operation is very inexpensive and requires a few seconds of central processing unit (CPU) time.

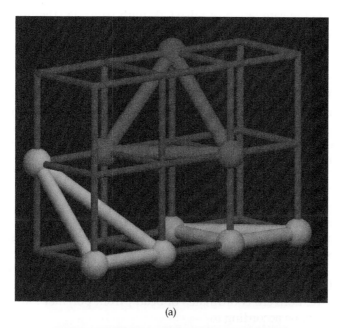

(a)

(b)

FIGURE 7.11
The result of annealing on different kinds of false cyclic polymers (CPs): (a) valid, (b) self-knotted (trefoil knot—lattice representation of the same knot in Figure 7.1d). After annealing, false CPs that are a trivial knot attain a contour length of at most $3\sqrt{2}$.

To examine the catenation of false CPs with true CPs, the linking number algorithm was used. The Gauss linking number of two CPs is an invariant (Gauss, 1833). Linking numbers of any two CPs are computed by counting the number of intersections in a two-dimensional projection. Two CPs (nearest periodic images) (Frenkel and Smit, 2002) isolated from the rest of the system are taken as inputs to the linking number algorithm. The CPs are then translated by equal amounts to the new coordinates $r_j^i = (x_{i,j}, y_{i,j}, z_{i,j})$, (where the index i is either 1 or 2, and identifies the CP to which a bead belongs; the index j denotes the bead number), satisfying the condition:

$$\min\{x_{i,j}, y_{i,j}, z_{i,j}\} = 1, \forall i \in [1,2], j \in [1,N] \qquad (7.28)$$

This is equivalent to translating (without rotating) the original coordinate system in such a way that for the two polymers under consideration, the smallest coordinate of any of their beads is unity (in lattice units). While the choice of unity is somewhat arbitrary, the translation of the coordinate system is done to avoid large numbers when computing the linking number.

The positive, nonzero coordinates of bead j from polymer i are then projected onto a plane according to

$$X_{i,j} = x_{i,j} + \frac{z_{i,j}}{L'+1}$$

$$Y_{i,j} = y_{i,j} + \frac{z_{i,j}}{2(L'+1)} \qquad (7.29)$$

where L' is the side of the cubic box drawn to contain both CPs and is defined as

$$L' = \max\{x_{i,j}, y_{i,j}, z_{i,j}\} + 1, \forall i \in [1,2], j \in [1,N] \qquad (7.30)$$

This transformation is a variant of the transformation used by Lua et al. (2004) and guarantees that (a) no two polymer beads are projected onto the same point and (b) no two projected bonds are collinear. Each projected chain is then assigned a direction. The direction of the projection was arbitrarily chosen to be in the direction of increasing bead numbers. The intersection points on the projection were identified and assigned either +1 or –1, based on the crossing type, according to the schematic shown in Figure 7.12. The number of intersections of each type was also counted, and the linking number was computed as $(n_a + n_b - n_c - n_d)/2$, where n_x is the number of crossings of type x.

In general, a linking number of zero implies that the two CPs in question are not catenated, although notable exceptions such as the Whitehead link

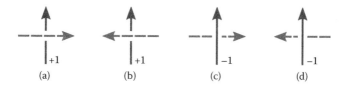

FIGURE 7.12
In (a)–(d), sign assignment for various types of crossings in the two-dimensional projection of two cyclic polymers (CPs).

(Whitehead, 1963) exist. For purposes of this study, the fraction of catenated CPs declared as noncatenated by the linking number algorithm was found to be negligible; thus, any two CPs with a linking number of zero were taken to be noncatenated. This implementation of the catenation constraint does not inspect catenation between two false CPs.

Based on the state of knotting and catenation, the false CPs thus obtained were classified as either of (a) valid, V, (b) only self-knotted, S, (c) only catenated, C, or (d) both self-knotted and catenated, SC. These categories correspond to the regions shown schematically in Figure 7.13. The superscripts V, S, C, and SC are used to distinguish each class of false CPs, and the superscript A is used to designate the superset of all the false CPs (i.e., $A = V \cup S \cup C \cup SC$, where V, S, C, and SC are disjoint sets). The classification along with the conformational information of each false CP were recorded.

The closing bond of the false CP was then removed, converting it back into the original LP, and the simulation was allowed to proceed. For each

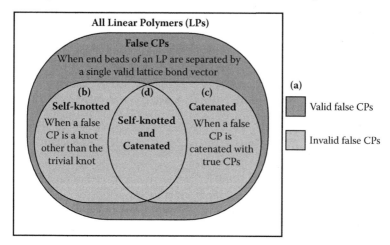

FIGURE 7.13
Venn diagram indicating the four categories of false cyclic polymers (CPs): (a) valid, (b) self-knotted, (c) catenated, and (d) self-knotted and catenated. The superscripts V, S, C, and SC are respectively used for each of these categories. The superscript A is used to denote the superset of all false CPs. In other words, $A = V \cup S \cup C \cup SC$, where V, S, C, and SC are disjoint sets.

class, the frequency with which LPs form false CPs leads to an estimate of $\lambda(N, \phi_C)$, and the conformational information of the false CPs collected over the course of the simulation provides an estimate of the probability density function $P(R_C)$.

7.6.3 Discussion

Systems with $N \in \{10, 30, 80, 150, 200\}$ were considered. For each value of N, the fraction of CPs considered was $f_C \in \{0.025, 0.250, 0.475\}$, mimicking different environments, ranging from mostly LPs to mostly CPs. For reference, for the S-BFM, the number of beads corresponding to an entanglement strand in LPs is $N_e \approx 30$ (Shanbhag and Larson, 2005; Subramanian and Shanbhag, 2008b). Thus, even the longest polymers simulated here are smaller than seven entanglements long. While longer polymers with approximately 20 entanglements and larger have been simulated with this model in earlier work (Shaffer, 1994; Shanbhag and Larson, 2005; Suzuki et al., 2008), simulations larger than the ones considered in this study become prohibitively expensive, because the probability with which an LP forms *any type of* false CP, λ^A, decreases rapidly with increasing molecular weight N. This is not surprising because as N increases, the probability with which the ends of an LP come within one lattice spacing of each other decreases, as the LP has a far greater number of conformational degrees of freedom to explore. When the normal distribution is used to determine the probability that a Gaussian LP has a vanishing end-to-end vector and forms a Gaussian CP, it is expected that $\lambda^A \sim N^{-3/2}$, which is observed in Figure 7.14. The actual fit obtained is $\lambda^A \approx 2.06N^{-1.56}$. The probability of forming a valid CP $\lambda^V < \lambda^A$, and unlike λ^A it decreases with increasing f_C.

As the fraction of true CPs in the melt f_C increases, an LP has a greater likelihood of threading through a true CP before its ends are within one lattice spacing of each other, thereby increasing the probability of catenation, and decreasing the probability of forming a valid false CP. It may be observed that at the lowest value of f_C, where an LP is surrounded largely by other LPs, the values of λ^V ($N, f_C = 0.025$) are close to λ^A (Gaussian CPs). This is because the strength of the catenation constraint that increases with f_C is weakest in this case.

The internal length scale of the CPs was also examined using the following procedure: if $\mathbf{r}_i$ denotes the coordinates of the ith bead, then the internal distribution of distances between beads (i.e., the internal length scale) is obtained by examining

$$r^2(n \,|\, N; \phi_C) = \langle (\mathbf{r}_i - \mathbf{r}_{i+n}) \cdot (\mathbf{r}_i - \mathbf{r}_{i+n}) \rangle \qquad (7.31)$$

where the average is computed over false CPs of a particular type formed during the course of the simulation. The analytical expression for the

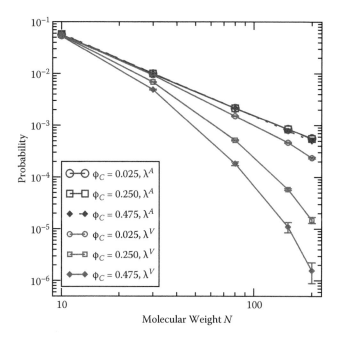

FIGURE 7.14
The probability of a linear polymer (LP) forming any type *of* false cyclic polymer (CP), λ^A decreases with increasing molecular weight N, and is independent of f_C, the fraction of CPs in the melt. On the other hand, the probability of a LP forming a *valid* false CP, λ^V, decreases as either of N or f_C increase. The fit obtained is $\lambda^A \sim 2.06N^{-1.56}$.

internal length scales of Gaussian CPs was given by Zimm and Stockmayer (1949) as

$$r^2(n\,|\,N;\phi_C) = C_\infty b^2 n\left(1 - \frac{n}{N}\right) \tag{7.32}$$

where the characteristic ratio, and mean square bond length have been previously reported as $C_\infty = 1.2$, and $b^2 = 2.05$ (Subramanian and Shanbhag, 2008b).

Figure 7.15 shows a representative plot of the quantity $r^2(n\,|\,N;f_C)n^{-1}(1 - n/N)^{-1}$ for $N = 150$ and $f_C = 0.250$. This molecular weight and concentration of CPs is very close to the threshold concentration that demarcates the dilute regime from the semidilute, and thus, the valid false CPs are nearly Gaussian. It is seen that on average, false CPs that are only self-knotted are the smallest. On the other hand, false CPs that are catenated with true CPs but are otherwise unknotted are the largest. The self-knotted and catenated false CPs are smaller than the purely catenated variety and larger than the purely self-knotted variety. These data, and other measurements published previously (Subramanian and Shanbhag, 2009) support the validity of these simulations.

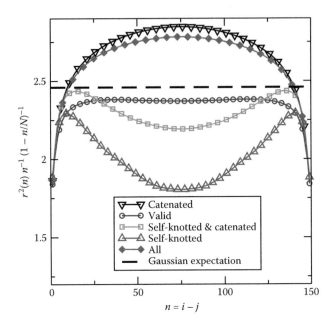

FIGURE 7.15
The average distance between beads of the four different classes of false cyclic polymers (CPs) for $N = 150$. As expected, the catenated, but otherwise unknotted, false CPs are the largest, and the self-knotted false CPs are the smallest. This plot is representative of the trend observed over all values of N and f_C.

We now turn our attention to the second term on the right-hand side of Equation 7.27, $F_{gauss}(R_C \mid N)$. Computing this quantity requires knowledge of the probability density function of R_C (or equivalently, the probability density function (PDF) of R_C^2) of Gaussian CPs, denoted here as $P_{gauss}(R_C)$. It may be computed in either of two ways:

1. Analytically, an expression for $P_{gauss}(R_C^2)$ in three dimensions has been derived as (Minato and Hatano, 1977)

$$P_{gauss}(R_C^2) = \frac{1}{2\pi} \int_{-\infty}^{\infty} K(s) \exp\left[-is\, R_C^2\right] ds$$

$$K(s) = \left[\frac{Y/2}{\sin(Y/2)}\right]^3 \tag{7.33}$$

$$Y = \sqrt{\frac{2is\, N_K}{3}}$$

where the number of Kuhn steps is N_K, and the Kuhn step length is assumed to be unity. It should be pointed out that this relationship does not require $N_K \gg 1$.

2. Alternatively, the CPs in the S-BFM are known to display Gaussian statistics, if bond uncrossability is switched off (Brown et al., 2001). By running a simulation consisting of pure CPs which are allowed to cross each other, $P_{gauss}(R_C^2)$ may be obtained.

The relationship between the molecular weight N and the number of Kuhn steps N_K is obtained using $(N_K - 1) = (N - 1)/C_\infty$. Using the relationship in Equation (7.33), the probability distribution function of R_C^2 is computed by direct numerical integration for the molecular weights considered in this study. These analytical results (method 1) are overlaid with histograms of the probability distribution function obtained from a direct simulation of crossing CPs (method 2) in Figure 7.16, and there is agreement between both methods.

Having obtained the free energy of Gaussian CPs and the free energy of valid CPs, the free energy of catenation can now be computed using Equation (7.27). In the expressions for $F(R_C)$ and $F_{gauss}(R_C)$, the natural logarithm of the partition function, $\log \mathbf{Z}_L$, appears as an additive constant, which cancels out when we take the difference, and thus does not play a role in determining $F_{cat}(R_C)$. Thus, we arbitrarily set $\mathbf{Z}_L = 1$. If we were to directly compare the free energies of two different molecular weights, it would be necessary to know the actual value of $\mathbf{Z}_L$.

Computing the free energy of catenation, $F_{cat}(R_C)$ in this manner, the quantity $F_{cat}N/\phi_C R_C^3$ is expected to be relatively constant, according to the modified Cates–Deutsch conjecture in Equation (7.22). However, as shown in Figure 7.17, this holds true only for large values of R_C. Performing a multiple linear regression, we find $F_{cat} = 0.066 \, \phi_C^{0.91} N^{0.51} R_C^{1.35}$. Using this regressed expression for F_{cat}, and the numerical procedure for computing F_{gauss}, we compute the total free energy F and thereby compute the average CP size as

$$\overline{R_C}(N,\phi_C) = \frac{\int_0^\infty R_C \exp\left[-F(R_C \mid N;\phi_C)\right]dR_C}{\int_0^\infty \exp[-F(R_C \mid N;\phi_C)]dR_C} \qquad (7.34)$$

Figures 7.18a and 7.18b depict the variation of $\overline{R_C}(N,\phi_C)$ with N and f_C, computed using the PDF of R_C of Gaussian CPs to obtain $F_{gauss}(R_C \mid N)$, and the regressed expression for $F_{cat}(R_C \mid N; f_C)$ over different molecular weights and fractions of CPs in the semidilute regime. Overlaid on these plots are the values of $\overline{R_C}$ obtained from the simulations in this section, and Section 7.3. While the agreement between the predicted $\overline{R_C}$ values and the observed $\overline{R_C}$ values seems reasonable, there are some ambiguities that should be pointed out.

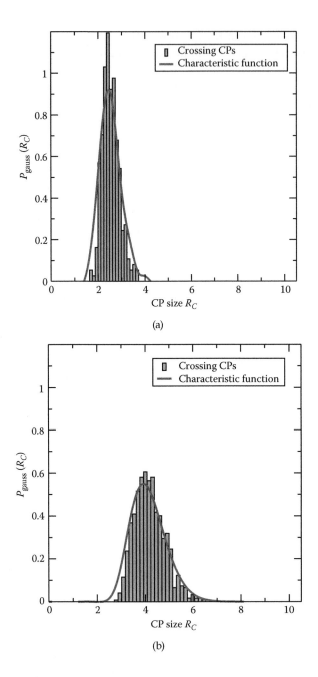

FIGURE 7.16

The PDF of R_C obtained by numerically integrating the characteristic function, and from crossing the bond fluctuation model proposed by Shaffer (Shaffer, J. S., 1994. Effects of chain topology on polymer dynamics—Bulk melts, *J. Chem. Phys.*, 101:4205–4213) (S-BFM) simulations for N = (a) 30, (b) 80, (c) 150, and (d) 200. The results of integrating Equation (7.33) have been rescaled to S-BFM lattice units. Also shown is the PDF of R_C of all false cyclic polymers (CPs).

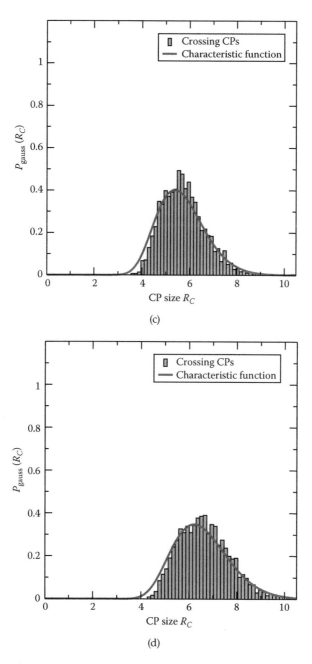

(c)

(d)

FIGURE 7.16 (continued)

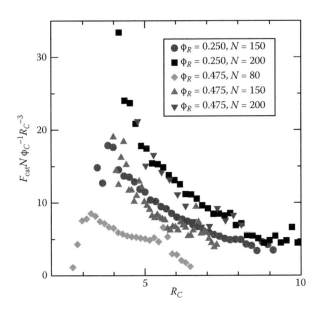

FIGURE 7.17

The free energy of catenation cast in the modified Cates–Deutsch form in Equation (7.22) seems to be valid only for larger values of R_C.

The values of $\overline{R}_C$ predicted using the regressed expression for $F_{cat}(R_C|N;f_C)$ seem to have the correct dependence on N. The observed dependence on f_C, however, seems stronger than predicted. In addition to this, the prediction overestimates $\overline{R}_C$ for $N = 80$, and underestimates $\overline{R}_C$ for $N = 300$. This error lies in estimating the probability of forming a valid CP, λ^V (Subramanian and Shanbhag, 2009), due to insufficient sampling at large f_C. In terms of computational time and effort, it becomes increasingly difficult to probe the semidilute regime with increasing N. This is because the chances of observing a valid false CP diminish rapidly as f_C and N increase. There are three reasons for this: (1) at large f_C, the number of LPs in the simulation box is small, thereby reducing the number of possible false CPs; (2) the probability with which an LP forms a false CP decreases rapidly with increasing N. Indeed, as seen in Figure 7.14, $\lambda^A \sim N^{-1.56}$, and (3) as f_C increases, the number of true CPs that an LP can thread through is far greater than at low f_C. Thus, in terms of obtaining the probability of forming a valid false CP, λ^V, probing the semidilute regime requires simulations at large N and f_C, which become prohibitively expensive. Because λ^V is one of the key pieces of information required to compute the free energy of catenation, the regressed expression for F_{cat} may be weighted less by the actual semidilute regime and more by the transition regime from dilute to semidilute, and might suffer in part from finite size effects.

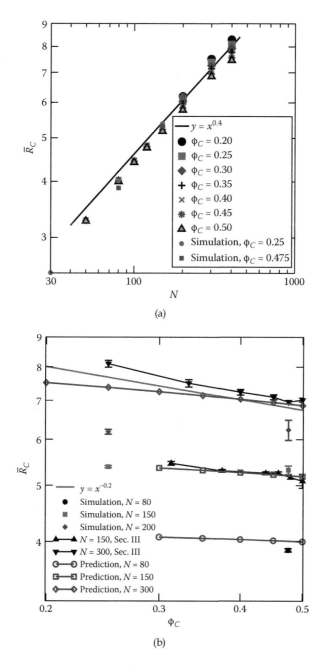

FIGURE 7.18

The average size $\overline{R}_C$ obtained using the regressed formula for $F_{cat}(R_C \mid N; f_C)$, and the analytical expression for $F_{gauss}(R_C \mid N)$. The average size $\overline{R}_C$ of valid false cyclic polymers (CPs), and results from Section 7.3 on size are plotted for comparison. (a) The scaling of $\overline{R}_C$ as a function of N seems to be captured correctly, but (b) its dependence on ϕ_C seems more like $\overline{R}_C \sim \phi_C^{-0.14}$.

7.7 Deviations from Theory at High Molecular Weights

The theory presented in Section 7.3 is a mean-field theory and has been shown to be correct for the range of molecular weights studied. Recent results on pure CP melts (Subramanian, 2010; Suzuki et al., 2009; Vettorel et al., 2009a, 2009b) indicate that the theory might require some modifications for higher molecular weights. In particular, while low and moderate molecular weight CPs display a scaling of size as $R_C \sim N^{0.4}$, at high molecular weights the scaling changes to $R_C \sim N^{1/3}$. This deviation has been observed by multiple researchers, using off-lattice molecular dynamics methods and the S-BFM. In the case of S-BFM, the deviation is independent of equilibration method. Suzuki et al. (2009) presented an alternate expression for the free energy due to the catenation constraint as

$$F_{cat}(R_C \mid N) = \frac{R_C^2}{N} \left[\frac{1}{R_C^2} \left(\frac{4\pi}{3} R_C^3 - N \right) \right]^2 \qquad (7.35)$$

and then showed that using this modified expression for F_{cat} leads to the correct scaling at high molecular weights. This asymptotic form of F_{cat} becomes fully operational only in the high molecular weight regime, which, as shown in Figure 7.19, is in the vicinity of $N \approx 2500$. For the S-BFM, which has a molecular weight between entanglements of $N_e = 30$, this corresponds to approximately $Z \approx 80$ entanglements. Experimentally accessible CPs, on the other hand, are limited to $Z \approx 20$ (Kapnistos et al., 2008). Based on the experimental materials that are currently accessible, the "moderate molecular weight" regime is perhaps of more importance.

The minimal constraint release model for dynamics that was presented in Section 7.5 is not expected to be greatly affected, as its realm of validity is the high ϕ_L (low ϕ_C) regime, where CPs are essentially immersed in a sea of LPs, and the effect of the noncatenation constraint is negligible. However, for moderate values of ϕ_L, the size of CPs enters the picture indirectly through the term $D_C(\phi_L = 0)$, and might affect, by a small amount, the asymptotic behavior. Thus, the implications of the asymptotic scaling of F_{cat} for the minimal constraint release model, at the time of writing, are mostly academic, as the high molecular weight regime is not really accessible via experiments. Probing the high molecular weight regime via simulations, on the other hand, can be prohibitively expensive in terms of computation time. In particular, while algorithms for rapid equilibration exist, as mentioned in Section 7.2, it is difficult to speed up the dynamics of polymers, which relax on a timescale of approximately one reptation time, which scales as $N^{3.4}$. While methods such as accelerated molecular dynamics (Voter et al., 2002) have been used for hard potentials, obtaining atomistic descriptions of diffusion dynamics in polymers remains a daunting task due to the inapplicability of these methods to soft materials.

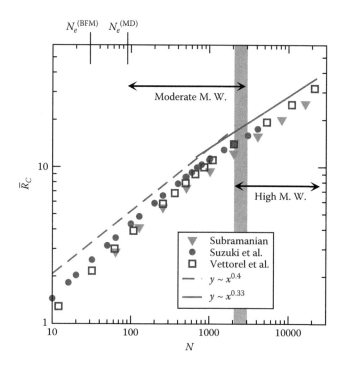

FIGURE 7.19
Deviations from the modified Cates–Deutsch scaling argument manifest themselves at high molecular weight. Data shown are obtained from three different simulation methods: Subramanian (Subramanian, G., 2010. A topology preserving method for generating equilibrated polymer melts in computer simulations, *J. Chem. Phys.*, 133:164902) used an off-lattice molecular dynamics method; Suzuki et al. (Suzuki, J., Takano, A., Deguchi, T., and Matsushita, Y., 2009. Dimension of ring polymers in bulk studied by Monte-Carlo simulation and self-consistent theory, *J. Chem. Phys.*, 131(14):144902+) used the S-BFM, and the equilibration method presented in Section 7.2; Vettorel et al. (Vettorel, T., Grosberg, A. Y., and Kremer, K., 2009a. Statistics of polymer rings in the melt: A numerical simulation study, *Phys. Biol.*, 6(2):025013+; Vettorel, T., Reigh, S. Y., Yoon, D. Y., and Kremer, K., 2009b. Monte-Carlo method for simulations of ring polymers in the melt, *Macromol. Rapid Commun.*, 30(4–5):345–351) used the S-BFM, and a combination of crankshaft rotation, wedge flip, and kink translocation moves to equilibrate the melt. Also indicated by N_e are the entanglement molecular weights for both models.

References

Adams, C. C. 2004. *The Knot Book*, Providence, RI: American Mathematical Society.
Alexiadis, O., Daoulas, K. C., and Mavrantzas, V.G. 2008. An efficient Monte Carlo algorithm for the fast equilibration and atomistic simulation of alkane-thiol self-assembled monolayers on a au(111) substrate, *The J. Phys. Chem. B*, 112(4):1198–1211.

Baig, C., Alexiadis, O., and Mavrantzas, V. G. 2010. Advanced Monte Carlo algorithm for the atomistic simulation of short- and long-chain branched polymers: Implementation for model h-shaped, a3aa3 multiarm (pom-pom), and short-chain branched polyethylene melts, *Macromolecules*, 43(2):986–1002.

Ball, R. C., and McLeish, T. C. B. 1989. Dynamic dilution and the viscosity of star polymer melts, *Macromolecules*, 22:1911–1913.

Brown, S., Lenczycki, T., and Szamel, G., 2001. Influence of topological constraints on the statics and dynamics of ring polymers, *Phys. Rev. E.*, 63(5):052801.

Brown, S., and Szamel, G., 1998a. Computer simulation study of the structure and dynamics of ring polymers, *J. Chem. Phys.*, 109:6184–6192.

Brown, S., and Szamel, G., 1998b. Structure and dynamics of ring polymers, *J. Chem. Phys.*, 108:4705–4708.

Carmesin, I., and Kremer, K., 1988. The bond fluctuation method—A new effective algorithm for the dynamics of polymers in all spatial dimensions, *Macromolecules*, 21:2819–2823.

Cates, M. E., and Deutsch, J. M., 1986. Conjectures on the statistics of ring polymers, *J. Phys-Paris.*, 47:2121–2128.

Daoulas, K. C., Harmandaris, V. A., and Mavrantzas, V. G., 2005. Detailed atomistic simulation of a polymer melt/solid interface: Structure, density, and conformation of a thin film of polyethylene melt adsorbed on graphite, *Macromolecules*, 38(13):5780–5795.

de Gennes, P. G., 1971. Reptation of a polymer chain in the presence of fixed obstacles, *J. Chem. Phys.*, 55(2):572–579.

de Pablo, J. J., Laso, M., and Suter, U. W., 1992. Simulation of polyethylene above and below the melting point, *J. Chem. Phys.*, 96(3):2395–2403.

Dean, F. B., Stasiak, A., Koller, T., and Cozzarelli, N. R., 1985. Duplex DNA knots produced by *Escherichia coli* topoisomerase I. Structure and requirements for formation, *J. Biol. Chem.*, 260(8):4975–4983.

Dodd, L. R., Boone, T. D., and Theodorou, D. N., 1993. A concerted rotation algorithm for atomistic Monte Carlo simulation of polymer melts and glasses, *Mol. Phys.: An Int. J. at Interface Between Chem. Phys.*, 78(4):961–996.

Doi, M., and Edwards, S. F., 1986. *The Theory of Polymer Dynamics*, Oxford: Clarendon Press.

Doi, M., and Kuzuu, N. Y., 1980. Rheology of star polymers in concentrated solutions and melts, *J. Polym. Sci. Polym. Lett. Ed.*, 18(12):775–780.

Everaers, R., Sukumaran, S. K., Grest, G. S., Svaneborg, C., Sivasubramanian, A., and Kremer, K., 2004. Rheology and microscopic topology of entangled polymeric liquids, *Science*, 303:823–826.

Frenkel, D., and Smit, B., 2002. *Understanding Molecular Simulation: From Algorithms to Applications*, 2nd ed. Orlando, FL: Academic Press.

Frischknecht, A. L., and Milner, S. T., 2000. Diffusion with contour length fluctuations in linear polymer melts, *Macromolecules*, 33:5273–5277.

Gauss, C., 1833. *Werke: Königliche Gesellschaft der Wissenschaften*, Göttingen: Teubner.

Geyler, S., and Pakula, T., 1988. Conformation of ring polymers in computer simulated melts, *Die Makromol. Chemie, Rapid Comm.*, 9(9):617–623.

Graessley, W. W., 1982. Entangled linear, branched and network polymer systems—Molecular theories, *Advan. Polym. Sci.*, 47:67–117.

Graessley, W. W., and Struglinski, M. J., 1986. Effects of polydispersity on the linear viscoelastic properties of entangled polymers. 2. Comparison of viscosity and recoverable compliance with tube model predictions, *Macromolecules*, 19(6):1754–1760.

Green, P. F., and Kramer, E. J., 1986. Matrix effects on the diffusion of long polymer chains, *Macromolecules*, 19(4):1108–1114.

Green, P. F., Mills, P. J., Palmström, C. J., Mayer, J. W., and Kramer, E. J., 1984. Limits of reptation in polymer melts, *Phys. Rev. Lett.*, 53(22):2145–2148.

Hur, K., Winkler, R. G., and Yoon, D. Y., 2006. Comparison of ring and linear polyethylene from molecular dynamics simulations, *Macromolecules*, 39:3975–3977.

Iyer, B. V. S., Lele, A. K., and Juvekar, V. A., 2006. Flexible ring polymers in an obstacle environment: Molecular theory of linear viscoelasticity, *Phys. Rev. E.*, 74(2):021805.

Iyer, B. V. S., Lele, A. K., and Shanbhag, S., 2007. What is the size of a ring polymer in a ring-linear blend?, *Macromolecules*, 40(16):5995–6000.

Jones, V., 1985. A polynomial invariant for knots via von Neumann algebras, *Bulletin of the American Mathematical Society*, 12:103–111.

Kapnistos, M., Lang, M., Vlassopoulos, D., Pyckhout-Hintzen, W., Richter, D., Cho, D., Chang, T., and Rubinstein, M., 2008. Unexpected power-law stress relaxation of entangled ring polymers, *Nat. Mater.*, 7(12):997–1002.

Karayiannis, N. C., Giannousaki, A. E., Mavrantzas, V. G., and Theodorou, D. N., 2002. Atomistic Monte Carlo simulation of strictly monodisperse long polyethylene melts through a generalized chain bridging algorithm, *J. Chem. Phys.*, 117(11):5465–5479.

Klein, J., 1986. Dynamics of entangled linear, branched, and cyclic polymers, *Macromol.*, 19(1):105–118.

Leontidis, E., de Pablo, J., Laso, M., and Suter, U., 1994. A critical evaluation of novel algorithms for the off-lattice Monte Carlo simulation of condensed polymer phases. In L. Monnerie and U. W. Suter, Eds., *Atomistic Modeling of Physical Properties*, Springer Berlin Heidelberg, *Advances in Polymer Science*, vol. 116, chap. 8, 283–318.

Liu, C. Y., Keunings, R., and Bailly, C., 2006. Do deviations from reptation scaling of entangled polymer melts result from single- or many-chain effects?, *Phys. Rev. Lett.*, 97:246001.

Lua, R. C., Borovinskiy, A. L., and Grosberg, A. Y., 2004. Fractal and statistical properties of large compact polymers: A computational study, *Polymers*, 45:717–731.

Mavrantzas, V. G., Boone, T. D., Zervopoulou, E., and Theodorou, D. N., 1999. End-bridging Monte Carlo: A fast algorithm for atomistic simulation of condensed phases of long polymer chains, *Macromolecules*, 32(15):5072–5096.

McKenna, G. B., Hadziioannou, G., Lutz, P., Hild, G., Strazielle, C., Straupe, C., Rempp, P., and Kovacs, A. J., 1987. Dilute-solution characterization of cyclic polystyrene molecules and their zero-shear viscosity in the melt, *Macromolecules*, 20:498–512.

McKenna, G. B., Hostetter, B. J., Hadjichristidis, N., Fetters, L. J., and Plazek, D. J., 1989. A study of the linear viscoelastic properties of cyclic polystyrenes using creep and recovery measurements, *Macromolecules*, 22:1834–1852.

McKenna, G. B., and Plazek, D. J., 1986. The viscosity of blends of linear and cyclic molecules of similar molecular mass, *Polym. Commun.*, 27(10):304–306.

Menasco, W., and Thistlethwaite, M., Eds., 2005. *Handbook of Knot Theory*, Amsterdam: Elsevier.

Micheletti, C., Marenduzzo, D., Orlandini, E., and Sumners, D. W., 2006. Knotting of random ring polymers in confined spaces, *J. Chem. Phys.*, 124:064903.

Milner, S. T., and McLeish, T. C. B., 1997. Parameter-free theory for stress relaxation in star polymer melts, *Macromolecules*, 30:2159–2166.

Minato, T., and Hatano, A., 1977. On the distribution function of the square radius of gyration of a ring chain, *J. Phys. Soc. Jpn.*, 42:1992–1996.

Müller, M., Wittmer, J. P., and Barrat, J. L., 2000a. On two intrinsic length scales in polymer physics: Topological constraints vs. entanglement length, *Eur. Lett.*, 52:406–412.

Müller, M., Wittmer, J. P., and Cates, M. E., 1996. Topological effects in ring polymers: A computer simulation study, *Phys. Rev. E.*, 53:5063–5074.

Müller, M., Wittmer, J. P., and Cates, M. E., 2000b. Topological effects in ring polymers. II. Influence of persistence length, *Phys. Rev. E.*, 61:4078–4089.

Obukhov, S. P., Rubinstein, M., and Duke, T., 1994. Dynamics of a ring polymer in a gel, *Phys. Rev. Lett.*, 73:1263–1266.

Ozisik, R., von Meerwall, E. D., and Mattice, W. L., 2002. Comparison of the diffusion coefficients of linear and cyclic alkanes, *Polymers*, 43:629–635.

Pant, P. V. K., and Theodorou, D. N., 1995. Variable connectivity method for the atomistic Monte Carlo simulation of polydisperse polymer melts, *Macromolecules*, 28(21):7224–7234.

Ramos, J., Peristeras, L. D., and Theodorou, D. N., 2007. Monte Carlo simulation of short chain branched polyolefins in the molten state, *Macromolecules*, 40(26):9640–9650.

Robertson, R. M., and Smith, D. E., 2007a. Self-diffusion of entangled linear and circular DNA molecules: Dependence on length and concentration, *Macromolecules*, 40(9):3373–3377.

Robertson, R. M., and Smith, D. E., 2007b. Strong effects of molecular topology on diffusion of entangled DNA molecules, *Proc. Natl. Acad. Sci.*, 104(12):4824–4827.

Roovers, J., 1985. Melt properties of ring polystyrenes, *Macromolecules*, 18:1359–1361.

Roovers, J., and Toporowski, P. M., 1988. Synthesis and characterization of ring polybutadienes, *J. Polym. Sci. B-Polym. Phys.*, 26:1251–1259.

Rouse, P. R., 1953. A theory of the linear viscoelastic properties of dilute solutions of coiling polymers, *J. Chem. Phys.*, 21:1272–1280.

Rubinstein, M., 1986. Dynamics of ring polymers in the presence of fixed obstacles, *Phys. Rev. Lett.*, 57:3023–3026.

Rubinstein, M., and Colby, R. H., 1988. Self-consistent theory of polydisperse entangled polymers—Linear viscoelasticity of binary blends, *J. Chem. Phys.*, 89:5291–5306.

Shaffer, J. S., 1994. Effects of chain topology on polymer dynamics—Bulk melts, *J. Chem. Phys.*, 101:4205–4213.

Shanbhag, S., and Kroger, M., 2007. Primitive path networks generated using annealing and geometrical methods: Insights into differences, *Macromolecules*, 40(8):2897–2903.

Shanbhag, S., and Larson, R. G., 2005. Chain retraction potential in a fixed entanglement network, *Phys. Rev. Lett.*, 94(7):076001.

Shanbhag, S., and Larson, R. G., 2006. Identification of topological constraints in entangled polymer melts using the bond-fluctuation model, *Macromolecules*, 39(6):2413–2417.

Shanbhag, S., Park, S. J., Zhou, Q., and Larson, R. G., 2007. Implications of microscopic simulations of polymer melts for mean-field tube theories, *Mol. Phys.*, 105(2):249–260.

Sides, S. W., Grest, G. S., Stevens, M. J., and Plimpton, S. J., 2004. Effect of end-tethered polymers on surface adhesion of glassy polymers, *J. Polym. Sci. Part B: Polym. Phys.*, 42(2):199–208.

Siepmann, J. I., and Frenkel, D., 1992. Configurational bias Monte Carlo: A new sampling scheme for flexible chains, *Mol. Physics: An Int. J. at Interface Between Chem. Phys.*, 75(1):59–70.

Subramanian, G., 2010. A topology preserving method for generating equilibrated polymer melts in computer simulations, *J. Chem. Phys.*, 133:164902.

Subramanian, G., and Shanbhag, S., 2008a. Conformational properties of blends of cyclic and linear polymer melts, *Phys. Rev. E.*, 77(1):11801.

Subramanian, G., and Shanbhag, S., 2008b. On the relationship between two popular lattice models for polymer melts, *J. Chem. Phys.*, 129(14):144904.

Subramanian, G., and Shanbhag, S., 2008c. Self-diffusion in binary blends of cyclic and linear polymers, *Macromolecules*, 41(19):7239–7242.

Subramanian, G., and Shanbhag, S., 2009. Conformational free energy of melts of ring-linear polymer blends, *Phys. Rev. E.*, 80(4):041806+.

Suzuki, J., Takano, A., Deguchi, T., and Matsushita, Y., 2009. Dimension of ring polymers in bulk studied by Monte-Carlo simulation and self-consistent theory, *J. Chem. Phys.*, 131(14):144902+.

Suzuki, J., Takano, A., and Matsushita, Y., 2008. Topological effect in ring polymers investigated with Monte Carlo simulation, *J. Chem. Phys.*, 129(3):034903.

Trigueros, S., Salceda, J., Bermudez, I., Fernandez, X., and Roca, J., 2004. Asymmetric removal of supercoils suggests how topoisomerase II simplifies DNA topology, *J. Mol. Biol.*, 335:723–731.

Uhlherr, A., Leak, S. J., Adam, N. E., Nyberg, P. E., Doxastakis, M., Mavrantzas, V. G., and Theodorou, D. N., 2002. Large scale atomistic polymer simulations using Monte Carlo methods for parallel vector processors, *Comput. Phys. Commun.*, 144(1):1–22.

Vasquez, R., and Shanbhag, S., 2011. Percolation of trace amounts of linear polymers in melts of cyclic polymers, *Macromol. Theory Simul.*, 20:205–211.

Vettorel, T., Grosberg, A. Y., and Kremer, K., 2009a. Statistics of polymer rings in the melt: A numerical simulation study, *Phys. Biol.*, 6(2):025013+.

Vettorel, T., Reigh, S. Y., Yoon, D. Y., and Kremer, K., 2009b. Monte-Carlo method for simulations of ring polymers in the melt, *Macromol. Rapid Commun.*, 30(4–5):345–351.

Viovy, J. L., Rubinstein, M., and Colby, R. H., 1991. Constraint release in polymer melts—Tube reorganization versus tube dilation, *Macromolecules*, 24:3587–3596.

Voter, A. F., Montalenti, F., and Germann, T. C., 2002. Extending the time scale in atomistic simulation of materials, *Annu. Rev. Mater. Res.*, 32(1):321–346.

Whitehead, J. H. C., 1963. *The Mathematical Works of J. H. C. Whitehead*, Oxford: Pergamon Press.

Yan, J., Magnasco, M. O., and Marko, J. F., 1999. A kinetic proofreading mechanism for disentanglement of DNA by topoisomerases, *Nat. (London)*, 401:932.

Zimm, B. H., and Stockmayer, W. H., 1949. The dimensions of chain molecules containing branches and rings, *J. Chem. Phys.*, 17(12):1301–1314.

8

Strain Hardening in Polymer Blends with Fibril Morphology

Joung Sook Hong

Soongsil University
Seoul, Korea

Hyung Tag Lim, Kyung Hyun Ahn, and Seung Jong Lee

Seoul National University
Seoul, Korea

CONTENTS

8.1 Introduction

The properties of a multiphase system are significantly affected by its morphology. The lack of thermodynamic interaction between phases results in immiscibility, and an immiscible multiphase system shows diverse morphology depending on the parameters such as composition, interfacial tension, viscoelasticity, shear rate, flow field, and so on (Han, 1981; Utracki, 1990; Sperling, 1997). Because blending of polymers with different characteristics is an economical way of developing new materials, it is important from an industrial point of view to understand the development of morphology

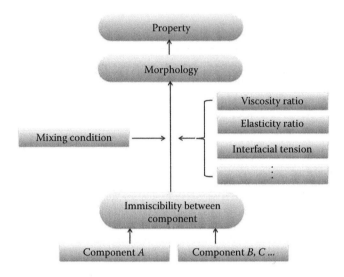

FIGURE 8.1
Generation of new materials by blending multicomponents.

during processing and to control the microstructure of the blend. As polymeric components experience complex deformation history of both shear and extension during mixing, it is difficult to understand how they develop morphology. Even with the same composition, the blend morphology can change from droplet to fibril structure depending on the parameters: flow rate, viscosity ratio, interfacial tension, and so on. Despite the critical influence of morphology on the properties of the blend, most research has been focused on droplet structure. Because blending typically aims at achieving good dispersion of the second phase, it is usually performed at maximal deformability of the dispersed phase, leading to droplet morphology over a wide range of blending conditions as a consequence of interfacial tension. However, there exists a process window that produces morphology other than droplet shape. As illustrated in Figure 8.1, the presence of immiscible components with diverse morphology produces different kinds of materials with unique rheological and mechanical performances. It has already been reported that blends with fibril morphology have unique and improved mechanical properties compared to droplet structure (Blizard and Baird, 1987; Yamamoto and Matsuoka, 1995; Cassagnau and Michel, 2001). Changes in morphology sometimes affect performance more than blend composition. Strain hardening behavior, as one example of a synergic morphological effect, is an abrupt increase in extensional viscosity with time, and it is of great importance in polymer processing in which a significant component of extensional deformation exists as in film blowing, blow molding, and thermoforming. As strain hardening provides strong resistance against fast stretching of polymer chains, the process becomes easier to control if the polymer melt shows strain hardening behavior, and the product can be produced in higher quality. There have been many studies on the extensional behavior of polyolefins with both linear and branch structures (Petrie, 1979; Münstedt and Laun, 1981; Linster and Meissner, 1986). Since the late 1970s much research has been carried out to induce strain hardening to linear polymers to expand their applications (Ishizuka and Koyama, 1980; Münstedt, 1980; Khan et al., 1987; Wagner et al., 2000; Yamaguchi, 2001; Gabriel and Münstedt, 2003; Stadler et al., 2007; Cheng et al., 2010). Ever since it was known that strain hardening originates from branch structure, many efforts have been made to induce strain hardening to linear polymers, by blending with a branch polymer that shows strain hardening behavior (Valenza et al., 1986; Utracki, 1990; Takahashi et al., 1999; Minegishi et al., 2001; Xian-Wu et al., 2010; Yamaguchi, 2001; Yamaguchi and Suzuki, 2002; Ajji et al., 2003; Joshi et al., 2003; Leonardi et al., 2005; Hong et al., 2005a, 2005b; Filipe et al., 2006; McCallum et al., 2007). However, even when low-density polyethylene (LDPE) that shows strain hardening behavior was dispersed in high-density polyethylene (HDPE), it was not easy to observe synergic hardening behavior. Despite the economic advantage of blending, it is still hard to obtain new materials with synergic performance, mainly because past attempts were restricted to droplet morphology and there was little consideration of the effect of morphology. It is still not possible to produce synergic effect of strain hardening with droplet morphology, due to immiscibility between the components and poor interface between the phases (Wool, 1995; Chaffin et al., 2000). There have also been trials to modify the microstructure in order to overcome the problems associated with poor interface. Long-chain branching by chemical or physical modification was introduced by making use of new catalysts (Yamaguchi and Suzuki, 2002), cross-linking agents (Takahashi et al., 1999; Yamaguchi, 2001; Minegishi et al., 2001; Leonardi et al., 2005; McCallum et al., 2007), electron beam irradiation, and grafting of functionalized polymers at interfaces (Münstedt, 1980; Hingmann and Marczinke, 1994; Kurzbeck et al., 1999; Yan et al., 1999; Lim et al., 2010). These techniques generate a few ultra-high molecular weight (UHMW) chains or long-chain branches in the polymer backbone. The existence of a few UHMW chains or long-chain branches in a linear polymer can lead to pronounced strain hardening behavior. Furthermore, the dependence of strain hardening behavior on strain rate can be controlled by diverse branch structures (Wagner et al., 2000; Gabriel and Münstedt, 2003). Strain hardening is known to originate from restricted stretching of the backbone between the branch points (Inkson et al., 1999). In other words, strain hardening behavior can be induced by a minimal amount of materials having strong resistance against stretching. In a multiphase system, strain hardening behavior can be induced by varying the morphology of the dispersed phase. Anisotropic structure such as a sheet or fibril is known to be more favorable for induction of strain hardening than the droplet structure (Hong et al., 2005a).

In the next section, fibrillation is explained in terms of deformability of the dispersed phase and morphology evolution (Section 8.2). The rheological properties of polymer blends exhibiting fibril morphology will be compared to those with droplet morphology (Section 8.3). The effect of anisotropic fibril morphology on strain hardening is also discussed. As a practical application, linear polyethylene blended with dispersed phase of fibril morphology is blow-molded and its superior inflation performance will be demonstrated (Section 8.4).

8.2 Fibrillation

Depending on processing conditions, diverse morphologies such as nodular, laminar, fibril, and droplet can be obtained in polymer blends even with the same composition. As morphology can be controlled as a function of deformability of the dispersed phase, it will be necessary to understand the deformability of the dispersed phase to control the morphology of the blend.

8.2.1 Deformation-Dependent Morphology Variation

When a blend flows, the dispersed phase or the domain is deformed to an ellipsoid or fibril, after which it is broken up into small droplets depending on the capillary number (Ca) and viscosity ratio (λ) (Taylor, 1932, 1934; Acrivos and Lo, 1978; Chin and Han, 1979; Grace, 1982). The capillary number is defined as the ratio between the viscous force and the interfacial force, wherein the former tends to break the droplet and the latter tends to keep it spherical. To induce deformation and breakup of a domain, an external flow field should be applied against interfacial energy. Beyond the critical flow condition (Ca_{cr}), the deformed domain is broken up into several small domains. The ratio of Ca and critical Ca_{cr} determines the deformation and breakup process. In Newtonian systems, the deformability of a domain is categorized as follows: for $Ca/Ca_{cr} < 0.1$, there is no deformation; for $0.1 < Ca/Ca_{cr} < 1$, droplet starts to deform, but there is no breakup; for $1 \leq Ca/Ca_{cr} \leq 4$, droplet breaks up; for $Ca/Ca_{cr} > 4$, large deformation and formation of long and stable fibril structure (Utracki, 1990). The difference in deformability under shear and extensional flow is reflected in Ca_{cr} (Acrivos and Lo, 1978; Chin and Han, 1979; Grace, 1982). Under extensional flow, the domain will deform and break up if Ca is larger than Ca_{cr}, while it is more complicated in shear flow because Ca_{cr} depends on the viscosity ratio as well. Under shear flow, Ca_{cr} quickly increases from 1 to infinity as λ increases from 1 to about 3.7. It is also known that for $\lambda < 0.2$, small drops are shed off from two tips of the original drop; for $0.2 < \lambda < 0.7$, Taylor's relation is obeyed;

for $0.7 < \lambda < 3.7$, the drops elongate to threads, which may break up by a capillary instability mechanism; and for $\lambda > 3.7$, the drops deform to prolate ellipsoids but not break up (Rumscheidt and Mason, 1961). Based on deformability, though restricted to a Newtonian system, the domain may form a nonspherical structure of high aspect ratio under certain conditions of λ and **Ca** (Champagne et al., 1996). Fibrillation can be obtained at high λ ($1 < \lambda < 3.7$) or at fast deformation (**Ca**/**Ca**$_{cr}$ > 4). Fibrillation under fast deformation (**Ca**/**Ca**$_{cr}$ > 4) typically takes place during capillary flow through a die (Min et al., 1984) or during the take-up process after extrusion (Evstatiev et al., 1998, 2000). The polybutylene terephthalate(PBT)/polyamide 6(PA 6) blend, for example, maintains a microfibrillar structure by cold and hot drawing and annealing after melt blending (Evstatiev et al., 1998, 2000). Application of dynamic thermal and deformation history during postextrusion maintains microfibrillar structure. In practice, it is hard to precisely predict the deformability of the dispersed phase due to unique viscoelastic response of each component, complex flow field, and coalescence. Though it is difficult to predict deformability, polymer blends of diverse morphology can be designed, provided that the viscosity ratio is controlled (Hong et al., 2005b). Three polypropylene (PP)/polyethylene (PE) blends with different viscosity ratios, 3.9, 1, and 0.5 (based on complex viscosity measured at 10 rad/s), were extruded under the same mixing condition. Figures 8.2 and 8.3 show the diverse morphologies of three PP/PE blends (PP/PE-1, PP/PE-2, and PP/PE-3). When the

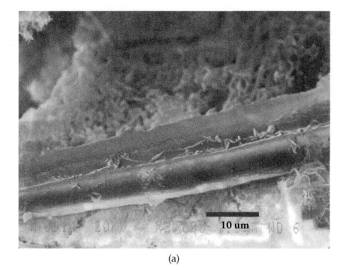

(a)

FIGURE 8.2
Morphology of three polypropylene (PP)/polyethylene (PE) blends with different viscosity ratios. The viscosity ratio is 3.9 for PP/PE-1 (a), 1 for PP/PE-2 (b), and 0.5 for PP/PE-3 (c). The surface of the blend was etched with xylene to improve morphology observation. (Reproduced from Hong, J. S., K. H. Ahn, and S. J. Lee. 2005a. Strain hardening behavior of polymer blends with fibril morphology. *Rheologica Acta* 45:202–208, with permission.)

(b)

(c)

FIGURE 8.2 (continued)

viscosity ratio is larger than one, the PP/PE blend shows fibril morphology. PP/PE-3 with the lowest viscosity ratio shows no fibril morphology. PP/PE-1 having a relatively high viscosity ratio (>3) shows a long straight fibril structure with a length of several hundred micrometers. When the viscosity ratio is low, in between 1.4 and 0.66, the fibril becomes shorter and thinner compared to PP/PE-1. Depending on the viscosity ratio, PP phase has different deformability, which is reflected on its morphology. The viscosity ratio of PP/PE-1 blend varies from 9.4 to 0.2 (based on complex viscosity measured in the range of 0.2 rad/s to 100 rad/s) due to shear thinning behavior, and the

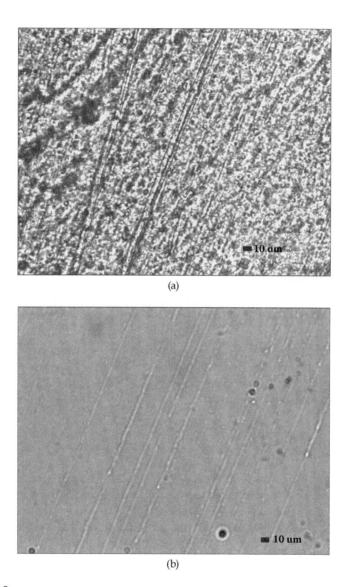

(a)

(b)

FIGURE 8.3
Optical microscope images of polypropylene (PP)/polyethylene (PE) blends; PP/PE-1 (a), PP/PE-2 (b), PP/PE-3 (c). The fibril structure is observed along the flow direction for PP/PE-1 and PP/PE-2 at a temperature between melting temperatures of PP and PE where PP remains in solid state and PE in melt. (Reproduced from Hong, J. S., K. H. Ahn, and S. J. Lee. 2005a. Strain hardening behavior of polymer blends with fibril morphology. *Rheologica Acta* 45:202–208, with permission.)

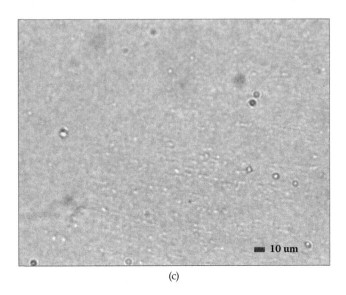

(c)

FIGURE 8.3 (continued)

morphology of PP can be controlled as long as the flow field in an extruder is defined under the given operation condition. The flow field in an extruder can be modeled depending on extruder configuration (Rauwendaal, 1986; Dealy and Wissbrun, 1989). The apparent shear rate in an extruder is a function of extrusion conditions such as screw rotation speed, barrel size, and channel depth. Then, the fibrillation can be induced by controlling the mixing condition. It is also important to understand the procedure of morphology evolution.

8.2.2 In Situ Fibrillation in an Extruder

In general, blending is performed either in an internal mixer or in an extruder. The flow field in an extruder is significantly dependent upon the processing condition and extruder configuration. The material experiences complex deformation of both shear and extension. During mixing in an extruder, a multiphase system experiences a significant reduction in the size and shape of the dispersed phase, from macroscale to microscale and from sheet to droplet. There have been several reports on morphology evolution during extrusion or mixing (Tadmor and Gogos, 1979; Min et al., 1984; Favis, 1990; Sundararaj et al., 1992; Scott and Macosko, 1995; Evstatiev et al., 1998). In extrusion, the morphology evolution of a polymer blend is generally accomplished in two steps. In the first step, significant reduction in the size of the dispersed phase is observed from a millimeter to a micrometer scale, and it takes place in a short axial length after solid conveying. In this stage, the morphology evolution is driven by the mechanical drag (Sundararaj et al., 1992; Covas et al., 2001). In the second step, morphology development to a smaller domain size is accompanied by a significant decrease in the drop size distribution via competition between coalescence and breakup. In a batch mixer, changing the mixing time from 2 to 20 min does not result in any significant reduction in domain size (Favis, 1990). It is expected that the changes in morphology have already occurred before observation, because the torque is already significantly reduced in just 2 min. Scott and Macosko (1991, 1995) studied morphology development in an intensive mixer. Sundararaj et al. (1992) studied morphology evolution of the polystyrene (PS)/polyamide (PA) and the PS/PP blends in an extruder and compared the result with that in an intensive mixer. Scott and Macosko (1991) reported that a major reduction in domain size occurs in an extruder in conjunction with melting or softening of the component, and the initial mechanism of morphology development involves the formation of a sheet or ribbon structure of the dispersed phase. However, the morphology like a sheet or ribbon was found to be unstable. The fibril morphology that appears as an intermediary during morphology evolution quickly changes into thermodynamically stable droplets (Champagne et al., 1996; Li et al., 1997; Monticciolo et al., 1998; Evstatiev et al., 1998, 2000; Covas et al., 2001; Cassagnau and Michel, 2001). These experiments were performed at a typical extrusion temperature, which is sufficiently higher than the melting temperature of the constituent polymer. Even though an intermediate structure may be maintained up to the die in the extruder, the dispersed phase quickly turns into a thermodynamically stable structure like a droplet during solidification. When a polymer blend comes out of a die, the flexible polymer relaxes and retracts against stress depending on the viscoelasticity of each component. And the blend keeps on developing its morphology due to heterogeneous stress distribution around the dispersed phase (Stone et al., 1986). As this process is fast and spontaneous, it is difficult to keep the intermediate morphology such as a sheet or fibril within the blend of the final product. However, if the deformability of the dispersed phase is controlled, the morphology observed at an initial stage of mixing can be maintained. Because the deformability depends on the viscosity ratio of the blend as well, it will change according to the extrusion temperature. Therefore, if a polymer blend consists of components having a large difference in melting temperature, it is possible to obtain a fibril or ellipsoidal structure depending on thermal and deformation history. In a PBT/PE blend, the morphology of PBT in PE can be changed from a sheet to droplet according to the extrusion condition (Hong et al., 2005b). If the extrusion temperature is set at 220°C, which is slightly lower than the melting temperature of PBT, PBT becomes deformable with restricted thermal energy, and the morphology evolution can be controlled by adjusting deformation history (Table 8.1). It is possible to keep a film or fibril structure in the final product if the dispersed phase is rigid enough to avoid relaxation related with instability originating from interfacial tension. This approach can generate fibril morphology in immiscible blends with multicomponents having a large difference in melting temperature such as PBT/PE, PBT/PS, PET/PE, PA/PE, and so on.

TABLE 8.1

Morphology Evolution of Polybutylene Terephthalate (PBT)/Polyethylene (PE) Blend (1 wt% PBT in PE Matrix)

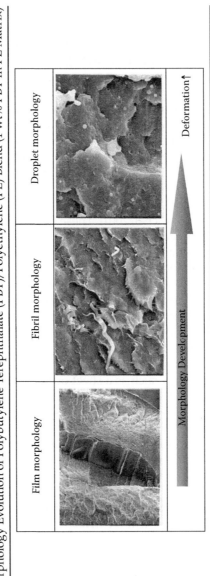

8.3 Rheological Properties of Blends with Fibril Morphology

8.3.1 Effect of Anisotropy on Rheological Properties

Anisotropic morphology has different effects on rheological properties compared to droplet morphology. When there is no interaction between the components in an immiscible multicomponent system, the rheological properties of a blend can be determined by a linear combination of each contribution:

$$\text{Viscosity, } \eta^*_{blend} = \sum \phi_i \eta_i^* \tag{8.1a}$$

$$\text{Modulus, } G^*_{blend} = \sum \phi_i G_i^* + G^*_{interface} \tag{8.1b}$$

where η^*_{blend}, G^*_{blend}, η_i^*, and G_i^* are the complex viscosity and complex modulus of the blend and each component, respectively; ϕ is the volume fraction of the droplet; and $G^*_{interface}$ is the contribution of interfacial tension to modulus. Even though the volume fraction of the dispersed phase is the same, there is a difference in interfacial contribution between droplet and fibril morphology, which leads to a different rheological effect. The effect of interface is reflected on extensional deformation more sensitively than shear deformation. However, when the volume fraction of the dispersed phase is low ($\phi < 0.1$), the effect of morphology on rheological properties is not clearly distinguishable between droplet and fibril (an example is shown in Section 8.3.3). When the dispersed phase has a higher melting temperature than the matrix, the polymer blend may be considered as a particle-suspension system in between two melting temperatures. The viscosity is increased in the presence of solid spherical particles due to an increase in hydrodynamic volume (i.e., $\eta_s = \eta_m(1 + 2.5\phi)$ for dilute suspension; η_m is the viscosity of the matrix, and η_s is the viscosity of the suspension). The viscosity of the blend would increase as the volume fraction of the dispersed phase increases. The effect of anisotropy is more complicated. When semidilute cylindrical fibers of appreciable size are suspended in a Newtonian medium ($\phi > (L/d)^{-2}$), the contribution of matrix (σ_m) and fiber (σ_d) to the stress of the suspension (σ_s) can be expressed by Equation (8.2) (Batchelor, 1970, 1971; Leal and Hinch, 1971; Larson, 1999):

$$\sigma_s = \sigma_m + \sigma_d = 2\eta_m D + \nu \xi_{str} \langle \mathbf{uuuu} \rangle : D \tag{8.2}$$

Here, ν is the number concentration of fiber (volume fraction $\phi = \nu \pi d^2 L/4$) and ξ_{str} is the viscous drag coefficient. The stress contribution by fiber (σ_d) is determined by ξ_{str} and tensor parameter ($\langle \mathbf{uuuu} \rangle$). ξ_{str} is dependent upon the anisotropy of the particle. For a fiber with a high aspect ratio (L/d, L is the length of the fiber and d is the diameter), Batchelor (1971) estimated $\xi_{str} \sim \pi\eta mL^3/3\ln(\pi/\phi)$ by introducing the concept of hydrodynamic screening when the average distance between the neighboring particles is between d and L. Under the simple shear flow, if hydrodynamic interaction between the fibers is neglected, the fiber quickly rotates until it aligns to the flow direction. The contribution of rotation to shear viscosity is proportional to the time average of ensemble, $\langle u_x^2 u_y^2 \rangle$, of fiber orientation to the flow direction (u_x) and to the gradient direction (u_y) (Larson, 1991). The stress tensor of Equation (8.2) gives the shear viscosity as follows:

$$\frac{\eta - \eta_s}{\eta_s} = v\xi_{str}\left\langle u_x^2 u_y^2 \right\rangle \tag{8.3}$$

where $\langle u_x^2 u_y^2 \rangle$ for non-Brownian fiber depends on L/d and reduced concentration (nL^3). For dilute concentration ($nL^3 < 6$), $\langle u_x^2 u_y^2 \rangle$ depends only on the aspect ratio and decreases as L/d increases. For example, $\langle u_x^2 u_y^2 \rangle$ is approximately 0.01 for the long fiber of L/d of 31.9, and is 0.03 for the short fiber of L/d of 10 (Leal and Hinch, 1971). For the uniaxial extensional flow of constant strain rate ($\dot{\varepsilon}$), the long fiber is perfectly aligned to the stretching direction when steady state is reached (<**uuuu**> $\cong$ 1). Then, Equation (8.2) can be simplified as follows:

$$\eta_e^+ = 3\eta_m\left[1 + \frac{4\phi(L/d)^2}{9\ln(\pi/\phi)}\right] \tag{8.4}$$

Though the contribution of fiber to viscosity in Equation (8.3) is weak under shear flow, the fiber contribution to extensional viscosity is significant.

8.3.2 Extensional Viscosity

When the dispersed phase has droplet morphology, there is no significant increase in viscosity and no indication of strain hardening behavior. As described in the previous section, the viscosity of a blend is proportionally contributed by each component. Under extensional flow, the characteristics of anisotropic morphology are more clearly differentiated compared to droplet morphology. The energy dissipation is maximized by highly anisotropic structure aligned along the extensional flow direction. In the case of fibril morphology, the long fibrils induce not only significant enhancement of the magnitude of the extensional viscosity but also strong strain hardening behavior even at a small volume fraction. Figure 8.4 shows the transient extensional viscosity of the three 1 wt% PP/PE blends introduced earlier under uniaxial extensional flow. The morphologies of the three PP/PE blends vary depending on the deformability of PP. The viscosity ratio is 3.9 for PP/PE-1,

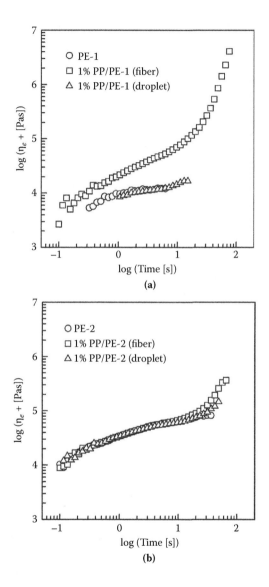

FIGURE 8.4

Transient extensional viscosity (η_e+) of polypropylene (PP)/polyethylene (PE) blends at extension rate of 0.05 [1/sec]. Three PP/PE blends have different viscosity ratio; (a) 3.9 (PP/PE-1), (b) 1 (PP/PE-2), and (c) 0.5 (PP/PE-3). (Reproduced from Hong, J. S., K. H. Ahn, and S. J. Lee. 2005a. Strain hardening behavior of polymer blends with fibril morphology. *Rheologica Acta* 45:202–208, with permission.)

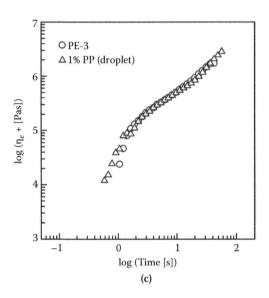

FIGURE 8.4 (continued)

1 for PP/PE-2, and 0.5 for PP/PE-3. The morphologies of the three blends are clearly differentiated in Figure 8.2. They show different extensional behavior according to the different morphologies. The PP/PE-1 blend shows not only significant enhancement in the magnitude of extensional viscosity but also strong strain hardening behavior due to the presence of long fibril structure. The PP/PE-2 blend shows a slight increase in viscosity due to a small aspect ratio, but the small fibrils also induce weak strain hardening behavior as more extension is applied. For PP/PE-3, 1 wt% PP with droplet morphology does not show an effective increase in extensional viscosity. As another example, Figure 8.5 shows the transient extensional viscosity of PE with both spherical and fibrillar PBT (5 wt%). PE with fibrillar structure shows a significant increase in extensional viscosity. The fibrillar structure of the blend is shown in Table 8.2. Because PBT threads entangle and form physical networks, they induce significant resistance against extension, leading to a significant enhancement of extensional viscosity. The effect of the filled long fiber on extensional viscosity can be determined from Equation (8.4). In Equation (8.4), the aspect ratio (L/d) was assumed to be constant (Mewis and Metzner, 1974). If the filled fiber is deformable like a thermoplastic polymer above the glass transition temperature, it will be stretched under extensional flow. The deformability of the fiber should be reflected on L/D. Local deformation of the fiber is mainly governed by global deformation, but there is a difference in deformation between the filled fiber and matrix because the viscosity ratio is very large ($\lambda \gg 1$) (Mighri et al., 1997). Just below the melting temperature of the polymer, it is rigid with low deformability. Under uniaxial extensional flow, the fiber is stretched depending on the rigidity ζ

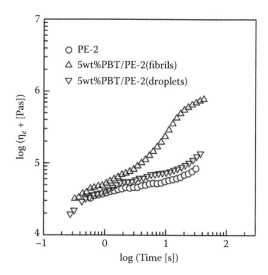

FIGURE 8.5
Transient extensional viscosity (η_e+) of PE-2, PE-2 embedded with polybutylene terephthalate (PBT) drops, and PE-2 embedded with PBT fibrils at extension rate of 0.05 [1/sec] and 150°C.

of the fiber ($0 < \zeta < 1$). If the fiber is very rigid, ζ is 1 (no deformation) while ζ approaches 0 as the fiber becomes more deformable (affine deformation). Between these two extremes, the strain rate of the fiber ($\dot{\varepsilon}_f$) can be described as $\dot{\varepsilon}_f = \dot{\varepsilon}(1 - \zeta)$. As the fiber is stretched, the length and diameter will change according to $L = L_o \exp(\dot{\varepsilon}_f t)$ and $d = d_o \exp(-1/2\dot{\varepsilon}_f t)$. Then, the aspect ratio (L/d) in Equation (8.4) can be described as follows:

$$(L/d)^2 = (L_o/d_o)^2 \exp(3\dot{\varepsilon}_f t) \tag{8.5}$$

where L_o and d_o are the initial length and initial diameter of the fiber, respectively. Rigid fiber ($\zeta = 1$) does not induce strain hardening because there is no deformation of fiber over stretching and it increases the transient extensional viscosity only (Mewis and Metzner, 1974). It seems that the deformable fiber acts like a macrochain that resists stretching similar to the long-chain branch polymer (Inkson et al., 1999). The restricted stretching of deformable fibers induces strong strain hardening behavior. For example, filled PP fibers in PP/PE blend induce significant strain hardening behavior at 150°C as seen in Figure 8.4. PP is deformable enough to be stretched when extension is performed at 150°C because the melting temperature of PP is 153°C. The rigidity of PP is 0.45 at 150°C, which is predicted from fitting the transient extensional viscosity with Equation (8.4). Figure 8.6 proves that the dispersed phase with fibril morphology can induce a significant strain hardening effect even with a very small amount (1 wt%) of the dispersed phase as long as it is deformable enough to be stretched. The initial aspect ratio of the fiber, L_o/d_o, was

TABLE 8.2

Fibrillation of PBT

1 wt% PBT in Polyethylene (PE)	5 wt% PBT in Polystyrene (PS)

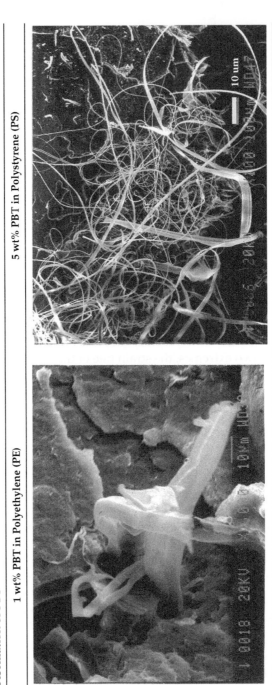

TABLE 8.3

Rigidity ζ of Fibers

Materials	Temperature	ζ
Polypropylene (PP)	150°C	0.45
Polybutylene terephthalate (PBT)	170°C	0.55
Glass	<300°C	1

Note: The rigidity of microfibril (ζ) indicates how much fiber is rigid against deformation. It decreases from one to zero as fiber is more deformable, and it was obtained by fitting η_{e^+} with Equation (8.4) and Equation (8.5).

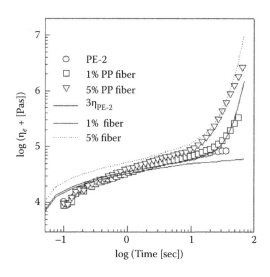

FIGURE 8.6

Transient extensional viscosity (η_e+) of polypropylene (PP)/polyethylene (PE)-2 blends with fibril morphology. Experimental data (o for PE-2, $\triangledown$ for 1 wt% PP/PE-2, and □ for 5 wt% PP/PE-2) are compared with the predictions of Equations (8.4) and (8.5) (— for 1 wt% PP/PE-2 and ---- for 5 wt% PP/PE-2). The extension rate is fixed at $\dot{\varepsilon}$ = 0.05 [1/sec], and the measurement temperature is 150°C. (Reproduced from Hong, J. S., K. H. Ahn, and S. J. Lee. 2005a. Strain hardening behavior of polymer blends with fibril morphology. *Rheologica Acta* 45:202–208, with permission.)

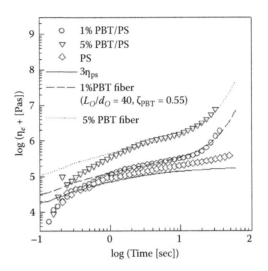

FIGURE 8.7

Transient extensional viscosity (η_{e+}) of PBT/polystyrene (PS) blends with fibril morphology. Experimental data ($\diamond$ for PS, $\circ$ for 1 wt% PBT/PS, and $\triangledown$ for 5 wt% PBT/PS) are compared with the predictions of Equations (8.4) and (8.5) (– – for 1 wt% PBT/PS and ---- for 5 wt% PBT/PS). The experimental condition is $\dot{\varepsilon} = 0.05$ s^{-1} and 170°C. (Reproduced from Hong, J. S., K. H. Ahn, and S. J. Lee. 2005a. Strain hardening behavior of polymer blends with fibril morphology. *Rheologica Acta* 45:202–208, with permission.)

obtained from Equation (8.4) with the transient extensional viscosity at an initial extension time of 1 sec. The I_o/d_o for PP/PE-2 blend is approximately 11. The strain hardening behavior of PP/PE-2 blend can be explained when the rigidity of PP at 150°C is considered. As the composition is increased to 5 wt%, the fiber induces strain hardening earlier, and hardening becomes more pronounced. Figure 8.7 shows the transient extensional viscosity of the PBT/PS blends with fibril morphology. The prediction with $\zeta_{PBT} = 0.55$ at 170°C is also provided for comparison. It proves that the addition of stretchable PBT fiber induces strain hardening behavior. The transient extensional viscosity is significantly increased as the composition of the anisotropic fiber is increased.

8.3.3 Shear Viscosity

The morphology may affect the rheological properties under shear and extension in different manners. If the dispersed phase is rigid but deformable, it more effectively contributes to the rheological properties of the blend. In Section 8.3.2, the transient extensional viscosity was measured at a lower temperature than the melting temperature of the dispersed phase. Rigid fibrils enhance extensional viscosity even with a small amount of the dispersed phase (1 wt%). Nevertheless, the morphological effect under shear flow is not

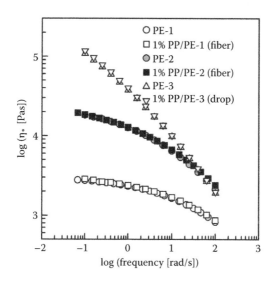

FIGURE 8.8
Complex viscosity (η^*) of polyethylene (PE) and 1 wt% polypropylene (PP)/PE blends at 150°C. Three blends have different morphology of long fiber (PP/PE-1, $L_o/d_o = 60$), short fiber (PP/ PE-2, $L_o/d_o = 11$), and droplet (PP/PE-3), but the increase in viscosity by the addition of 1 wt% PP is very slight (symbol □ for 1 wt% PP/PE-1, ■ for 1 wt% PP/PE-2, and ▽ for 1 wt% PP/PE-3). (Reproduced from Hong, J. S., K. H. Ahn, and S. J. Lee. 2005a. Strain hardening behavior of polymer blends with fibril morphology. *Rheologica Acta* 45:202–208, with permission.)

as strong as under extensional flow. Under shear flow, a small amount of the dispersed phase with either fibril or droplet structure has virtually no effect on viscosity. A suspension with 1 wt% noninteracting drops ($\phi_{volume} = 0.0087$) shows an increase in viscosity of only 2.2% compared to that of the matrix viscosity (η_m). For a noninteracting anisotropic particle, the stress contributed by the dispersed phase (σ_d) depends on the particle aspect ratio (L/d) and the degree of orientation. However, due to a composition as small as 0.87 vol%, it does not contribute to the viscosity significantly even though the dispersed phase has a high aspect ratio. Figure 8.8 shows the complex viscosity of the PP/PE blends with both fibril and droplet structures at 150°C. It is worthwhile to note that PP/PE-1 and PP/PE-2 have a fibril structure while PP/PE-3 has a droplet structure. Compared to the significant morphological effect on extensional viscosity in Figure 8.4, it is difficult to discriminate between droplet and fibril morphology with 1 wt% PP under shear flow. When the composition of PP increases to 5 wt%, the complex viscosity of the PP/PE-2 blend increases only 14% compared to that of PE-2 (Figure 8.9a). Even with 5 wt% PP, the effect of morphology on viscosity is not significant under the shear flow of small deformation. Figure 8.9 shows the dependence of composition on complex viscosity at 150°C. The 1 wt% and 5 wt% PBT/PE-2 with both fibril and droplet structures are compared. There is no increase in shear

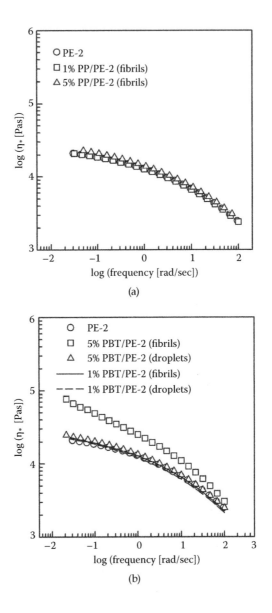

FIGURE 8.9
Complex viscosity (η^*) of polypropylene (PP)/polyethylene (PE)-2 and PBT/PE-2. For PP/PE-2 (a), the effect of fibril concentration is compared. For PBT/PE-2 (b), the effect of morphology is compared. The dependence of concentration is also compared. (Reproduced from Hong, J. S., J. L. Kim, K. H. Ahn, and S. J. Lee. 2005b. Morphology development of PBT/PE blends during extrusion and its reflection on the rheological properties. *Journal of Applied Polymer Science* 97:1702–1709, with permission.)

viscosity with 1 wt% PBT regardless of the structure. However, as the composition is increased to 5 wt%, a difference between the fibril and droplet structure becomes exposed in the rheological properties. Different from 5 wt% PP/PE-2 blend, the PBT fiber increases the viscosity of the blend more than PP fiber due to larger L/d of PBT fiber (L_o/d_o for PBT = 40 and L_o/d_o for PP = 11; L_o/d_o is obtained from Equation (8.4)). Although the blend morphology is reflected on the shear viscosity as the aspect ratio of fibril increases or its concentration increases, the morphological influence on shear viscosity is far less than that of extensional viscosity. According to Equation (8.3), the shear viscosity slightly increases with the aspect ratio of the fiber because the tensor parameter, $<u_x^2u_y^2>$, is less than one for the fiber under shear flow. The fibril structure enhances viscosity over the entire frequency range because PBT maintains a high aspect ratio. In the case of droplet structure, however, it is hard to expect any significant increase in viscosity. In 1 wt% and 5 wt% PBT/PE blends with droplet structure ($\phi_{v\ 1wt\%PBT}$ = 0.005 and $\phi_{v\ 5wt\%PBT}$ = 0.026), the viscosity of the blend (η_s) can be increased by 1.25% and 6.5%, respectively, over the matrix viscosity (η_m). However, as the anisotropy of the dispersed phase increases, they can be oriented by a flow field and become more dissipative than the dispersion of droplets. Then the fibril structure significantly enhances viscosity compared to droplet structure.

8.4 Applications

Because blending is an economic and promising way to develop new materials, understanding the structure formation of the blend (fibril or droplet morphology) is important for better processing and performance of the products, especially in processing with large and rapid deformation (Micic et al., 1998; La Mantia et al., 2005; Münstedt et al., 2005, 2006). For example, thermoforming and blow molding are widely used as primary industrial processes for forming large, thin-walled plastic parts. During processing, polymers undergo large deformation, high strain rate, contact with cold mold, and sometimes instability during inflation. The problems frequently encountered in this processing are mainly related with large variations of wall thickness and instability during inflation—rupture of the sheet or parison, and shrinkage and warpage exhibited in the final product (Covas et al., 2001). All of these problems cause detrimental effects on both performance and manufacturing cost. Nonuniform wall thickness variation and parison rupture are thought to arise in part due to mold geometry and process conditions (Isayev, 1991). Other factors affecting wall thickness variation include initial dimension of the sheet or parison, temperature distribution, extensional viscosity of the material, inflation dynamics, and cooling and solidification that occur in the mold cavity. In order to address these problems

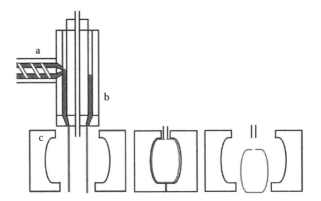

FIGURE 8.10
Extrusion blow molding process: (a) single-screw extruder, (b) die with a gas supplier, and (c) mold. The blow molding condition of high-density polyethylene (HDPE) was set as typically used in industry. The maximum extrusion temperature of a single-screw extruder was set at 220°C for HDPE and 250°C for HDPE/polyethylene terephthalate (PET) blend.

effectively, it is necessary to better understand the extensional behavior of the material and deformation characteristics of forming. If the material shows strain hardening behavior under fast extension, for example, there are many advantages both in operability and in product performance—uniform thickness distribution and less waste. Now the effect of strain hardening behavior on the blow molding process is to be discussed as an example.

Blow molding is a process that generates hollow plastic parts such as bottles and containers. It was initially used in the packaging industry but recently expanded its application to automotive industry to produce large parts such as fuel tank, bumper, dashboard, and seatbacks (Baird and Collias, 1995). Provided that strain hardening can be induced by an economical blending method, it is cost-effective and it can improve the quality of the products and performance. The extrusion blow molding process, as schematically illustrated in Figure 8.10, involves a single-screw extruder (a) fitted with a die (b) and mold (c). The procedure from parison to bottle formation is sequentially shown in Figure 8.10. When the parison becomes soft, it is pinched off between the two halves of a mold, and the soft parison is blown against the cooled mold surface by air that is injected through a blow pin. After the polymer is solidified, the mold is opened and the product is pulled out. The typical circumferential blow-up ratio is in the range of 3 to 4. Under large and rapid deformation, the parison requires strain hardening during inflation for easy process control and good performance. In the previous section, it was noted that the anisotropic secondary phase can induce strain hardening provided that it has restricted rigidity. For example, when polyethylene terephthalate (PET)/high-density polyethylene (HDPE) blend and HDPE were subjected to an extrusion blow molding process (500 ml detergent bottle), the PET/HDPE blend more easily finds the blow molding condition than HDPE (melt index = 1 g/10 min, ASTM1238). The PET/HDPE blend was blown at 250°C, which is close to the melting temperature of PET (257°C). When the PET/HDPE blend with fibril morphology passes through the extruder and die, its fibril morphology is maintained. This anisotropic morphology significantly improves the blowing performance compared to HDPE. Figure 8.11 compares the bottles produced with HDPE (Figure 8.11a) and the HDPE/PET blend (Figure 8.11b). In the case of HDPE, the thickness at the curved corner is thin and ruptures are easily produced, whereas the thickness of the bottle made of the PET/HDPE blend is thicker than HDPE. The PET/HDPE blend shows good performance and produces a good shaped bottle as seen in Figure 8.11b. When the morphology of the bottle made with the HDPE/PET blend is observed in Figure 8.12, the ellipsoidal microstructure (aspect ratio >2) is homogeneously dispersed without alignment. The anisotropic structure of PET is expected to induce strain hardening. It increases the resistance against fast inflation and maintains uniform thickness, which results in good performance and uniform thickness of the bottle.

8.5 Summary

Strain hardening in an immiscible polymer blend plays an important role in polymer processing, in which large and rapid deformation is dominant. Strain hardening is induced by a nonlinear molecular structure such as chain branching. In polymer blends, strain hardening can be induced by controlling the morphology such that the fibrillation of the dispersed phase is maintained. If an anisotropic fiber restrictedly deforms under extensional flow, it induces strong strain hardening even though the content of the dispersed phase is very low. The fibrillar morphology can be obtained by controlling the deformability of the dispersed phase when $\lambda > 1$ and $Ca/Ca_{cr} > 4$. The outstanding performance of the blend with fibril morphology can be exploited to design new materials and to improve the process.

Acknowledgment

This work was supported by the National Research Foundation of Korea (NRF) grant funded by the Korean government (MEST) (No. 20100027746). J. S. Hong acknowledges the support from Basic Science Researcher Program through NRF funded by the MEST, Korea (No. 20100005545).

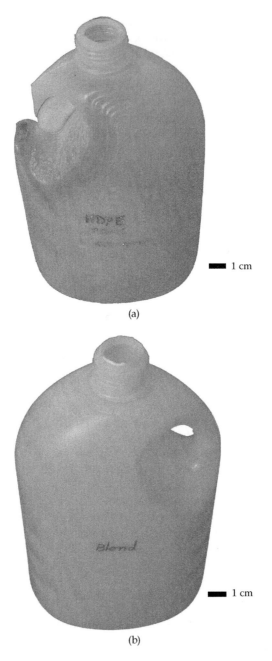

(a)

(b)

FIGURE 8.11
Blown bottles (500 ml detergent bottle) by extrusion blow molding: (a) high-density polyethylene (HDPE) and (b) HDPE/polyethylene terephthalate (PET) blend.

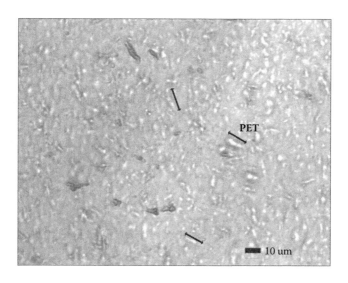

FIGURE 8.12
Morphology of bottle (high-density polyethylene [HDPE]/polyethylene terephthalate (PET) blend) taken through an optical microscope at magnification of 200×.

References

Acrivos, A., and T. S. Lo. 1978. Deformation and breakup of a single slender drop in an extensional flow. *Journal of Fluid Mechanics* 86:641–672.

Ajji, A., P. Sammut, and M. A. Huneault. 2003. Elongational rheology of LLDPE/LDPE blends. *Journal of Applied Polymer Science* 88:3070–3077.

Baird, D. G., and D. I. Collias. 1995. *Polymer Proceesing: Principles and Design.* New York: Wiley.

Batchelor, G. K. 1970. Slender-body theory for particles of arbitrary cross-section in Stokes flow. *Journal of Fluid Mechanics* 44:419–440.

Batchelor, G. K. 1971. The stress generated in a non-dilute suspension of elongated particles by pure straining motion. *Journal of Fluid Mechanics* 46:813–829.

Blizard, K. G., and D. G. Baird. 1987. The morphology and rheology of polymer blends containing a liquid crystalline copolyester. *Polymer Engineering and Science* 27:653–662.

Cassagnau, P., and A. Michel. 2001. New morphologies in immiscible polymer blends generated by a dynamic quenching process. *Polymer* 42:3139–3152.

Chaffin, K. A., J. S. Knutsen, P. Brant, and F. S. Bates. 2000. High-strength welds in metallocene polypropylene/polyethylene laminates. *Science* 288:2187–2190.

Champagne, M. F., M. M. Dumoulin, L. A. Utracki, and J. P. Szabo. 1996. Generation of fibrillar morphology in blends of block copolyetheresteramide and liquid crystal polyester. *Polymer Engineering and Science* 36:1636–1646.

Cheng, S., E. Phillips, and L. Parks. 2010. Processability improvement of polyolefins through radiation-induced branching. *Radiation Physics and Chemistry* 79:329–334.

Chin, H. B., and C. D. Han. 1979. Studies on droplet deformation and breakup. I. Droplet deformation in extensional flow. *Journal of Rheology* 23:557–590.

Covas, J. A., O. S. Carneiro, and J. M. Maia. 2001. Monitoring the evolution of morphology of polymer blends upon manufacturing of microfibrillar reinforced composites. *International Journal of Polymeric Materials* 50:445–467.

Dealy, J. M., and K. F. Wissbrun. 1989. *Melt Rheology and Its Role in Plastics Processing: Theory and Applications*. New York: Van Nostrand Reinhold.

Evstatiev, M., S. Fakirov, G. Bechtold, and K. Friedrich. 2000. Structure-property relationships of injection- and compression-molded microfibrillar-reinforced PET/PA-6 composites. *Advances in Polymer Technology* 19:249–259.

Evstatiev, M., J. M. Schultz, S. Petrovich, G. Georgiev, S. Fakirov, and K. Friedrich. 1998. In situ polymer/polymer composites from poly(ethylene terephthalate), polyamide-6, and polyamide-66 blends. *Journal of Applied Polymer Science* 67:723–737.

Favis, B. D. 1990. The effect of processing parameters on the morphology of an immiscible binary blend. *Journal of Applied Polymer Science* 39:285–300.

Filipe, S., M. Cidade, and J. Maia. 2006. Uniaxial extensional flow behavior of immiscible and compatibilized polypropylene/liquid crystalline polymer blends. *Rheologica Acta* 45:281–289.

Gabriel, C., and H. Münstedt. 2003. Strain hardening of various polyolefins in uniaxial elongational flow. *Journal of Rheology* 47:619–630.

Grace, H. P. 1982. Dispersion phenomena in high viscosity immiscible fluid systems and application of static mixers as dispersion devices in such systems. *Chemical Engineering Communications* 14:225–277.

Han, C. D. 1981. *Multiphase Flow in Polymer Processing*. New York: Academic Press.

Hingmann, R., and B. L. Marczinke. 1994. Shear and elongational flow properties of polypropylene melts[sup a]. *Journal of Rheology* 38:573–587.

Hong, J. S., K. H. Ahn, and S. J. Lee. 2005a. Strain hardening behavior of polymer blends with fibril morphology. *Rheologica Acta* 45:202–208.

Hong, J. S., J. L. Kim, K. H. Ahn, and S. J. Lee. 2005b. Morphology development of PBT/PE blends during extrusion and its reflection on the rheological properties. *Journal of Applied Polymer Science* 97:1702–1709.

Inkson, N. J., T. C. B. McLeish, O. G. Harlen, and D. J. Groves. 1999. Predicting low density polyethylene melt rheology in elongational and shear flows with "pom-pom" constitutive equations. *Journal of Rheology* 43:873–896.

Isayev, A. I. 1991. *Modeling of Polymer Processing: Recent Developments*. New York: Hanser.

Ishizuka, O., and K. Koyama. 1980. Elongational viscosity at a constant elongational strain rate of polypropylene melt. *Polymer* 21:164–170.

Joshi, S. C., Y. C. Lam, C. Y. Yue, K. C. Tam, L. Li, and X. Hu. 2003. Energy-based predictive criterion for LCP fibrillation in LCP/thermoplastic polymer blends under shear. *Journal of Applied Polymer Science* 90:3314–3324.

Khan, S. A., R. K. Prud'homme, and R. G. Larson. 1987. Comparison of the rheology of polymer melts in shear, and biaxial and uniaxial extensions. *Rheologica Acta* 26:144–151.

Kurzbeck, S., F. Oster, H. Munstedt, T. Q. Nguyen, and R. Gensler. 1999. Rheological properties of two polypropylenes with different molecular structure. *Journal of Rheology* 43:359–374.

La Mantia, F. P., R. Scaffaro, G. Carianni, and P. Mariani. 2005. Rheological properties of different film blowing polyethylene samples under shear and elongational flow. *Macromolecular Materials and Engineering* 290:159–164.

Larson, R. G. 1999. *The Structure and Rheology of Complex Fluids*. New York: Oxford University Press.

Leal, L. G., and E. J. Hinch. 1971. The effect of weak Brownian rotations on particles in shear flow. *Journal of Fluid Mechanics* 46:685–703.

Leonardi, F., C. Derail, and G. Marin. 2005. Some applications of molecular rheology: Polymer formulation and molecular design. *Journal of Non-Newtonian Fluid Mechanics* 128:50–61.

Li, X., M. Chen, Y. Huang, G. Lin, S. Zhao, B. Liao, C. Wang, and G. Cong. 1997. In-situ composite based on polypropylene and nylon 6. *Advances in Polymer Technology* 16:331–336.

Lim, H. T., H. Liu, K. H. Ahn, S. J. Lee, and J. S. Hong. 2010. Effect of added ionomer on morphology and properties of PP/organoclay nanocomposites. *Korean Journal of Chemical Engineering* 27:705–715.

Linster, J. J., and J. Meissner. 1986. Melt elongation and structure of linear polyethylene (HDPE). *Polymer Bulletin* 16:187–194.

McCallum, T. J., M. Kontopoulou, C. B. Park, E. B. Muliawan, and S. G. Hatzikiriakos. 2007. The rheological and physical properties of linear and branched polypropylene blends. *Polymer Engineering and Science* 47:1133–1140.

Mewis, J., and A. B. Metzner. 1974. The rheological properties of suspensions of fibres in Newtonian fluids subjected to extensional deformations. *Journal of Fluid Mechanics* 62:593–600.

Micic, P., S. N. Bhattacharya, and G. Field. 1998. Transient elongational viscosity of LLDPE/LDPE blends and its relevance to bubble stability in the film blowing process. *Polymer Engineering and Science* 38:1685–1693.

Mighri, F., A. Ajji, and P. J. Carreau. 1997. Influence of elastic properties on drop deformation in elongational flow. *Journal of Rheology* 41:1183–1201.

Min, K., J. L. White, and J. F. Fellers. 1984. Development of phase morphology in incompatible polymer blends during mixing and its variation in extrusion. *Polymer Engineering and Science* 24:1327–1336.

Minegishi, A., A. Nishioka, T. Takahashi, Y. Masubuchi, J.-i. Takimoto, and K. Koyama. 2001. Uniaxial elongational viscosity of PS/a small amount of UHMW-PS blends. *Rheologica Acta* 40:329–338.

Monticciolo, A., P. Cassagnau, and A. Michel. 1998. Fibrillar morphology development of PE/PBT blends: Rheology and solvent permeability. *Polymer Engineering and Science* 38:1882–1889.

Münstedt, H. 1980. Dependence of the elongational behavior of polystyrene melts on molecular weight and molecular weight distribution. *Journal of Rheology* 24:847–867.

Münstedt, H., S. Kurzbeck, and J. Stange. 2006. Importance of elongational properties of polymer melts for film blowing and thermoforming. *Polymer Engineering and Science* 46:1190–1195.

Münstedt, H., and H. M. Laun. 1981. Elongational properties and molecular structure of polyethylene melts. *Rheologica Acta* 20:211–221.

Münstedt, H., T. Steffl, and A. Malmberg. 2005. Correlation between rheological behaviour in uniaxial elongation and film blowing properties of various polyethylenes. *Rheologica Acta* 45:14–22.

Petrie, C. J. S. 1979. *Elongational flows: Aspects of the behavior of model elasticoviscous fluids*. London: Pitman.

Rauwendaal, C. 1986. *Polymer Extrusion*. New York: Hanser.

Rumscheidt, F. D., and S. G. Mason. 1961. Particle motions in sheared suspensions XII. Deformation and burst of fluid drops in shear and hyperbolic flow. *Journal of Colloid Science* 16:238–261.

Scott, C. E., and C. W. Macosko. 1991. Model experiments concerning morphology development during the initial stages of polymer blending. *Polymer Bulletin* 26:341–348.

Scott, C. E., and C. W. Macosko. 1995. Morphology development during the initial stages of polymer-polymer blending. *Polymer* 36:461–470.

Sperling, L. H. 1997. *Polymeric Multicomponent Materials: An Introduction*. New York: Wiley.

Stadler, F., A. Nishioka, J. Stange, K. Koyama, and H. Münstedt. 2007. Comparison of the elongational behavior of various polyolefins in uniaxial and equibiaxial flows. *Rheologica Acta* 46:1003–1012.

Stone, H. A., B. J. Bentley, and L. G. Leal. 1986. An experimental study of transient effects in the breakup of viscous drops. *Journal of Fluid Mechanics* 173:131–158.

Sundararaj, U., C. W. Macosko, R. J. Rolando, and H. T. Chan. 1992. Morphology development in polymer blends. *Polymer Engineering and Science* 32:1814–1823.

Tadmor, Z., and C. G. Gogos. 1979. *Principles of Polymer Processing*. New York: Wiley-Interscience.

Takahashi, T., J.-I. Takimoto, and K. Koyama. 1999. Elongational viscosity for miscible and immiscible polymer blends. II. Blends with a small amount of UHMW polymer. *Journal of Applied Polymer Science* 72:961–969.

Taylor, G. I. 1932. The viscosity of a fluid containing small drops of another fluid. *Proceedings of the Royal Society of London Series A* 138:41–48.

Taylor, G. I. 1934. The formation of emulsions in definable fields of flow. *Proceedings of the Royal Society of London Series A* 146:501–523.

Utracki, L. A. 1990. *Polymer Alloys and Blends: Thermodynamics and Rheology*. New York: Hanser.

Valenza, A., F. P. La Mantia, and D. Acierno. 1986. The rheological behavior of HDPE/LDPE blends. V. Isothermal elongation at constant stretching rate. *Journal of Rheology* 30:1085–1092.

Wagner, M. H., H. Bastian, P. Hachmann, J. Meissner, S. Kurzbeck, H. Münstedt, and F. Langouche. 2000. The strain-hardening behaviour of linear and long-chain-branched polyolefin melts in extensional flows. *Rheologica Acta* 39:97–109.

Wool, R. P. 1995. *Polymer Interface: Structure and Strength*. New York: Hanser.

Xian-Wu, C., Z. Zi-Cong, X. Ying, and Q. Jin-Ping. 2010. The effect of polypropylene/polyamide 66 blending modification on melt strength and rheologic behaviors of polypropylene. *Polymer Bulletin* 64:197–207.

Yamaguchi, M. 2001. Rheological properties of linear and crosslinked polymer blends: Relation between crosslink density and enhancement of elongational viscosity. *Journal of Polymer Science Part B: Polymer Physics* 39:228–235.

Yamaguchi, M., and K.-I. Suzuki. 2002. Enhanced strain hardening in elongational viscosity for HDPE/crosslinked HDPE blend. II. Processability of thermoforming. *Journal of Applied Polymer Science* 86:79–83.

Yamamoto, S., and T. Matsuoka. 1995. Dynamic simulation of fiber suspensions in shear flow. *The Journal of Chemical Physics* 102:2254–2260.

Yan, D., W. J. Wang, and S. Zhu. 1999. Effect of long chain branching on rheological properties of metallocene polyethylene. *Polymer* 40:1737–1744.

9

Modification of Polymer Blends by E-Beam and γ-Irradiation

Rodolphe Sonnier and Aurélie Taguet

Centre de Recherche CMGD
Ecole des Mines d'Alès, France

Sophie Rouif

IONISOS SA
Dagneux, France

CONTENTS

9.1 Introduction

Radiation processing of polymers can be operated with several types of ionizing radiations: e-beam (accelerated electrons generated by an accelerator), gamma rays (photons emitted by a cobalt 60 source), and X-rays (since recently). It acts by a spontaneous ionizing process, in two steps: formation of ions that decomposes into free radicals. There is no need to add an initiator like peroxides or photoinitiators. Then, the free radicals can induce some chemical reactions, depending on different parameters that will be considered hereafter.

Generally, the effects of ionizing radiation are a balance between ruptures and molecular rearrangements, such as cross-linking or cyclizations. There is no general rule of behavior, but trends can be identified by considering the polymer structure and conditions of radiation. For the first aspect, the presence of a quaternary carbon atom on a polymer chain generally promotes β-scission, phenyl groups promote stability, double bonds (vinyl) and ethylene promote cross-linking. For the second aspect, radiolysis reactions and scissions are generally favored by the presence of oxygen, less by temperature, and very little by the dose rate.

But the dose is the main factor of influence. The dose is the amount of energy received per kilogram of material, its unit is the kiloGray (1 kGy = 1 kJ/kg). Radiation processing of a polymer is generally operated at doses between 15 kGy and 200 kGy. Most polymers are stable or cross-link on this dose range, except those that can degrade. This is the case of polypropylene (PP), polyvinyl chloride (PVC), polyoxymethalene (POM), butyl rubber, and cellulose. Polytetrafluoroethylene (PTFE) degrades at a lower dose. Cross-linking of polyethylene (PE) begins at about 15 kGy and increases up to 200 kGy, according to an asymptotic curve.

In polymer blends, these phenomena are complicated by the possible interactions between both polymers. Interfacial cross-linking could take place and allows the compatibilization of the polymers. A more stable polymer like polystyrene (PS) could protect a less stable polymer and limit its degradation. The net effect of irradiation on a polymer blend could not be determined from the phenomena induced by irradiation on pure polymers.

This chapter is divided into three parts. The first part gathers the four first sections and discusses several specific characteristics of irradiation, with a particular emphasis on the specific phenomena concerning the polymer blends. The four following sections deal with different processes involving irradiation (polymerization, preirradiation, in situ compatibilization, and cross-linking). The last part exposes some applications of irradiation to design materials in various fields (recycling, shape memory materials, preparation of hydrogels, conductive polymer blends, synthesis of silicon carbide fibers, etc.). Finally, new trends in irradiation are evoked in relation with polymer blending.

9.2 Comparisons of Different Processes

Irradiation (e-beam or γ-irradiation) is classified among the high-energy solid-state processes. Some similarities between mechanical milling and irradiation to compatibilize nonmiscible blends have been observed by different authors.[1–3] Smith et al. considered that these processes lead to the same phenomena (chain scission, cross-linking, amorphization), and they claimed that the factors influencing these phenomena (polymer chemical structure and temperature) are not dependent on the process.[3]

Nevertheless, it is quite difficult to compare different processes while their proper parameters are different. E-beam and γ-irradiations could be easily compared because the main differences concern the penetration depth and the dose rate. Penetration depth is significantly higher for γ-irradiation. This point has a technical consequence. γ-Irradiation could be performed on very large samples (big-bags) while e-beam could be applied only on thin materials (few centimeters).

Dose rate is of few kGy per hour for γ-irradiation against several hundreds of kGy per minute for e-beam irradiation. Three consequences are expected from this difference. First, in e-beam irradiation, a significant increase in temperature could occur[4–6] that could induce some important changes (e.g., the increase in temperature could enhance the mobility of generated radicals and then the rate of their decays should increase). Second, a high dose rate leads to a high concentration of radicals and the probability of their recombination is higher. Gottlieb et al.[7] compared e-beam and γ-irradiation to prepare PVME-PVP hydrogels. Radiation doses between 40 and 120 kGy were applied. From Charlesby-Pinner plots, the authors showed that the gel fraction is higher for γ-irradiation (Figure 9.1). Gelation dose was also lower for γ-irradiation: 7.1 kGy versus 14.6 kGy for e-beam, in the case of

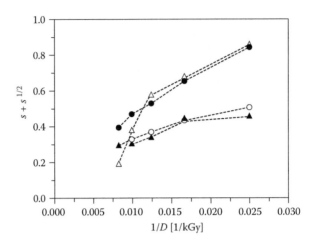

FIGURE 9.1

Charlesby-Pinner plot of pure poly(vinyl methyl ether) (PVME) (white triangle), pure poly(vinylpyrrolidinone) (PVP) (black triangle), and a 1:1 mixture (black circle: electron beam; white circle: γ-irradiation); *s* corresponds to the sol fraction and *D* to the radiation dose. (From Gottlieb, R., Schmidt, T., Arndt, K., Synthesis of temperature-sensitive hydrogel blends by high-energy irradiation. *Nuclear Instruments and Methods in Physics Research Section B: Beam Interactions with Materials and Atoms* 2005, 236, 371–376. With permission.)

a 50/50 PVME/PVP hydrogel. The authors explained that the dose rate is lower in γ-irradiation than the concentration of radicals and their recombination are limited, leading to a higher probability to react with macromolecules and to form cross-links. Finally, a low dose rate in γ-irradiation needs a high time of treatment. If irradiation is performed in air, oxidation could occur. Balance between oxidation and other phenomena (e.g., cross-linking) could be modified and could induce changes in mechanical properties.[8]

Ali et al.[9] have irradiated blends of virgin and scrapped polyethylenes using e-beam and γ-irradiations. Trimethylol propane triacrylate (TMPTA) was added as a cross-linking additive. The authors observed that maximum gel fraction is higher for e-beam than for γ-irradiation: 80% to 90% against 50% to 70% according to the formulations. Moreover the thermal stability of the blends is higher for e-beam irradiation. On the contrary, the gelation dose is lower for γ-irradiation: 5 to 7 kGy against 7 to 9 kGy for e-beam irradiation. These differences could be explained by the relative importance of various phenomena (cross-linking, oxidation) for both irradiation methods.

Raj et al.[10–13] have compared the efficiency of microwave and e-beam irradiations to stabilize the interface of various partially miscible or nonmiscible blends: polystyrene (PS)/polymethyl methacrylate (PMMA), polyvinyl chloride (PVC)/ethylene vinyl acetate (EVA), PP/acrylonitrile butadiene rubber (NBR), and polyvinyl chloride (PVC)/poly(styrene acrylonitrile) (SAN). For this purpose, they used positron annihilation lifetime measurements, and they considered particularly a hydrodynamic interaction parameter α. This parameter is close to 0 when no friction is generated at the interface. When interfacial modifications occur, α are negative values. The authors show that microwave irradiation is more efficient to stabilize the interfaces when the polymers contain some polar groups. Polar groups (e.g., carbonyl groups of PMMA) are able to absorb microwaves and thus become more reactive. On the contrary, for nonpolar or weakly polar polymers (like PVC/SAN), e-beam irradiation is more efficient. Nevertheless, the authors do not explain the choice of irradiation conditions: 100 kGy with an electron accelerator of energy 8 MeV and 60s with a 30 Watt microwave facility. Therefore, it is not sure that the comparison is correct.

Lee et al.[14] prepared high-density polyethylene/ethylene ethacrylate copolymer blends filled with carbon black. They cross-linked the blends using e-beam irradiation (dose 100 kGy) or using vinyltrimethoxysilane during blending. Reactive vinylsilane groups were grafted on the polymer backbone. Further hydrolysis of silane led to the formation of intermolecular link by condensation reactions. The authors studied the electrical properties of the materials and in particular the positive temperature coefficient (PTC). The PTC is the temperature for which a material shows a sudden increase in electrical resistance. While gel fraction of the radiation-cross-linked blend is higher than those of silane-cross-linked blends (70 wt% against 50 to 63 wt%), the stability of the volume resistivity and PTC after several heat cycles is much better for silane-cross-linked blends. This is due to the difference in the cross-linking process. While radiation-cross-linking is performed in the solid state (and preferentially in amorphous domains), silane-cross-linking is carried out in the melting state. According to the authors, silane condensation reaction allows multiple network formations; these multiple networks could restrict the movements of carbon black particles in the materials better than radiation cross-linking. The cross-linking structure is more stable through heat cycles.

9.3 Lifetime of Free Radicals

The presence of long lifetime free radicals is one of the main differences between the irradiation processing and the incorporation of peroxide in the melting state. In the latter, free radicals react faster due to the high processing temperature. Moreover, the dispersion of these radicals in the material is homogeneous. In the former, even if radicals are generated homogeneously, the reactions take place mainly in the amorphous phase, and many free radicals are "frozen" in the crystalline phase. Consequently, a slow postprocessing degradation could be expected.[15] In a polymer blend, both polymers have a different chemical structure and crystallinity; therefore, the stability and the lifetime of radicals generated under irradiation should be different according to their localization (in amorphous or crystalline phases, in one polymer or in the second one, or at the interface).

The interactions between two polymers could limit the free-radical decay after irradiation. Miklesova and Szocs[2] carried out γ-irradiation of miscible poly(methylmethacrylate) (PMMA)/polyethyleneoxyde (PEO) blends at liquid N_2 temperature with a total dose of 10 kGy. The blends were prepared by mixing on a Brabender plasticorder during 6 or 12 min. The authors observed that the free-radical decay is slower when the time of mixing is longer because the transfer of the radical center is controlled by molecular motions. These motions are affected when more interactions occur between both polymers and consequently the decay is slowed.

Przybytniak et al.[16] have drawn similar conclusions in the case of a non-miscible PE/styrene-butadiene-styrene (SBS) blend. The authors assumed that both polymers are characterized by different initial amounts of radicals and different rates of their decays. Moreover, the whole process (creation of radicals and rates of decay) is not a simple sum of the processes occurring in PE and SBS, even if these polymers are phase separated.

If the irradiation dose is reasonably chosen, a polymer material could be extruded or molded-injected after irradiation. In this case, the final properties of the material are different according to the order of the processing steps. Sonnier et al.[17] carried out two processing sequences on 80/20 PP-PE blends. In the first sequence, the blends were γ-irradiated after mold injection. In the second sequence, the mold injection was performed after γ-irradiation. The ESR (electron spin resonance) method was used to quantify the significant presence of residual free radicals even 1 month at room temperature after irradiation. The mechanical properties are worse for the blends prepared according to the second sequence because more free radicals were able to promote the chain scission of PP. On the contrary, for PE/ground tire rubber (GTR) blends, this second sequence allows an increase of mechanical properties because interfacial cross-linking is enhanced due to the "thawing" of free radicals trapped in the PE crystalline phase (Figure 9.2).

9.4 Influence of Additives

Before irradiating a polymer, two kinds of additives could be incorporated. The first kind is that of radical scavengers (like hydroquinone, and more generally aromatic additives). Chen et al.[18] studied the effect of four different aromatic additives on the radiation resistance of 60/40 styrene-ethylene/butylene-styrene copolymer (SEBS)-PS blends. Considering the changes in mechanical properties, thermal stability and volume resistivity, pyrene was found the most efficient additive and diphenylacetylene (DPA) the worst. The authors concluded that the best protection effect could be attributed to

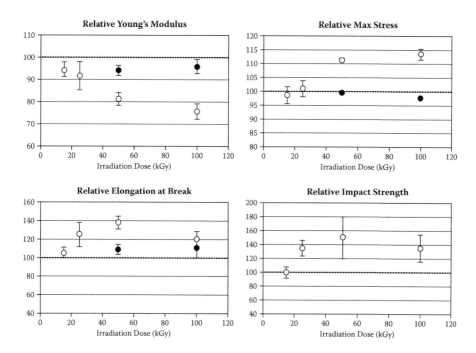

FIGURE 9.2

Mechanical properties of 50/50 high density polyethylene (HDPE)-GTR (100% for noinirradi-ated blend). Black circles correspond to irradiation after injection molding; white circles correspond to irradiation before injection molding. (From Sonnier, R., Massardier, V., Clerc, L., Lopez-Cuesta, J. M., Bergeret, A., Reactive compatibilization of polymer blends by gamma-irradiation: Influence of the order of processing steps. *Journal of Applied Polymer Science* 2010, 115(3), 1710–1717. With permission.)

the compound with the highest degree of aromaticity. Moreover, the unsaturated bonds of DPA could react with the transient intermediates caused by irradiation. Then DPA could be grafted on the polymer macromolecules contrarily to other additives. While DPA is the less efficient protective compound, the authors considered that the protective effect is more dependent on the excess energy release via excitation energy and charge transfer than on the grafting reaction with macromolecules.

The second kind is that of polyfunctional agents that could promote cross-linking (against chain scissioning). Trimethylol propane trimethacrylate (TMPTMA), trimethylol propane triacrylate (TMPTA) or triallyl isocyanurate (TAIC) belong to this category (Figure 9.3).

Dubey et al.[19] compared the efficiency of different additives to promote the cross-linking of a nonmiscible ethylene propylene diene monomer rubber (EPDM)/styrene-ethlene/butylene-styrene copolymer (SEBS) blend γ-irradiated between 50 and 500 kGy: ethyleneglycol dimethacrylate styrene-ethylene/butylene-styrene copolymer (EGDMA), tri(propylene glycol) diacrylate (TRPGDA), trimethylol propane triacrylate (TMPTA), and trimethylol propane

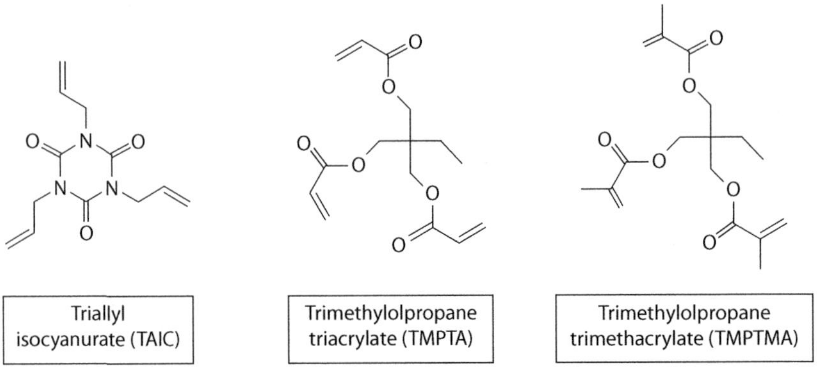

| Triallyl isocyanurate (TAIC) | Trimethylolpropane triacrylate (TMPTA) | Trimethylolpropane trimethacrylate (TMPTMA) |

FIGURE 9.3

Some polyfunctional cross-linking additives.

trimethacrylate (TMPTMA). They established that the functionality is the first criterion to assess the efficiency of the additives. EGDMA and TRPGDA with a functionality equal to 2 are less efficient than TMPTA and TMPTMA (functionality = 3) based on gel fraction results. Moreover, acrylate monomers are more efficient than methacrylate monomers. The authors explain this observation by the rate constants for propagation that are significantly lower for methacrylates than for acrylates. Finally, it could be noticed that the mechanical properties do not follow the same hierarchy as gel fraction results. For example, elongation at break of the blends irradiated under 25 kGy is higher with EGDMA and TMPTMA in comparison to TRPGDA and TMPTA while at such a dose, and the gel fraction of the blend cross-linked with TRPGDA is lower than those of blends cross-linked with TMPTMA and TMPTA.

The previous work has studied the efficiency of additives on the gel fraction of the whole blend. However, in polymer blends, the system is quite more complicated. First, one polyfunctional agent could be efficient for one polymer and not for the other. Second, the dispersion of the additive could be nonhomogeneous in both phases (in the case of a nonmiscible blend). If two agents are used (in principle one for each phase), the system becomes very complex. In our knowledge, up to now, few studies were devoted to these parameters (differential efficiency of an additive toward both polymers in blending, interactions between additives, heterogeneous dispersion, etc.).

9.5 Role of Miscibility—Stabilizing Effect from Aromatic Rings

Different miscible blends under irradiation were studied: PS-PVME,[20] PS-PMMA,[21] PVOH-PEG,[22] PVOH-PAM,[23] PVAc-PMMA.[24] While such polymers are highly compatible, no compatibilization could be expected by irradiation. Studies are carried out to estimate the interactions between both polymers.

Main results about McHerron and Wilkes' studies will be discussed below.[20] Grafting between PS and PVME occurs under irradiation when these polymers are miscible. When phase separation is carried out by annealing at 160°C, gel fraction decreases in a large range of composition because few PS macromolecules are co-cross-linked with PVME chains. Interestingly, for PS-rich blends (85% of PS), an inverse tendency is observed, and higher gel fraction is obtained for phase-separated blends.

El-Din et al.[23] studied PVOH/PAM blends under irradiation. While pure PVOH and PAM are brittle polymers with and without irradiation, a 50/50 PVOH-PAM blend exhibits a yielding behavior. Moreover, the strain at break of this blend is higher than for pure polymers, and it increases under irradiation while strain at break of PVOH significantly falls (see Table 9.1). Even if this behavior is not clearly explained, these results illustrate quite well that irradiation could induce some changes in polymer blends that are not expected according to the radiation behavior of pure polymers.

Another effect is strongly influenced by the miscibility level of a polymer blend: it is the stabilizing effect from aromatic rings. Polymers containing aromatic rings are well known to be very stable under irradiation due to the ability of phenyl rings to absorb energy without undergoing bond rupture. Such a polymer could protect another polymer in blending against degradation induced by irradiation due to a shielding effect.

Torikai et al.[25] have γ-irradiated poly(styrene-*co*-methacrylate) copolymers and blends of polystyrene and polymethacrylate. They investigated the influence of irradiation on PMMA by ultraviolet and Fourier transform infrared spectroscopies and by viscosity measurements. In the case of the blends, the degradation of the PMMA is similar to that awaited. No shielding effect

TABLE 9.1

Tensile Mechanical Properties of poly(vinyl alcohol) (PVOH) and PAM Homopolymers and Their Blends at Equal Contents before and after Gamma Irradiation at a Dose of 100 kGy (Standard Deviations Are Omitted)

Polymer Blend Composition (%)	Irradiation Dose (kGy)	Yield Stress (%)	Yield Strain (%)	Break Stress (MPa)	Break Strain (%)
PVA	0	None	None	4.31	70
	100	None	None	4.55	21
50-50 PVA-PAM	0	0.29	70	3.68	89
	100	0.11	61	3.33	115
PAM	0	None	None	2.86	20
	100	None	None	3.63	58

Source: El-Din, H. M. N., El-Naggar, A. W. M., Ali, F. I., Miscibility of poly(vinyl alcohol)/polyacrylamide blends before and after gamma irradiation. *Polymer International* 2003, 52, 225–234. With permission.

of PS is noticed. On the contrary, a protective effect of PS is observed for the copolymers. Thus, a chemically bonded structure is needed for energy transfer between both polymers. Moreover, the authors calculated the spatial extent of protection of a styrene unit. Assuming a random distribution of methyl methacrylate (MMA) and styrene in the copolymers, they concluded that the styrene unit could protect only the immediate neighboring MMA unit in the macromolecular backbone.

Other works confirm that the radiation protection is not efficient when aromatic rings are not located into the macromolecular backbone. Babanalbandi and Hill[26] studied immiscible blends of arylpolyesters (bisphenol-A polycarbonate and poly(bisphenol-A-co-phtalate) and poly(3-hydroxybutyrate-co-3-valerate). γ-irradiation was performed at 77 K and blends were analyzed by ESR (electron spin resonance). Aromatic polyesters had a greater contribution to the spectrum than the aliphatic one, and this result was explained by the fact that ejected electrons were scavenged more efficiently by the aromatic polymers. Nevertheless, the radiation chemical yields for radical formation in the blends were close to that expected according to the linear additive model. The authors concluded that both polymers are not miscible at the level required for effective radiation protection of poly(3-hydroxybutyrate-co-3-valerate).

McHerron and Wilkes[20] have shown that PS does not offer any protection against radiation cross-linking into PS/PVME blends while these blends are miscible. On the contrary, a significant amount of radiation grafting occurs between both polymers.

Lee et al.[27] have irradiated different immiscible blends of poly(phenylene oxide) and poly(2-vinylnaphtalene). The objective of this work was to check if the latter polymer could improve the radiation resistance of the former. On the basis of thermal stability, mechanical properties and light resistance, no protective effect of poly(2-vinylnaphtalene) could be confirmed.

9.6 Radiation Polymerization in Polymer Blends

One alternative to physical blending consists in polymerizing one component (at least) in the presence of the other one. This is called reactive blending. Generally the blend is homogeneous at the beginning but during the polymerization a phase separation occurs. Such a method could lead to a fine morphology and possibly to a compatibilization between both components by creation of covalent bondings at the interface.

Spadaro et al.[28,29] polymerized methyl methacrylate (MMA) monomers in the presence of acrylonitrile-butadiene rubber by γ-irradiation at a temperature of 70°C. For pure MMA, a total dose of 4 kGy is enough to complete polymerization and further irradiation (6.3 kGy) leads to a degradation of PMMA macromolecules. On the contrary, for PMMA/ABN blends, a higher dose

TABLE 9.2

Tensile Properties of Pure Polymethyl Methacrylate (PMMA) and PMMA-Acrylonitrile Butadiene Rubber (ABN) Blends (Standard Deviations Are Omitted)

System	Dose (kGy)	Young's Modulus (MPa)	Stress at Break (MPa)	Elongation at Break (%)
Pure PMMA	4	1790	61	4.0
Pure PMMA	6.3	1650	50	4.1
PMMA-ABN blend	4	1350	37	5.6
PMMA-ABN blend	6.3	1710	61	7.4
PMMA-ABN blend	8	1520	48	4.8
PMMA-ABN blend; annealed at 120°C, 120 min	4	1280	35	5.0
PMMA-ABN blend; annealed at 120°C, 120 min	8	1540	50	4.7

Source: Spadaro, G., Dispenza, C., Mc Grail, P. T., Valenza, A., Cangialosi, D., Submicron structured polymethyl methacrylate/acrylonitrile-butadiene rubber blends obtained via gamma radiation induced in situ polymerization. *Advances in Polymer Technology* 2004, 23, 211–221. With permission.

favors the grafting reactions between both polymers. Percentage of PMMA grafted onto ABN reaches 20% at a total dose of 8 kGy. ABN forms very small nodules (less than 100 nm) in the PMMA matrix. Mechanical properties have been measured and showed maximum values at 6.3 kGy (Table 9.2). A further dose probably degrades the PMMA. It should be noticed that the cross-linking of ABN under radiation was not taken into account by the authors.

The same researchers' team studied also the cationic curing of difunctional epoxy resin in the presence of a thermoplastic polysulfone as a toughening agent and an onium salt as initiator.[4–6] Irradiation presents some interesting advantages to induce curing of a thermoset material. In particular, this is a cold process that does not need thermal activation or chemical additives (in most cases). Moreover, the curing could be very quick and localized to a part of the sample that could have a specific shape. Different total doses (80 and 150 kGy) and dose rate (90 and 840 kGy/h) have been tested. At a low dose rate, a heterogeneous structure was achieved consisting of two epoxy fractions with different cross-linking degrees and nodules of polysulfone. A nodular morphology could be observed. Thermally postcuring at 100°C during 2 h improves the cross-linking of the less cross-linked epoxy fraction while polysulfone nodules size increases significantly.

At a high dose rate, the temperature increases strongly during processing (up to 200°C) and then the curing is both radiation and thermally induced. Epoxy fraction is homogeneous but the cross-linking degree is lower. Probably the high speed of the curing causes the material to quickly reach a glass transition temperature higher than the process temperature. Consequently, a vitrification phenomenon occurs and inhibits the cross-linking. In the same time, the phase separation between polysulfone and epoxy is inhibited leading to a very fine morphology.

In another work, the same authors cure a blend of two epoxy monomers (difunctional and trifunctional) by e-beam irradiation and in the presence of 10 wt% of polysulfone.[6] The temperature is measured during the curing, and the start of curing corresponds to a sharp increase of temperature. For the pure difunctional epoxy resin, the adding of polysulfone leads to a higher induction dose probably due to a dilution effect and to an increase in system viscosity. On the contrary, when the blend of difunctional and trifunctional epoxy monomers is cured in the presence of polysulfone, the induction dose decreases. The authors explain this point by the dilution effect of polysulfone causing a lower extent of the precuring reactions. These precuring reactions lead to a very high viscosity for the system without polysulfone. Consequently, the polysulfone allows the curing to start after a lower absorbed dose. Finally, the dynamic mechanical analysis reveals that the presence of thermoplastic polysulfone rises to a less cross-linked structure according to the lower glass transition temperature and maximum tan δ value.

9.7 Preirradiation of Polymer before Blending

Preirradiation could be used to modify a polymer before blending it with another one. The preirradiation could be performed for different purposes (cross-linking of a rubber, grafting, functionalization, or oxidation). Peng et al.[30] prepared ultrafine carboxylated styrene-butadiene-rubber (CSBR) powders by γ-irradiation. The control of irradiation conditions (dose and presence of cross-linking sensitizer, 2-ethyl hexyl acrylate (2-EHA)) allows a desired cross-linking density to be obtained. Moreover, ultrafine particles (150 nm) could be obtained easily when the cross-linking density is high enough. The powder was then incorporated into PA6 to enhance its impact strength. The content of CSBR was fixed at 5 wt%. The authors determined the optimum cross-linking density (for 25 kGy with 3 wt% of 2-EHA) leading to the highest mechanical properties and considered that irradiation could be a useful method to prepare rubber powders as a toughening agent for thermoplastics.

The grafting functionalization of a poly(vinylidene fluoride) powder by γ-irradiation was achieved by Valenza et al.[31] The amount of grafted methacrylic acid onto poly(vinylidene fluoride) (PVDF) powder was 19.7 w%. The grafted polymer was then blended at different ratios with an ionomer based on ethylene-methacrylic acid copolymer, partially neutralized (Surlyn 9970). Nongrafted PVDF and this ionomer are highly immiscible. The functionalization of the PVDF with methacrylic acid allows to compatibilize both polymers resulting in a finer morphology. The best compatibilization was found for a combination of the radiation grafting and the addition of zinc acetyl acetonate dihydrate to neutralize the blend.

Irradiation of polyethylenes (PEs) could lead to graft some oxidized species on the polyethylene backbone.[32] This treatment was performed in order to improve the compatibility of polyethylenes toward polar polymers as poly(ethylene terephthalate)[33] or polyamide 6.[31,34–36] Irradiation in air of polyethylene causes oxidative degradation with formation of carbonyl and hydroxyl groups.[32,36] Valenza et al.[31] studied the influence of irradiation on several polyethylenes differing from their structure (two high- and low-density polyethylenes and one linear low-density polyethylene, high density polyethylene (HDPE), low density polyethylene (LDPE), and linear low density polyethylene (LLDPE), respectively). To promote oxidation (compared to cross-linking and chain scission or branching), irradiation was carried out on thin films of PE in air at room temperature. A finer morphology and some improvement of mechanical properties were obtained and attributed to some interactions between oxidized groups on irradiated polyethylene and functional groups of the polar polymer (amid functions or end-chain carboxylic or amine groups). For example, the elongation at break and the Izod impact strength of a 10/90 LLDPE-PA6 blend were increased from 8% to 21% and 39 to 65 J.m^{-1}, respectively when LLDPE was preirradiated at 50 kGy. Nevertheless, no cocrystallization phenomenon between PE and PA6 was observed, but crystallization of polyethylene was modified according to the compatibilization with PA6.

Some differences were noted in relation to the structure of the polyethylene. In particular, the morphology of HDPE-PA6 was less homogeneous. This result could be attributed to a less extent of oxidation of HDPE because of its crystallinity, which limits the oxygen diffusion into the bulk. Moreover, the presence of cross-linked chains (the gel fraction was measured at about 10%) contributed to a less regular distribution of HDPE droplets. The dose and the dose rate are also influent parameters. For 10/90 LLDPE-PA6 blends, the total dose was optimized at 50 kGy.[36] A higher dose (135 kGy) led to the cross-linking of LLDPE that was detrimental for mechanical properties. In another study,[31] Valenza et al. considered the influence of the dose rate on the final properties of a LDPE/PA6 blend. The content of LDPE was fixed at 10 or 25 wt%. Mechanical properties of LDPE-PA6 were slightly improved when LDPE was preirradiated at 0.03 Gy/s in comparison to the unirradiated LDPE/PA6 blend. On the contrary, when LDPE was preirradiated at 1 Gy/s, the mechanical properties of the blends were the worst. This result was ascribed to the ratio between oxidation and cross-linking. The former phenomenon was promoted by a low dose rate. On the contrary, for a dose rate of 1 Gy/s, the oxidation was limited and cross-linking was improved (the gel fraction increased to 30% at 50 kGy).

For the same objectives as polyethylene, polypropylene (PP) was preirradiated before being mixed with PA6 in 70/30 PA6-PP blends.[37] The best properties were obtained at a dose of 100 kGy, either by electron or gamma beam irradiation; higher doses increase compatibility, but there is extensive degradation and it produced breakable films.

Other works were performed by using irradiation to increase the compatibility of a PE/polar polymer blend. For example, Senna et al.[38] performed electron beam irradiation in air at a total dose of 10 kGy on low-density polyethylene (LDPE)/plasticized starch (PLST) blends. By Fourier transform infrared (FTIR), they showed an increase of bands assigned to O-H and C=O bonds stretching, and the formation of hydrogen bonding between LDPE and PLST. Hence, irradiation created an interconnection between starch molecules and LDPE through the formation of free radicals. The interconnection at the interface between LDPE and PLST led to higher yield stress for increasing irradiation dose concomitant with lower yield strain. Then, irradiation of a polar/apolar blend can lead to a better compatibility.

9.8 In Situ Compatibilization of Polymer Blends via γ or E-Beam Radiation

Many polymers are immiscible and form heterogeneous systems when blended. Then, irradiation was commonly used to improve the compatibility between immiscible polymers in a blend.[33] In comparison with other methods of compatibilization based on the reactivity of functional groups grafted on the polymer backbone, the changes are not limited to the interface. Irradiation leads to changes not only in the interphase (interfacial cross-linking) but also in the bulk of both polymers (chain scission, cross-linking, etc.). Therefore, it is not always easy to determine if the macroscopic properties change due to the compatibilizing effect of the irradiation or due to the modification of the polymers in bulk. There are hundreds of authors working on irradiated immiscible polymer blends that claim, in their articles, to have increased the compatibility between the two polymers just by regarding the mechanical properties. In a concern of quality of results regarding the compatibilization of polymer blends by irradiation, we chose to depict here only the articles that clearly evidenced an increase in the interfacial adhesion between the two polymers, either by covalent bonding between the two components or by physical linkages.

9.8.1 Chemical Reaction between the Two Polymers Induced by Irradiation

Upon electron beam irradiation, blends based on polyesters undergo transesterification reactions. Kim et al.[39] investigated the transesterification reactions

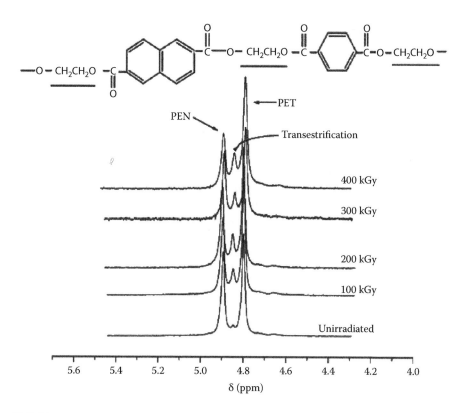

FIGURE 9.4
Proton nuclear magnetic resonance (¹H NMR) spectra of the thylene unit in poly(ethylene 2,6-hapthalate) (PEN)/poly(ethylene terephthalate) (PET) blends at various dose rates. (From Kim, J. U. N. Y., Kim, O. H. S., Kim, S. H. U. N., Jeon, H. A. N. Y., Effects of electron beam irradiation poly(ethylene terephthalate) blends. *Polymer Engineering and Science* 2004, 44(2), 395–405. With permission.)

between PEN and poly(ethylene terephthalate) (PET) by proton nuclear magnetic resonance (¹H NMR) spectroscopy. They identified that a new peak, with a chemical shift of 4.85 ppm appeared when crescent irradiation dose rates were applied. This peak was assigned to the protons of ethylene units that exist between terephthalate and naphthalate units in PEN-PET copolymers. The results are presented in Figure 9.4. This result permitted calculation of a transesterification percentage. It was shown that the percentage of transesterification increased with the dose rate. Moreover, transesterification reactions led to the formation of a single phase proved by a single glass transition temperature between that of the two pure polyester components. Finally, the melting temperature of irradiated PEN/PET blends decreased with the increase in the degree of transesterification. It was explained by the fact that the degree of transesterification increased with dose rate and led to the formation of more random chain structures in the irradiated blend.

Concerning water-soluble-based polymer blends, Abd Alla et al.[22] observed a limited compatibility for poly(vinyl alcohol) (PVA)-poly(ethylene glycol) (PEG) blends over the range of 0% to 30% of PEG. Exposure of PVA-PEG blends (with a %PEG up to 30%) to gamma irradiation (with doses up to 100 kGy) gave rise to an improvement in the tensile mechanical properties. It was suggested that the free radicals formed during irradiation of PVA was involved in the formation of covalent bonding along the boundaries of the polymers, and hence improved the compatibility of the blend.

In the same manner, blends containing (100% to 90%) polystyrene and (0% to 10%) styrene-butadiene rubber (SBR) exhibited improved impact properties after gamma irradiation at a dose of 100 kGy.[40] FTIR provided evidence that irradiation produced a radical in the benzene ring of PS that could react with the double bond of polybutadiene producing a metasubstituted benzene (Figure 9.5). Hence, this chemical link between the two polymers gave rise to the increase in Izod impact strength particularly for 100 kGy γ-irradiated 90/10 PS-SBR blend.

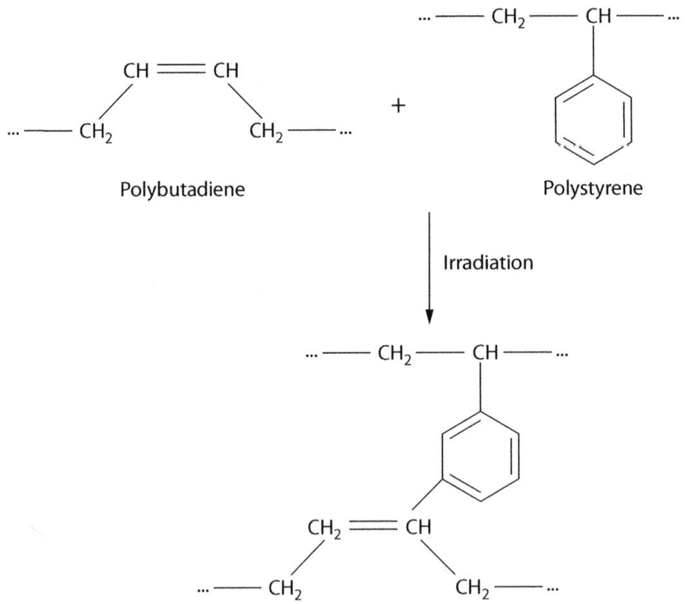

FIGURE 9.5
Mechanism of the chemical bonding between the benzene ring of polystyrene (PS) and the double bond of the polybutadiene. (From Martínez-Barrera, G., Studies on the rubber phase stability in gamma irradiated polystyrene-SBR blends by using FT-IR and Raman spectroscopy. *Radiation Physics and Chemistry* 2004, 69, 155–162. WIth permission.)

9.8.2 Incompatibility Reduction by Cross-Linking between the Two Phases

Incompatibility between two polymers in a blend can be reduced by both chain scissions in one phase and cross-linking between the two phases. This was greatly studied in the case of thermoplastic/elastomer blends.[41–45] Usually, when thermoplastic/elastomer blends are exposed to irradiation, the elastomeric phase is cross-linked (under air), whereas the thermoplastic phase is largely degraded.[43,45] For example, Sonnier et al.[42] performed in situ scanning electron microscopy observations under microtensile strain on a 50% recycled HDPE blended with 50% GTR (ground tire rubber). They clearly evidenced by this technique a better cohesion between HDPE matrix and the GTR dispersed phase after blend irradiation. In irradiated thermoplastic/elastomer or thermoplastic/thermoplastic elastomer (TPE) blends, the radiation cross-linking and degradation reactions mainly occurred in the amorphous regions of semicrystalline polymers, whereas higher amorphous content is favorable to the radiation cross-linking of polymers, especially for higher dose rate irradiation. For example, in LDPE/EVA blends, where the amorphous phases of both polymers are completely miscible, it was shown that EVA had an enhancement effect on radiation cross-linking because EVA made the blend more amorphous in nature, which in turn increase its efficiency toward cross-linking.[46–48] Hence, increasing EVA content gave rise to the increase in tensile strength and the decrease in heat shrinkability.[49] However, the HDPE/EVA blends are partly compatible in the amorphous region, and this was unfavorable for the enhancement effect of EVA on the radiation cross-linking of these blends.[50] Dalai et al. also worked on PP/EVA blends in which PP and EVA are compatible in the amorphous phase. They obtained the same results as for LDPE/EVA blends (i.e., the complete compatibility in the amorphous region between LDPE or PP and EVA favored the enhancement of radiation cross-linking and prevented LDPE or PP degradation).[51] Mihailova et al.[52] studied irradiated PP/EVA blends by using wide angle X-ray scattering (WAXS). At the interface between PP and EVA, they described a zone where amorphous chains of the two components interpenetrated each other, whereas PP and EVA macromolecules crystallize separately. Those results on HDPE or LDPE or PP blended with EVA evidenced that a good compatibility is a prerequisite for the enhancement effect of EVA on the radiation cross-linking of the polyolefin/EVA blend.

More generally, in a polyblend system containing both a radiation-cross-linkable polymer and a radiation-degradable polymer, the last one can penetrate into the cross-linking network of the first one by graft reaction on the interface. This was proved by Li et al.[53] for poly(ethylene-co-vinyl alcohol) (EVOH)/low-density polyethylene (LDPE) systems by regarding the gel content and FTIR spectra. They also suggested a "graft" structure formed on the interface when EVOH/LDPE blends were irradiated. This mechanism was based on recombination reactions between the two stable macromolecular

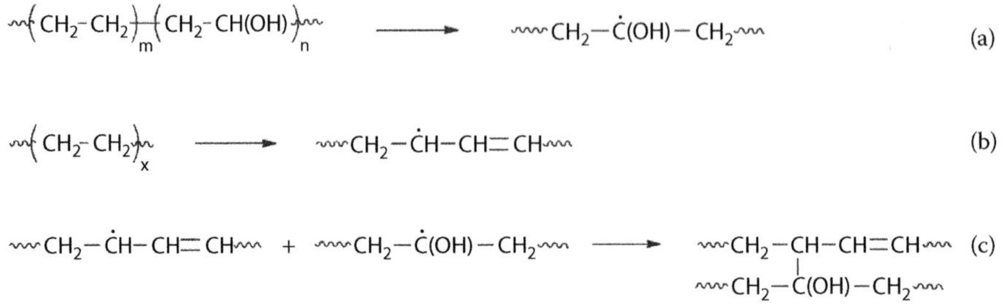

$$\sim\!\!\left(CH_2\text{-}CH_2\right)_m\!\!\left(CH_2\text{-}CH(OH)\right)_n\!\!\sim \quad\longrightarrow\quad \sim\!\!CH_2\text{-}\overset{\cdot}{C}(OH)\text{-}CH_2\!\!\sim \qquad\text{(a)}$$

$$\sim\!\!\left(CH_2\text{-}CH_2\right)_x\!\!\sim \quad\longrightarrow\quad \sim\!\!CH_2\text{-}\overset{\cdot}{C}H\text{-}CH\!\!=\!\!CH\!\!\sim \qquad\text{(b)}$$

$$\sim\!\!CH_2\text{-}\overset{\cdot}{C}H\text{-}CH\!\!=\!\!CH\!\!\sim \;+\; \sim\!\!CH_2\text{-}\overset{\cdot}{C}(OH)\text{-}CH_2\!\!\sim \quad\longrightarrow\quad \begin{array}{l}\sim\!\!CH_2\text{-}CH\text{-}CH\!\!=\!\!CH\!\!\sim \\ \qquad\qquad| \\ \sim\!\!CH_2\text{-}C(OH)\text{-}CH_2\!\!\sim\end{array} \;\text{(c)}$$

FIGURE 9.6
Stable macromolecular radical and possible new graft structure in the polyblend of EVOH/LDPE: (a) hydroxyl radical, (b) alkyl radical, and (c) graft structure. (From Li, H. H., Yin, Y., Liu, M. H., Deng, P. Y., Zhang, W. X., Sun, J. Z., Improved compatibility of EVOH/LDPE blends by γ-ray irradiation. *Advances in Polymer Technology* 2009, 28(3), 192–198. With permission.)

radicals (i.e., hydroxyl radicals for EVOH and allyl radicals for LDPE). This is presented in Figure 9.6.

9.8.3 Irradiation of Blends Containing a Coagent

Gamma and electron beam radiation were also used in the presence of coagents to improve polymer blend miscibility and properties. Table 9.3 summarizes all the formulations of irradiated blends containing a coagent.

9.8.3.1 Radiation-Induced Radicals from Coagents that Can Further React with the Unsaturation of the Polymeric Chains

Irradiation doses were commonly used in the presence of cross-linking agent, such as trimethylol propane trimethylacrylate (TMPTMA) or ditrimethylol propane tetraacrylate (DTMPTA), in blends containing isotactic (i)PP/LDPE[54] or LDPE/EVA,[55,56] respectively. In such blends, irradiation in the presence of trimethylol propane acrylates was particularly efficient, regarding tensile strength, when the LDPE content was high.

Coagents based on bismaleimide, multifunctional acrylate, and methacrylate esters, in the presence of irradiation, can form very reactive radicals that can homopolymerize or graft onto the unsaturation of the polymeric chains through an "ene" reaction mechanism. Hence, the created network can be enhanced through the grafting of such coagents in between the polymeric chains. It increases the cross-linking density. Zurina et al.[57] published several works dealing with electron beam radiation of epoxidized natural rubber (ENR) blended with ethylene vinyl acetate (EVA) in the presence of such coagent (*N,N′-m*-phenylenbismaleimide, HVA-2[57] and trimethylolpropane triacrylate, TMPTA[58]). They reported the gel content versus irradiation dose for blends without and with HVA-2 or TMPTA. The gel content was higher in the samples containing HVA-2 or TMPTA, indicating clearly that those coagents enhanced the cross-linking density in irradiated

TABLE 9.3

Formulations Irradiated with a Coagent

Polymer Blends	Proportions	Compatibilizer	Irradiation	Reference
rPP/wfibers-PAN rPP/wfibers-viscose rPP/wfibers-hemp-viscose rPP/wfibers-chopped glass fiber	70PP/30 fibers	Mixture of epoxy acrylate (EA) and tri-propyleneglycol diacrylate (TPGDA) (1%)	Electron beam at 8 kGy	1
LDPE/HDPE/ PP/ PS/PET	24LDPE/ 23HDPE /21PP/15PS/ 17PET	SEBS-g-MA (10%); TMPTMA (1%)	Electron beam 300 kGy	60,94
iPP/LDPE	100/0; 80/20; 50/50; 20/80; 0/100	TMPTMA (1, 3, and 5%)	Electron beam 5 to 50 kGy	9
LDPE/EVA	100/0; 70/30; 60/40; 50/50; 40/60; 30/70	DTMPTA (1 to 5 parts)	Electron beam (20 to 500 kGy)	55,56
Popropylene-coethylene/ polybutylene succinate	66/33; 50/50; 33/66	PP-g-MA (15%)	Electron beam (10 kGy/pass)	62,63
NR/LLDPE	60/40	Liquid natural rubber (LNR6 and LNR16)	3MeV e-beam; 40 to 240 kGy	128
wHDPE/wPS/ wPVC	70/15/15	EVA; SEBS (5, 7.5, and 15)	γ-irradiation 50 to 200 kGy (after blending)	92
ENR/EVA	50/50	HVA-2 (*N,N'-m-*phenylenbismaleimide)	Electron beam 0 to 100 kGy; 20 kGy/pass	57
ENR/EVA	50/50	TMPTA	Electron beam 0 to 100 kGy; 20 kGy/pass	58
NBR/SBR	50/50	NBR-g-CA; NBR-g-MMA	49 rad/s	61
wLDPE/wbutyl rubber	50/50	DEGDMA; DVB; TEGDMA; GMA; MAW[a]	γ-irr; 100 to 400 kGy; 5 to 7 kGy/h	59
SBR/EVA	0/100 to 100/0	SEBS-g-MA	γ-irradiation	64

[a] Diethylene glycol dimethacrylate (DEGDMA); divinyl benzene (DVB); tetraethylene glycol dimethacrylate (TEGDMA); glycidyl methacrylate (GMA); maleic anhydride (MA).

ENR/EVA blends. Irradiation produced the initiating free radicals that then reacted with a bismaleimide or a multifunctional acrylate molecule through unsaturation to produce cross-linking. The mechanism is given in Figure 9.7. The effective cross-linking by HVA-2 or TMPTA was confirmed by dynamic mechanical analyzer (DMA).

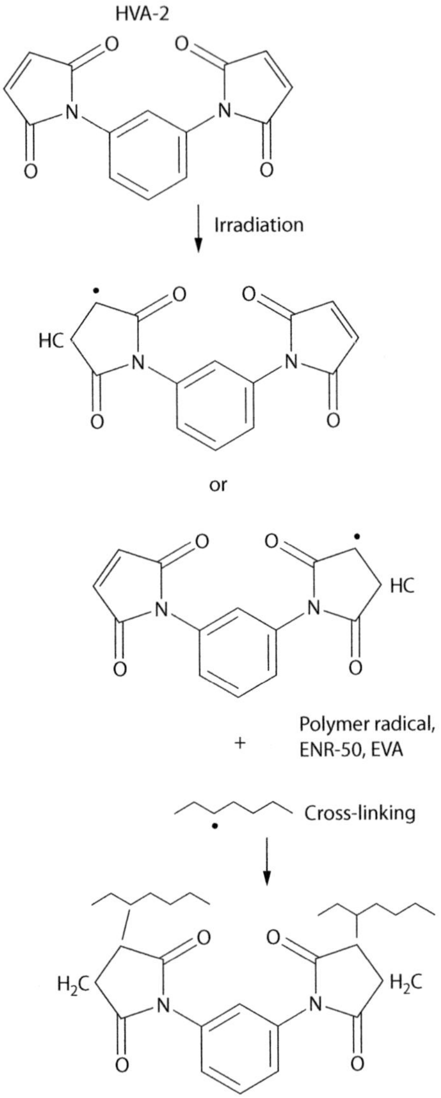

FIGURE 9.7
Proposed mechanism of crosslinking for epoxidized natural rubber (ENR) and ethylene vinyl acetate (EVA) with HVA-2. (From Zurina, M., Ismail, H., Ratnam, C., The effect of HVA-2 on properties of irradiated epoxidized natural rubber (ENR-50), EVA, and ENR-50/EVA blend. *Polymer Testing* 2008, 27, 480–490. With permission.)

Maziad et al.[59] compared the efficiency of the different compatibilizers toward the gel fraction of waste LDPE/waste butyl rubber irradiated blends. As expected, the gel fraction increased with irradiation dose; however, over the whole irradiation range, each compatibilizer had a different capacity regarding the enhancement of radiation-induced cross-linking and the order is as follows: diethylene glycol dimethacrylate (DEGDMA) > divinyl benzene (DVB) > tetraethylene glycol dimethacrylate (TEGDMA) > glycidyle methacrylate (GMA) > maleic anhydride (MA) > blank. The authors concluded that the efficiency of the compatibilizing agents is linked to its functionality (number of double bonds).

In some cases, lack of noticeable influence of electron beam radiation on the values of tensile and Charpy impact strengths of 24-23-21-15-17 LDPE-HDPE-PP-PS-PET blends (based on recycled polymers) compatibilized with 1% trimethylol propane trimethylacrylate (TMPTMA) were found. Zenkiewicz et al.[60] explained this lack of influence by the protective action of aromatic rings of PS and PET that hindered cross-linking. In the same article, the addition of 10% of styrene-ethylene/butylene-styrene elastomer grafted with maleic anhydride (SEBS-g-MA) led to the great increase of both tensile and Charpy impact strengths.

9.8.3.2 Irradiation to Create a Copolymer that Can Compatibilize an Immiscible Blend

Block and graft copolymers have been used successfully as interfacial agents to control the morphology and reinforce the interface between two immiscible polymers in a blend. Graft copolymerization can be carried out by irradiation. For example, copolymers were prepared from acrylonitrile butadiene copolymer (NBR) grafted with methylmethacrylate (MMA) or cellulose acetate (CA) using gamma radiation and incorporated in a 50/50 acrylonitrile butadiene rubber-styrene butadiene rubber (NBR-SBR) blend.[61] As expected, the presence of grafted copolymers (NBR-g-MMA or NBR-g-CA) improved the interfacial adhesion between NBR and SBR. This was proved by regarding the evolution of phase morphology between uncompatibilized and compatibilized blends by scanning electron micrography (SEM) and by the improvement of mechanical properties when adding a compatibilizer.

9.8.3.3 Further Irradiation of Compatibilized Blends

The addition of a polypropylene grafted maleic anhydride (PP-g-MA) compatibilizer into polypropylene-*co*-ethylene/polybutylene succinate blends and the subsequent irradiation allowed prevention of degradation mechanisms leading to mechanical stability.[62,63] The cyclic anhydride group of the PP-g-MA first permitted compatibilizion of both polypropylene-*co*-ethylene and polybutylene succinate and a second action as an "energy sink."

Moreover, it was shown that irradiation of the compatibilized blend was able to enhance the biodegradation slightly.

Styrene butadiene rubber (SBR)/poly(ethylene-*co*-vinyl acetate (EVA) blends were compatibilized with SEBS-g-MA and further gamma irradiated to evaluate their aging resistance toward radiation. The resistance of the blends to gamma irradiation is better for those that contain higher proportions of EVA.[64]

9.9 Cross-Linking

Elastomeric phases (El) (or thermoplastic elastomers, TPE) are commonly added to thermoplastic polymers (TP) to increase their impact strength and their toughness. And those blends consisting of TP/El or TP/TPE are commonly cross-linked to further increase the thermal and mechanical properties. This is particularly the case when using recycled polymers.[65] For that purpose, irradiation is often used in polymer blend systems to induce cross-linking inside the different phases, or between them, resulting in the improvement of physical properties. However, cross-linking and chain scission are processes always in competition, and experimental and environmental conditions can influence the balance between both. To compare the evolution of the cross-linking reaction versus irradiation doses, gel fraction is often measured by weighting the sample before and after extraction in an appropriate solvent. Dong et al.[66] described three different morphologies in irradiated immiscible binary blends. In the first, cross-linking occurs in the bulk and in the dispersed phase. In the second, degradation occurs in the bulk and cross-linking in the dispersed phase. Finally in the third, cross-linking occurs in the bulk and degradation in the dispersed phase. The same authors performed irradiation on 75/25 Nylon1010-HIPS blends and noted chain scission in high impact polystyrene (HIPS) and chain cross-link in Nylon1010, corresponding to the third morphology.[67]

9.9.1 Cross-Linking of Polyolefin/Elastomer or Polyolefin/Thermoplastic Elastomer (TPE)

To improve the impact strength of PP and PE, they were commonly blended with elastomeric phases such as EPDM,[68] SEBS,[69] SBS,[69] or thermoplastic elastomers such as EVA.[52,70]

Irradiation changes the chemical structure of polyolefins, especially in the amorphous phase, while the crystalline phase is more resistant to its action.[69] As mentioned earlier, all polyolefin/El or polyolefin/TPE irradiated blends can be classified in three different morphologies depending on the balance between chain scission and cross-linking.

9.9.1.1 Enhancement of Polymer Cross-Linking by Adding a Second Polymer

The cross-linking of a polymer could be enhanced by the incorporation of a second polymer easily cross-linkable. In PP/EVA blends containing 70% of PP and 30% of EVA, EVA was described to cross-link under low doses of irradiation, whereas PP underwent chain scissions.[71] Minkova et al.[72,73] found that higher amounts of EVA in PP matrix favored cross-linking and increased the cross-linking density. EVA in PP slows the oxidation of alkyl radicals of PP.[70]

Spenadel,[74] to increase the flexibility of wire insulation, had irradiated LDPE, EPDM rubber, and blends of these two polymers. In the case of LDPE, the gel fraction attains 70% to 80% for a dose of between 100 and 150 kGy. The gel fraction for EPDM rubber (Vistalon 3708) attains 90% for 50 kGy. The incorporation of a moderate amount of EPDM rubber (20 wt%) in the LDPE matrix allowed a gel fraction of 70% to 80% for 50 kGy to be obtained. Therefore, the gel fraction of the blend does not match a rule of mixture, and a strong decrease of the needed dose could be achieved due to the presence of EPDM rubber.

McHerron and Wilkes'[20] e-beam irradiated blends of polystyrene and poly(vinyl methyl ether) (PVME). The phase diagram of these two polymers displays a lower critical solution temperature (LCST) behavior, revealing that they are compatible over the entire composition range at room temperature. While PS is radiation resistant, PVME is cross-linked under radiation. Hence, the addition of small amounts of PVME to pure PS increases the gel fraction obtained after irradiation. When the polymers are miscible, irradiation allows not only the cross-linking of PVME but also a significant grafting between both polymers. On the contrary, when the morphology is biphasic, the gel fraction is closely related to the PVME content. No or little grafting between PS and PVME occurs in this case.

Dalai and Wenxiu draw similar conclusions about PE/EVA blends irradiated by γ-irradiation in air.[48,50] EVA could enhance efficiently the cross-linking of LDPE while the amorphous phases of both polymers are miscible.[50] Zhang et al.[75] showed that the radiation cross-linking behavior of LDPE/EVA blends fits an equation between radiation dose and gel fraction developed initially for pure polymers. They concluded that LDPE and EVA are miscible in amorphous phase. On the contrary, the amorphous phases of EVA and HDPE are only partially miscible. Therefore, when EVA is incorporated into HDPE, poor enhancement of PE cross-linking is obtained under irradiation.[48]

9.9.1.2 Influence of the Process

The process of irradiation has an influence on the final properties of the blend. For example, the SBS elastomeric phase could be preirradiated before incorporation in the polyolefin PP matrix.[76] In that case, PP/SBS blends exhibit lower Young's modulus but higher elongation at break.

Van Gisbergen et al.[77] used electron beam radiation to fixate the morphology of PP/EPDM blends. An optimal morphology for toughening was obtained via extrusion of a high molecular weight PP (matrix) blended with EPDM (dispersed phase). This morphology was fixed by electron beam radiation before subsequent processing of the blend (injection molding). Hence, the non-cross-linked phase flowed, while the cross-linked phase largely maintained its shape. At low dose, the dispersed phases of EPDM rubber were cross-linked by irradiation, whereas main chain scission occurred in PP. This method of morphology fixation was possible because the degrading-type polymer constituted the continuous phase.

Van Gisbergen[68] also studied the influence of cross-links, induced by electron beam (EB) irradiation, on the two major processes that play an important role in the formation of the morphology of a model system (i.e., thread break-up and coalescence). A 80/20 PS-LDPE model system was blended via a corotating twin-screw extruder and strands were quenched in water and exposed to EB irradiation (dose of 472 kGy) that induced cross-links in the LDPE dispersed phase. The irradiated and nonirradiated strands were then heated at 200°C for different periods and under no shear deformation to evaluate by SEM and microrheology the stability of the induced morphology. They showed that even in irradiated blends, the fixation of the morphology is prevented by the absence of a yield stress that would prohibit the flow of the dispersed phase completely and prevent from thread break-up. Hence, when heating of the blend lasted for a long period, both break-up and coalescence occurred.

9.9.1.3 Interphase Formation Consisting of Interfacial Cross-Links between the Phases

Irradiated PP/SEBS and PP/SBS blends were compared.[69] It was shown that SEBS and SBS exhibited different behavior toward irradiation; granulated blends were irradiated with electron beam 1 week after they were injection molded. SEBS-based systems were more resistant to irradiation than SBS. Hence, the property changes in the SEBS/PP system are generally smaller in comparison with the SBS-based blends. Moreover, irradiated PP/SBS blends led to structure changes of crystalline and amorphous PP and elastic SBS phase with the creation of new interphase consisting of a part of elastic SBS phase that was cross-linked and grafted by the PP radiolysis products. Perera et al.[78] showed the influence of the amount of SBS. With low amounts of SBS, chain scission predominated in PP, whereas for high amounts of SBS (>30%), cross-linking in the butadiene block of SBS compensated the chain scission in PP. Hence, high amounts of SBS copolymers in a PP matrix led to a decrease in their melt flow index (MFI) because the presence of SBS decreased the PP sensitivity to radiation effects.

LLDPE/polydirethylsibxane (PDMS) blends with various rates of PMDS (0, 10, 20, 30, 40, 50, 60, 70, and 100) were electron beam irradiated.[79] In the blends where LLDPE consisted in the matrix and PDMS was the dispersed phase, Giri et al.[79] evidenced not only intermolecular cross-links within the LLDPE matrix and within the PDMS rubber, but also interfacial cross-links between LLDPE and PDMS rubber. The optimum radiation doses were identified to range from 100 to 300 kGy for obtaining the best balance in physicomechanical properties depending on the composition.

9.9.2 Blends of Two Immiscible Elastomers

Elastomers are classically blended to improve the physical and mechanical properties of the final material. For example, Dubey et al.[19] incorporated suitable amounts of EPDM in the SBR matrix because it was expected to impart significant heat and ozone resistance to the SBR matrix. The irradiation-induced vulcanization of SBR/EPDM blends was proved to combine desired properties and high mechanical strength. For example, the formation of a cross-linked network led to a decrease in elongation at break. Cross-linking restricts the movement of the polymer chain against the applied force. Moreover, to decrease the radiation dose (and hence, the cost of radiation), multifunctional acrylates (MFAs) and allylic reactive molecules were used.

Natural rubber (NR)/SBR blends exhibit improved oxidative stability compared to pure component and their mechanical properties could be improved by vulcanization. Manshaie et al.[80] compared NR/SBR cured blends either by electron beam irradiation or by sulfur vulcanization. The irradiated blends have better mechanical properties and better heat stability than those cured by a sulfur system. The irradiated blends exhibited higher tensile strength, hardness, and abrasion resistance than nonirradiated ones. However, cross-linking provokes the decrease in elongation at break and resilience.

In the same manner as blends of thermoplastics or thermoplastic/elastomer, irradiated blends of elastomers can undergo chain scission due to degradation or cross-linking, depending especially on the dose range. For example, Zurina et al.[81] measured tanδ versus temperature for 50/50 epoxidized natural rubber (ENR-50)-EVA blends by DMA. At 60 kGy, the irradiation-induced cross-link enhanced the T_g of the blend, whereas at higher dose (100 kGy), the T_g decreased due to the occurrence of oxidative degradation that broke the cross-link structure.

In EPDM/SBR blends, if EPDM is the rich phase, 15 kGy irradiation decreases the tensile strength indicating that chain scission and decreasing in cross-linking are the most evident mechanisms. The increase in tensile strength of SBR-rich blends at low irradiation levels can be explained by the presence of aromatic rings in the polymer, which are relatively resistant toward degradation. However, higher dosages of radiations enhance cross-linking and increases the tensile strength but reduced the elongation at break.

Three different compositions of EPDM-butyl rubber (IIR) blends (3:1, 1:1, and 1:3) were γ-irradiated. The evolution of the gel fraction versus irradiation

doses for the three compositions showed that butyl rubber is more susceptible than EPDM to radiolytic degradation. Hence, the radicals provided by
IIR would be grafted on the macromolecules of EPDM or they would recom-
bine to restore the initial structure. Thus, the gel content of irradiated 75/25
EPDM-IIR is always higher (whatever the dose rate) than that of 50/50 or
25/75 composition. Finally, a greater content of EPDM in its blend with butyl
rubber provides a certain resistance against oxidation.

9.10 Irradiation in Some Application Fields

Polymer blends could be irradiated for the same purpose as polymers.[82]
For example, polymer devices for medical applications have been steril-
ized by irradiation for many years. Irradiation could also be used to obtain
a polymer material suitable to foam by modifying its structural parameters
(macromolecular mass, branch content).[83–85] For such an application, even if the
irradiation of polymer blends is more complex than the irradiation of a pure
polymer, the main objectives are the same in both cases. Thus, the following
part will consider only the specific applications of irradiated polymer blends.

9.10.1 Irradiation of Recycled Polymer Blends

It is possible to recycle polymers (mainly thermoplastics) to give them a second life. Generally, some treatments are needed to upgrade their properties,
because they were degraded or polluted in some extents during their first
life or their reprocessing. Irradiation was recognized as one of the methods used in this purpose.[33] Most often, a fraction of a recycled polymer is
incorporated into a virgin polymer with the same nature, and irradiation
is performed to improve mechanical (or other) properties. Such dilution was
studied in many cases for PE[9,86–90] or PET.[91] In some cases, the recycling of
two polymers leads to combining them to obtain a new material. For these
blends, the problem with compatibilization is the same as for virgin polymer blends.

However, in most cases, the presence of a second polymer (or more) is
not desired. According to the sorting process, two, three, or more polymers
could be blended in various ratios. Moreover, the compatibilizing treatment
should be robust to take into account the changes in composition of wastes
during time. Different researchers' teams studied if irradiation could be
such a robust treatment.[60,92–94]

Zenkiewicz et al.[60,93,94] blended five common polymers: LDPE, HDPE, PP,
PS, and PET. The content of each polymer was 24, 23, 21, 15, and 17 wt%,
respectively. This composition was considered close to that of packaging
plastic waste being dumped in Poland. E-beam irradiation was combined

with the incorporation of compatibilizers (10 wt% of SEBS-g-MAH or 1 wt% of TMPTMA) to improve the mechanical properties of the blend. Dose range was 0 to 300 kGy under ambient conditions. In any cases (with or without compatibilizers), irradiation did not lead to an improvement in mechanical properties. Charpy and tensile-impact strengths hardly changed even at a dose of 300 kGy for the noncompatibilized blend and the TMPTMA compatibilized blend. The addition of SEBS-g-maleic anhydride (MAH) allowed an increase in impact strength, but irradiation made this property decrease.

Elmaghor et al.[92] blended three common plastics in the packaging industry: HDPE, PVC, and PS. Compositions were changed but HDPE was always the major polymer. Two compatibilizers were used at different contents: EVA (to compatibilize HDPE and PVC) and SEBS (to compatibilize HDPE and PS). As in the previous study, γ-irradiation (up to 200 kGy) was also performed to compatibilize the polymers. In most cases, irradiation could not improve the mechanical properties. Tensile strength hardly changed while elongation at break decreased. However, in some cases, impact strength could increase, in particular for moderate doses (50 to 100 kGy) (Table 9.4). The highest increase

TABLE 9.4

Impact Strength (J.m^{-1}) of Various Systems

	Irradiation dose (kGy)				
	0	50	100	150	200
HDPE	14	17.21	23.44	NB	23.62
90-10 PS-SEBS	1.92	2.24	3.07	1.33	1.26
PS	0.66	0.66	0.66	0.66	0.66
90-10 HDPE-PVC	6.30	6.45	6.25	5.93	3.44
90-10 HDPE-PS	7.11	7.62	9.03	10.32	8.03
HDPE-PS-PVC-EVA-SEBS wt ratios					
70-15-15-0-0	2.39	2.43	3.34	2.29	2.15
70-15-15-2.5-2.5	3.09	3.33	3.65	2.36	2.22
70-15-15-5-5	3.27	3.57	4.38	2.47	2.54
70-15-15-7.5-7.5	3.38	3.87	4.63	3.69	2.77
70-15-15-10-10	3.96	5.00	5.38	7.22	4.86
70-15-15-15-15	4.44	5.82	8.18	NB	7.54
90-5-5-7.5-7.5	21.27	23.74	NB	NB	NB
80-10-10-7.5-7.5	6.75	7.94	10.95	8.61	6.62
70-15-15-7.5-7.5	3.38	3.87	4.63	3.69	2.77
60-20-20-7.5-7.5	2.06	2.15	2.82	2.20	1.94
50-25-25-7.5-7.5	1.88	1.97	2.35	2.08	1.70

Source: Elmaghor, F., Zhang, L., Li, H., Recycling of high density polyethylene/poly(vinyl chloride)/polystyrene ternary mixture with the aid of high energy radiation and compatibilizers. *Journal of Applied Polymer Science* 2003, 88, 2756–2762. With permission.

Note: NB, no break.

was obtained for a HDPE-rich blend with low contents of PVC and PS (5 wt% for each polymer) and higher contents of compatibilizers (7.5 wt% for each one). While the effect of irradiation depends strongly on the composition, this method could not be considered as a robust method overcoming the changes in waste compositions.

Another specific problem concerns the recycling of cross-linked macro-molecular structures as thermosets or rubbers. These materials could not be reprocessed by melting and molding. One solution is their grinding and incorporation into a new polymer matrix as fillers. Thus, several articles carried out the irradiation of a thermoplastic matrix containing ground rubber particles. No entanglement between rubber and thermoplastic macro-molecules could be expected due to the cross-linked structure of rubber. Moreover, particle size is generally very high (more than one hundred microns). Consequently, the incorporation of ground rubber decreases dras-tically the mechanical properties of the material.

R. Sonnier et al.[42] γ-irradiated some recycled HDPE/ground tire rubber (GTR). GTR content was varied between 0 and 70 wt%. The irradiation dose ranged between 0 and 100 kGy. Elongation at break and Charpy impact strength of pure polyethylene decreased while Young's modulus and yield stress increased with increasing irradiation dose. These changes are related to the cross-linking of polyethylene under irradiation. On the contrary, for the 50/50 HDPE-GTR blend, elongation at break and Charpy impact strength increase significantly with increasing dose up to 50 kGy. Such an improve-ment of ductility was ascribed to the cross-linking at the interface between the polyethylene and the rubber, which are two cross-linkable materials by irradiation. It must be noticed that this phenomenon is observed only when the irradiation is performed before injection molding.

M.M. Hassan et al.[59,65,95] incorporated waste rubber into different matrices (PA6/PA66, PE, NR) and irradiated the blends sometimes in the presence of additives. For PE/waste rubber blends, tensile strength increased and elonga-tion at break decreased with irradiation due to the cross-linking of the matrix. All mechanical properties of polyamide (PA)/waste rubber blends decreased with increasing irradiation dose probably because of a strong degradation of the matrix. Finally, a slight increase of tensile strength and elongation at break was reported for NR/GTR blends with a low dose of 30 kGy. Nevertheless, the same tendency was observed for pure NR. Abou Zeid et al.[41] have irradiated EPDM/HDPE blends. EPDM was partially substituted by GTR. Irradiation promotes the cross-linking of the material and thus increases in tensile strength and decrease to the elongation at break. The partial substitution of EPDM by GTR leads to the decrease of all mechanical properties.

In the above studies, no interfacial cross-linking as previously reported by Sonnier et al.[42] was observed. It is possible that this difference is ascribed to differences in processing. While blends were injection molded after irradia-tion in the work of Sonnier et al.,[42] irradiation was performed on standard specimens in all other studies.

9.10.2 Shape Memory Effect

Under specific stimulus, shape memory materials could move from a temporary shape to their original shape.[96] The stimulus could be light, pH, or electric or magnetic field, but the most common stimulus is heat. In this case, a shape memory polymer (SMP) possesses a switch transition temperature. When the SMP is subject to deformation, its cross-linking structure could store internal stress if it is cooled below this switch temperature. When the polymer is heated above this temperature, it returns to its original shape. Shape memory polymer blends could be achieved using irradiation.

Zhu et al.[97] studied a biocompatible shape memory polymer blend based on poly(ε-caprolactone) (PCL) and polymethylvinylsiloxane (PVMS). Pure PCL was subject to scission rather than cross-linking under irradiation. In the presence of a small amount of PVMS (<20 wt%), both polymers are miscible in the amorphous phase and the radiation cross-linking of PCL is enhanced. Mechanical properties were improved, and a strong shape memory behavior was achieved. Above the melting point of PCL, the blend exhibited a rubber-like state and could be deformed. The switch temperature was the melting temperature of PCL. With 5 to 15 wt% of PVMS and under 100 kGy γ-irradiation, the deformation fixation ratio and the deformation recovery ratio were 100%.

Irradiation of LDPE/EVA blends has been extensively studied because of their particular interest as heat-shrinkable materials for the wire and cable industry.[46–50,55,56,75,98–104] Heat-shrinkable materials are a kind of shape memory materials with heat as their stimulus.

Chattopadhyay et al.[49] prepared LDPE/EVA films irradiated using an electron beam accelerator. A sensitizer (ditrimethylol propane tetraacrylate, DTMPTA) was also added. They studied heat shrinkability after optimizing the conditions of the test. The films with an initial lengh L_0 (2.5 cm) were stretched to a particular percent stretching (40%) and at a definite temperature (343 K) for 120 s. After any elastic recovery, the lengh L_1 was measured. The films were placed in an oven at a given temperature (353 K) for a certain time (60 s). The new lengh L_2 was then measured. Heat shrinkage was defined as $100(L_1 - L_2)/L_1$. The amnesia rating was defined as $100(L_2 - L_0)/L_0$.

Their results were explained on the basis of elastic recovery (it means cross-linking density) and crystallinity. EVA increased the amorphous fraction of the blends and enhanced its radiation cross-linking. Heat shrinkability increased with the increase in crystallinity and decreased with the increase in gel fraction. DMPTMA level, EVA content, and radiation dose decreased the heat shrinkability. The amnesia rating also decreased with increasing radiation dose and EVA content.

Viksne et al.[103] studied the heat shrinkability of films based on polyolefin waste. Films based on LDPE, HDPE, and their blends were subjected to γ-irradiation (up to 250 kGy). They were drawn at 130°C and cooled under tension. By further heating under relaxed conditions, the films exhibit a "memory effect." The authors concluded that the extent of shrinkage depended on the degree of degradation of PE wastes and on the irradiation dose. They showed that PE wastes needed higher doses than virgin PE to exhibit the "memory effect" (150 kGy rather than 100 kGy).

9.10.3 Preparation of Hydrogels

Polymer hydrogels are cross-linked hydrophilic polymer networks that are able to swell considerably in an aqueous medium. The development of hydrogels is a dynamic field of research, and these materials could be used in various applications as agriculture, wastewater treatment, and of course biomedicine and pharmacy (drug delivery systems, wound dressing, soft contact lenses, etc.).

Irradiation is one of the best methods to prepare polymer hydrogels while it avoids the use of potentially toxic additives. Moreover, sterilization for biomedical applications could be performed in the same time. Irradiation of polymers in aqueous solution provokes a radiolysis of water, leading to the formation of reactive radicals, as OH⁻ radicals. The radicals could react with macromolecules and transfer the radical center to chains.

Polymer blend hydrogels could be prepared to combine properties of both polymers or to obtain a good balance between some properties. In particular, hydrogels based on a synthetic polymer and a polysaccharide are extensively studied. In aqueous solution, many polysaccharides like chitosan could not cross-link under irradiation. Macromolecules are not enough close to each other to cross-link. Therefore, degradation occurs. On the contrary, when polysaccharide is blended with a synthetic cross-linkable polymer like poly(vinyl alcohol) (PVA), polyethylene oxide (PEO), or polyvinyl pyrolidone (PVP), stable hydrogels could be prepared even as the addition of polysaccharide decreases the gel fraction and the cross-linking density. Yan et al.[105] studied hydrogels based on poly(vinylmethylether) (PVME) and carboxymethylchitosan (CM-chitosan or CMCS). The ratio between these two polymers was 50/50, and the e-beam radiation dose ranged from 20 to 60 kGy. The authors showed that irradiation led simultaneously to the cross-linking of PVME, the grafting of some CM-chitosan chains onto PVME macromolecules and the degradation of CM-chitosan chains to shorter chains (Figure 9.8). Similar results were obtained by Abad et al.[106] on PVP-kappa carrageenan (KC) hydrogels. KC was mainly entrapped into cross-linked PVP gel, leading to a semi-interpenetrating polymer network (Semi-IPN). Some KC chains were also grafted on PVP networks. Grafting of KC increased with increasing KC content.

Zhao et al. prepared antibacterial hydrogels based on PVA and carboxymethylated chitosan (CM-chitosan) with e-beam irradiation.[107] PVA was cross-linkable under irradiation, but this polymer had no antibacterial activity. A hydrogel based on 10 wt% of PVA and not more than 3 wt% of CM-chitosan showed an antibacterial activity against *Escherichia coli*.

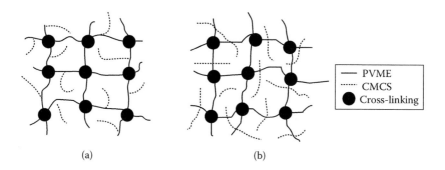

(a)   (b)

FIGURE 9.8
Structure of (a) poly(vinyl methyl ether) (PVME)-cn-chitosan graft hydrogel, (b) semi-inter-penetrating polymer network (IPN) hydrogel cross-linked using EB. (From Yan, S. F., Yin, J. B., Yu, Y., Luo, K., Chen, X. S., Thermo- and pH-sensitive poly(vinylmethyl ether)/carboxymethyl-chitosan hydrogels crosslinked using electron beam irradiation or using glutaraldehyde as a crosslinker. *Polymer International* 2009, 58, 1246–1251. With permission.)

Thermosensitive,[7] pH-sensitive,[108] and even thermo- and pH-sensitive[105] hydrogels were prepared by several research teams using irradiation. In the latter reference, PVME was used due to its thermoresponsive behavior in the temperature range studied (20 to 40°C). Indeed, this polymer presents a lower critical solution temperature close to 37°C. Below this temperature, there is hydration of PVME chains leading to the swelling of the PVME network. Above 37°C, dehydration and further deswelling occur. Although nonionic polymers like PVME do not exhibit pH sensitivity, CM-chitosan was used in this purpose. CM-chitosan is an amphoteric polymer containing amino and carboxyl groups, forming a network with oppositely charged structures. At low or high pH, swelling occurs due to the electrostatic repulsion between protonated amino groups and dissociated carboxylic groups. But at pH = 3, there is an equivalent amount of these groups and minimum swelling was obtained.

A generally increasing radiation dose leads to increasing gel strength, increasing gel fraction, but decreasing swelling behavior. Moreover, above a specific dose, degradation of macromolecules occurs. Thus, an optimal dose should be determined. For a hydrogel based on kappa-carrageenan and polyethylene oxide, Tranquilan-Aranilla et al.[109] obtained an optimal dose of 10 kGy to optimize the properties of their hydrogel.

Zhai et al.[110] studied PVA/starch hydrogels. Starch is composed of amylopectin and amylose and the effect of irradiation on both components is very different. Gel fraction decreases with increasing starch content. One starch fraction is physically entrapped in PVA gel while another fraction is chemically grafted on PVA chains. Further analysis showed that amylopectin is not intermiscible with PVA and could not react with PVA. On the contrary, amylose could be blended homogeneously with PVA and grafting reaction could occur.

Yang et al.[111,112] prepared hydrogels based on PVA and water-soluble chitosan using various processes: γ-irradiation alone (30 kGy), freeze-thawing alone (three cycles between –20 and 25°C—freeze-thawing is another common method to prepare hydrogels without chemical cross-linked polymer networks) and a combination of both processes (irradiation after freeze-thawing and freeze-thawing after irradiation). While hydrogels prepared by irradiation alone have too low mechanical properties for wound dressing application, hydrogels prepared by irradiation and freeze-thawing are suitable. The authors found that the structure of hydrogels is mainly determined by the first processing step. The gel fraction does not change significantly with the process. When irradiation is performed before freeze-thawing, hydrogels show better mechanical strength, higher swelling capacity and thermal stability, and are less turbid because irradiation leads to chemical cross-linking of chains from both polymers and then limits the phase separation occurring in the freeze-thawing process.

Nurkeeva et al.[113] studied the influence of pH on the structure of a poly(acrylic acid)-poly(vinyl alcohol) hydrogel. They found that hydrogel films could be prepared only between two critical pH values, pH_1 and pH_2. Below the lower value (pH_1 = 2.75 to 3), the formation of hydrophobic interpolymer complexes led to their aggregation into particles of a colloidal dispersion. Above the second critical pH (pH_2 = 4.5), both polymers were not miscible. Therefore, γ-irradiation did not lead to the cross-linking. The products were fully soluble. High gel fraction (>70%) was only obtained for pH ranging between pH_1 and pH_2.

9.10.4 Enhanced Electrical Conductivity of Polymer Blends by Irradiation

Under irradiation, some ionic species are generated that could enhance the conductivity. Nevertheless, this irradiation-induced conductivity is not stable and decreases faster. However, irradiation could be used indirectly to make some polymer blends durably conductive.

Polyaniline (PANI) is a nonconductive polymer that could be rendered conductive by HCl doping. Several methods exist to accomplish such a doping. One of them is the incorporation of a chlorine-containing polymer into a polyaniline matrix and the irradiation of the resultant blend. Ionizing radiation (e-beam or γ-irradiation) leads to the dehydrochlorination of the second polymer and the subsequent doping of the PANI (Figure 9.9). Several chlorine-containing polymers were tested in this approach, PVC, chlorinated polyisoprene, poly(vinylidene chloride-*co*-vinyl acetate), poly(vinylidene chloride-*co*-vinyl chloride).[114–116]

Bodügoz-Sentürk and Güven[114] have prepared by solution casting and γ-irradiated some PANI/P(VDC-*co*-VAc) films. A strong increase of 9 orders of magnitude in the conductivity of these films was obtained for a dose of 500 kGy (from 10^{-9} to almost 10^{-1} S/cm). The total dose, the temperature, the stretching, and the thickness of the films had an effect on the conductivity.

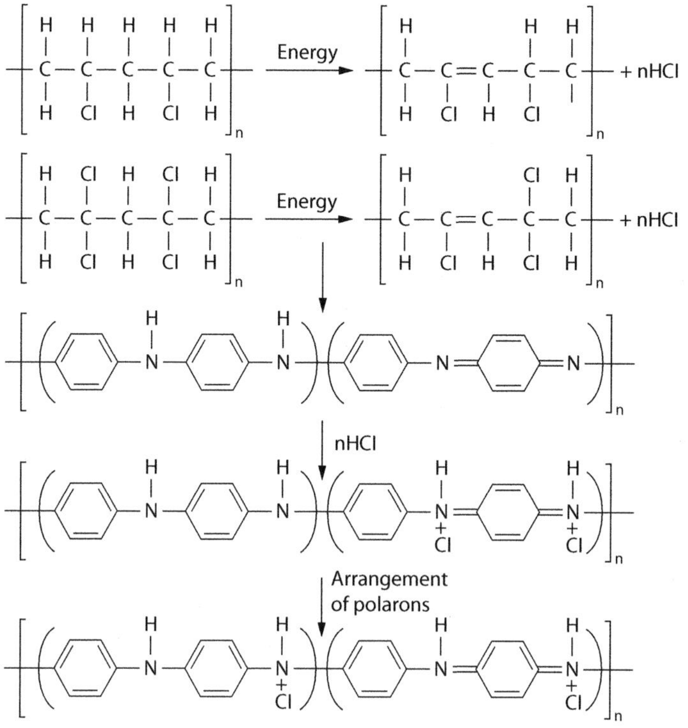

FIGURE 9.9
Radiation-induced dehydrochlorination of polyvinyl chloride (PVC) and poly(vinylidene chloride (PVDC) and concomitant doping of polyaniline (PANI)-base with released HCl. (From Güven, O., Radiation-induced conductivity control in polyaniline blends/composites. *Radiation Physics and Chemistry* 2007, 76, 1302–1307. With permission.)

However, the most influent parameter was the content of the P(VDC-*co*-VAc). An optimal P(VDC-*co*-VAc)/PANI molar ratio of 0.5 allowed to obtain a conductivity close to 0.1 S/cm at 500 kGy. If this ratio was fixed to 2, the conductivity decreased to less than 10^{-4} S/cm for the same dose. Moreover, the authors claimed that the conductivity of the films was stable for at least 1 year under ambient conditions.

Another interest of irradiation in the field of the conductivity of polymer blends was shown by Faez et al.[117] These authors incorporated EPDM to PANI in order to improve mechanical properties of the material. PANI was doped not with irradiation, but with dodecylbenzene sulfonic acid by reactive processing in an internal blender. To cross-link the EPDM, a phenolic resin and e-beam irradation (75 and 150 kGy) were compared. Contrary to the cross-linking by phenolic resin, the cross-linking by e-beam irradiation did not interfere with the presence of the acid necessary for doping PANI. Consequently, better mechanical properties were obtained. Moreover, irradiation-induced additional conductivity led to a slightly more conductive material.

9.10.5 Synthesis of Silicon Carbide Fibers

Silicon carbide fibers are difficult to obtain by conventional powder processing methods. One way to synthesize such fibers is to pyrolyze organosili-
con polymers like polycarbosilane after melt-spinning and cross-linking.
Polycarbosilane fibers should be cross-linked to avoid their melting during
pyrolysis. Two processes are used for this purpose: oxidation curing in hot
air and irradiation in an inert atmosphere.[118,119] The radiation dose is very
high in this process (1 to 10 MGy). Even though the fibers prepared from oxidation curing have low thermal resistance, irradiation allows us to achieve
fibers with low content of oxygen and consequently with an increasing service temperature (1300°C and more).[120]

Nevertheless, a polycarbosilane precursor is not easy to spin, and
researchers tried to incorporate a polymer additive to improve its spinning ability. Su et al.[120] incorporated a small amount of polypropylene to
polycarbosilane. Both polymers are compatible, because of the proximity
of their structures. Spinning ability is significantly improved and average
continuous fiber length increases from 4800 to 9312 m with 5 wt% of PP.
Moreover, the tensile strength of the fibers increases from 8.5 to 17 MPa
while the diameter remains unchanged (14.6 microns). Polypropylene
also has a great effect on electron-beam radiation curing. Under an inert
atmosphere, polypropylene is able to cross-link. A gel fraction of 80% is
achieved for only 2 MGy while polycarbosilane attains such a gel fraction
for a dose higher than 6 MGy. The incorporation of PP into polycarbosilane (PCS) leads to the cross-linking at lower doses. This result is very
important, because the dose rate is usually low to avoid the sticking of the
fibers due to the heat generated by radiation. Then the curing is costly. The
incorporation of a small amount of PP allows for reduction of the curing
time. The authors show also that there are chemical interactions between
PP and PCS under radiation. Finally there is no detrimental effect of PP on
the properties of the silicon carbide fibers.

Polyvinylsilane (PVS) has been also incorporated into PCS to improve the
spinning ability of the silicon carbide fibers.[121,122] PVS plasticizes PVC and
decreases the spinning temperature: 490 K for a 80/20 PCS-PVS blend versus
600K for PCS.[121] Thinner diameters are obtained with an addition of 10 or
20 wt% of PVS. After γ-irradiation and pyrolysis, tensile strength is higher
for fibers from PCS/PVS blends at the same dose.[122] Moreover, fibers (without fusing) could be achieved at a lower dose when PVS is added to PCS.
Nevertheless, the addition of PVS increases the oxygen content in the fibers
and therefore decreases their heat stability. Cross-linking under inert atmosphere has also been carried out to avoid this problem, but in this case, a
higher dose is needed for curing PCS/PVS blends than for pure PCS.

Wach et al.[123] used a precursor polymer blend to develop a silicon carbide
ceramic coating on porous alumina substrate as a gas separation membrane. A small amount (5 to 20 wt%) of polyvinylsilane was blended with

polycarbosilane in order to obtain SiC film without defects. Such defects (cracks or pinholes formed during pyrolysis) reduced drastically the selectivity of the membranes.

9.11 Conclusions and New Trends

Irradiation is an efficient method to modify the structure of a polymer. Various phenomena could be promoted by irradiation like cross-linking, grafting, chain scissioning, and oxidation. The net effect of irradiation depends mainly on the chemical structure of the polymer, on the radiation dose, on the presence of some additives (radicals scavengers or cross-linking promoters) and on some other conditions (like atmosphere).

In the case of polymer blends, phenomena could occur not only in the bulk of each polymer but also at the interface. Therefore, another parameter could influence the phenomena induced by irradiation: the morphology of the blend. In the previous pages, some works have described the role of morphology: the stabilizing effect from aromatic rings is allowed only for the closest chemical groups, and grafting between two polymers is promoted in the case of miscibility. Miscibility seems to be a strong condition to favor interactions between both constituents, but some works show also that the behavior under irradiation of nonmiscible blends does not follow a simple rule of mixtures.

Irradiation is distinguished from other methods of compatibilization by its decoupling with other processing steps like melt blending and injection molding. Moreover, irradiation is performed at room temperature, and macromolecules have low mobility and radicals could have a high lifetime, in particularly in crystalline domains. Therefore, postprocessing phenomena could be occurring. These differences allow irradiation leading to different materials according to their position during the processing of the material (before blending, before or after molding). From this point of view, irradiation could be considered as a very versatile method.

Recently, the Leibniz Institute of Polymer Research Dresden developed a process that allows the coupling between e-beam irradiation and melt-mixing.[124-126] A 1.5 MeV electron accelerator has been coupled to a Banbury mixing chamber. In this process, polymers are fully melted and then there is no crystallinity. Irradiation is performed at high temperatures and macromolecules are highly mobile. Moreover, contrary to classic static processes, the radiation dose is delivered during intensive mixing. First studies are carried out on polyolefins (PP and PE) filled with a high content of magnesium hydroxyde (MH) (60 wt%) and tryallyl cyanurate (TAC) (2 wt%) as a grafting agent.[126] Results show that better mechanical properties (elongation at break and toughness) are obtained with this in-line electron-induced reactive processing method, in comparison with the static process (irradiation after melt blending). Charpy toughness of PE/MH/TAC exceeds 35 kJ/m² in the former process against 12 kJ/m² in the latter.

Rooj et al.[125] prepared a polypropylene/epoxidized natural rubber 50/50 PP-ENR blend in the presence of triallyl cyanurate (TAC) (2 wt%). In a first step, they irradiated PP/TAC during melt mixing at 50 kGy. In a second step, ENR was added in the mixing chamber. This blend was compared to PP/TAC/ENR blend prepared in one step (without irradiation). Irradiation generated polar carboxylic acid groups on PP, and degradation was prevented by TAC. Carboxylic acid groups could react with epoxy groups on ENR resulting in finer morphology and better mechanical properties.

50/50 PP/EPDM thermoplastic vulcanizates were processed using in-line electron-induced reactive processing method.[124] The authors studied the influence of some parameters like radiation dose (25 to 100 kGy), electron treatment time (15 to 60 s), and electron energy (0.6 and 1.5 MeV). Two phenomena are promoted by irradiation: in situ compatibilization of PP and EPDM and cross-linking of EPDM. The absorbed dose has a great effect on final properties and morphology like in the static irradiation process. But electron treatment time and electron energy also have a significant effect. It should be noticed that in this process the penetration depth of electrons is limited to a part of the mixing chamber. Thus, contrary to static irradiation processes, these parameters control the ratio of radical generation rate to mixing rate (dose per rotation) and the ratio of modified volume to total mixing volume. Best mechanical properties (tensile strength and elongation at break) were obtained with a dose of 100 kGy delivered during 15 s (the lowest time applied) and an electron energy of 1.5 MeV. EPDM nodule size was as low as 0.04 µm.

References

1. Singh, A., Bahari, K., Use of high-energy radiation in polymer blends technology. *Polymer Blends Handbook*, 11, 757–859.
2. Miklesova, K., Szocs, F., Free-radical decay in a polymer blend PMMA/PEO studied by the EPR method. *European Polymer Journal* 1992, 28, 553–554.
3. Smith, A. P., Spontak, R. J., Ade, H., On the similarity of macromolecular responses to high-energy processes: mechanical milling vs. irradiation. *Polymer Degradation and Stability* 2001, 72(3), 519–524.
4. Alessi, S., Conduruta, D., Pitarresi, G., Dispenza, C., Spadaro, G., Hydrothermal ageing of radiation cured epoxy resin-polyether sulfone blends as matrices for structural composites. *Polymer Degradation and Stability* 2010, 95, 677–683.
5. Alessi, S., Dispenza, C., Fuochi, P., Corda, U., Lavalle, M., Spadaro, G., E-beam curing of epoxy-based blends in order to produce high-performance composites. *Radiation Physics and Chemistry* 2007, 76, 1308–1311.

6. Alessi, S., Dispenza, C., Spadaro, G., Thermal properties of e-beam cured epoxy/thermoplastic matrices for advanced composite materials. *Macromolecular Symposia* 2007, 247, 238–243.

7. Gottlieb, R., Schmidt, T., Arndt, K., Synthesis of temperature-sensitive hydrogel blends by high-energy irradiation. *Nuclear Instruments and Methods in Physics Research Section B: Beam Interactions with Materials and Atoms* 2005, 236, 371–376.

8. Valenza, A., Gallo, L., Spadaro, G., Calderaro, E., Acierno, D., Mechanical properties-structure relationship in blends of polyamide 6 with γ-irradiated LDPE. *Polymer Engineering and Science* 1993, 33, 1336–1340.

9. Ali, Z., Badr, Y., Eisa, W., Gel, Thermal, and x-ray diffraction characterization of virgin, scrapped polyethylene and its blends. *Polymer Composites* 2006, 27(6), 709.

10. Raj, J. M., Kumaraswamy, G. N., Ranganathaiah, C., Interfacial stabilization of binary polymer blends through radiation treatment: A free volume approach. *Physica Status Solidi (C)* 2009, 6, 2404–2406.

11. Raj, J. M., Ranganathaiah, C., A new method of stabilization and characterization of the interface in binary polymer blends by irradiation: A positron annihilation study. *Journal of Polymer Science Part B: Polymer Physics* 2009, 47, 619–632.

12. Raj, J. M., Ranganathaiah, C., A free-volume study on the phase modifications brought out by e-beam and microwave irradiations in PP/NBR and PVC/SAN blends. *Polymer Degradation and Stability* 2009, 94, 397–403.

13. Raj, J. M., Ranganathaiah, C., Ganesh, S., Interfacial modifications in PS/PMMA and PVC/EVA blends by e-beam and microwave irradiation: A free volume study. *Polymer Engineering and Science* 2008, 48(8), 1495–1503.

14. Lee, G. J., Han, M. G., Chung, S. C., Suh, K. D., Im, S. S., Effect of crosslinking on the positive temperature coefficient stability of carbon black-filled HDPE/ethylene-ethylacrylate copolymer blend system. *Polymer Engineering and Science* 2002, 42, 1740–1747.

15. Rivaton, A., Lalande, D., Gardette, J., Influence of the structure on the γ-irradiation of polypropylene and on the post-irradiation effects. *Nuclear Instruments and Methods in Physics Research Section B: Beam Interactions with Materials and Atoms* 2004, 222, 187–200.

16. Przybytniak, G. K., Zagorski, Z. P., Zuchowska, D., Free radicals in electron beam irradiated blends of polyethylene and butadiene–styrene block copolymer. *Radiation Physics and Chemistry* 1999, 55(5–6), 655–658.

17. Sonnier, R., Massardier, V., Clerc, L., Lopez-Cuesta, J. M., Bergeret, A., Reactive compatibilization of polymer blends by gamma-irradiation: Influence of the order of processing steps. *Journal of Applied Polymer Science* 2010, 115(3), 1710–1717.

18. Chen, J., Huang, X., Jiang, P., Wang, G., Protection of SEBS/PS blends against gamma radiation by aromatic compounds. *Journal of Applied Polymer Science* 2009, 112, 1076–1081.

19. Dubey, K. A., Bhardwaj, Y. K., Chaudhari, C. V., Bhattacharya, S., Gupta, S. K., Sabharwal, S., Radiation effects on SBR–EPDM blends: A correlation with blend morphology. *Journal of Polymer Science Part B: Polymer Physics* 2006, 44, 1676–1689.

20. McHerron, D., Wilkes, G., Electron beam irradiation of polystyrene-poly(vinyl methyl ether) blends. *Polymer* 1993, 34, 3976–3985.

21. El-Salmawi, K., Abu Zeid, M. M., El-Naggar, A. M., Mamdouh, M., Structure–property behavior of gamma–irradiated poly(styrene) and poly(methyl methacrylate) miscible blends. *Journal of Applied Polymer Science* 1999, 72, 509–520.

22. Abd Alla, S. G., Said, H. M., El-Naggar, A. W. M., Structural properties of γ-irradiated poly(vinyl alcohol)/poly(ethylene glycol) polymer blends. *Journal of Applied Polymer Science* 2004, 94, 167–176.

23. El-Din, H. M. N., El-Naggar, A. W. M., Ali, F. I., Miscibility of poly(vinyl alcohol)/polyacrylamide blends before and after gamma irradiation. *Polymer International* 2003, 52, 225–234.

24. El-Din, H. M. N., El-Naggar, A. W. M., Ali, F. I., Thermal decomposition behavior of γ-irradiated poly(vinyl acetate)/poly(methyl methacrylate) miscible blends. *Journal of Applied Polymer Science* 2006, 99, 1773–1780.

25. Torikai, A., Harayama, K., Hayashi, N., Mitsuoka, T., Fueki, K., Radiation-induced degradation of poly(styrene-co-methylmethacrylate) and blends of polystyrene and polymethylmethacrylate. *Radiation Physics and Chemistry* 1994, 43(5), 493–496.

26. Babanalbandi, A., Hill, D. J. T., Radiation chemistry of arylpolyester blends with a polyalkanoate. *Polymer International* 1999, 48(10), 963–970.

27. Lee, S. J., Kwon, S.-K., Cho, W.-J., Ha, C.-S., Properties and radiation resistance of aromatic polymer-based polyblends. *Journal of Applied Polymer Science* 1999, 73, 1697–1705.

28. Cangialosi, D., McGrail, P., Emmerson, G., Valenza, A., Calderaro, E., Spadaro, G., Properties and morphology of PMMA/ABN blends obtained via MMA in situ polymerisation through γ-rays. *Nuclear Instruments and Methods in Physics Research Section B: Beam Interactions with Materials and Atoms* 2001, 185, 262–266.

29. Spadaro, G., Dispenza, C., McGrail, P. T., Valenza, A., Cangialosi, D., Submicron structured polymethyl methacrylate/acrylonitrile-butadiene rubber blends obtained via gamma radiation induced in situ polymerization. *Advances in Polymer Technology* 2004, 23, 211–221.

30. Peng, J., Zhang, X., Qiao, J., Wei, G., Radiation preparation of ultrafine carboxylated styrene-butadiene rubber powders and application for nylon 6 as an impact modifier. *Journal of Applied Polymer Science* 2002, 86, 3040–3046.

31. Valenza, A., Carianni, G., Mascia, L., Radiation grafting functionalization of poly(vinylidene fluoride) to compatibilize its blends with polyolefin ionomers. *Polymer Engineering and Science* 1998, 38(3), 452–460.

32. Singh, A., Irradiation of polymer blends containing a polyolefin. *Radiation Physics and Chemistry* 2001, 60, 453–459.

33. Burillo, G., Clough, R. L., Czvikovszky, T., Guven, O., Le Moel, A., Liu, W. W., Singh, A., Yang, J. T., Zaharescu, T., Polymer recycling: Potential application of radiation technology. *Radiation Physics and Chemistry* 2002, 64(1), 41–51.

34. Spadaro, G., Acierno, D., Dispenza, C., Calderaro, E., Valenza, A., Physical and structural characterization of blends made with polyamide 6 and gamma-irradiated polyethylenes. *Radiation Physics and Chemistry* 1996, 48(2), 207–216.

35. Spadaro, G., Acierno, D., Dispenza, C., Valenza, A., Thermal analysis of blends made with polyamide 6 and gamma-irradiated polyethylenes. *Thermochimica Acta* 1995, 269/270, 261–272.

36. Valenza, A., Spadaro, G., Calderaro, E., Acierno, D., Structure and properties of nylon-6 modified by gamma-irradiated linear low-density polyethylene. *Polymer Engineering and Science* 1993, 33(13), 845–850.

37. Adem, E., Burillo, G., Avalos-borja, M., Carreon, M., Radiation compatibilization of polyamide-6/polypropylene blends, enhanced by the presence of compatibilizing agent. *Nuclear Instruments and Methods in Physics Research Section B: Beam Interactions with Materials and Atoms* 2005, 236, 295–300.
38. Senna, M. M., Hossam, F. M., El-naggar, A. W. M., Compatibilization of low-density polyethylene/plasticized starch blends by reactive compounds and electron beam irradiation. *Polymer Composites* 2008, 29(10), 1137–1144.
39. Kim, J. U. N. Y., Kim, O. H. S., Kim, S. H. U. N., Jeon, H. A. N. Y., Effects of electron beam irradiation on poly(ethylene2,6-naphtalate)/poly(ethylene terephthalate) blends. *Polymer Engineering and Science* 2004, 44(2), 395–405.
40. Martínez-Barrera, G., Lopêz, H., Castano, V. M., Rodriguez, R., Studies on the rubber phase stability in gamma irradiated polystyrene-SBR blends by using FT-IR and Raman spectroscopy. *Radiation Physics and Chemistry* 2004, 69, 155–162.
41. Abou Zeid, M. M., Rabie, S. T., Nada, A. A., Khalil, A. M., Hilal, R. H., Effect of gamma irradiation on ethylene propylene diene terpolymer rubber composites. *Nuclear Instruments and Methods in Physics Research Section B: Beam Interactions with Materials and Atoms* 2008, 266, 111–116.
42. Sonnier, R., Leroy, E., Clerc, L., Bergeret, A., Lopez-cuesta, J., Compatibilisation of polyethylene/ground tyre rubber blends by γ irradiation. *Polymer Degradation and Stability* 2006, 91, 2375–2379.
43. Zaharescu, T., Chipara, M., Postolache, M., Radiation processing of polyolefin blends. II. Mechanical properties of EPDM-PP blends. *Polymer Degradation and Stability* 1999, 66, 5–8.
44. Zaharescu, T., Assessment of compatibility of ethylene–propylene–diene terpolymer and polypropylene. *Polymer Degradation and Stability* 2001, 73, 113–118.
45. Zaharescu, T., Setnescu, R., Jipa, S., Setnescu, T., Radiation processing of polyolefin blends. I. Crosslinking of EPDM-PP blends. *Journal of Applied Polymer Science* 2000, 77, 982–987.
46. Ali, Z. I., Effect of electron beam irradiation and vinyl acetate content on the physicochemical properties of LDPE/EVA blends. *Journal of Applied Polymer Science* 2007, 104(5), 2886–2895.
47. Dadbin, S., Frounchi, M., Sabet, M., Studies on the properties and structure of electron-beam crosslinked low-density polyethylene/poly[ethylene-co-(vinyl acetate)] blends. *Polymer International* 2005, 54, 686–691.
48. Dalai, S., Wenxiu, C., Radiation effects on LDPE/EVA blends. *Journal of Applied Polymer Science* 2002, 86, 1296–1302.
49. Chattopadhyay, S., Chaki, T. K., Bhowmick, A. K., Heat shrinkability of electron-beam-modified thermoplastic elastomeric films from blends of ethylene-vinylacetate copolymer and polyethylene. *Radiation Physics and Chemistry* 2000, 59(5–6), 501–510.
50. Dalai, S., Wenxiu, C., Radiation effects on HDPE/EVA blends. *Journal of Applied Polymer Science* 2002, 86, 553–558.
51. Dalai, S., Wenxiu, C., Radiation effects on poly(propylene) (PP)/ethylene-vinyl acetate copolymer (EVA) blends. *Journal of Applied Polymer Science* 2002, 86, 3420–3424.
52. Mihailova, M., Kresteva, M., Aivazova, N., Krestev, V., Nedkov, E., X-ray investigation of polypropylene and poly(ethylene-co-vinyl acetate) blends irradiated with fast electrons WAXS investigation of irradiated i-PP/EVA blends. *Radiation Physics and Chemistry* 1999, 56(5–6), 581–589.

53. Li, H. H., Yin, Y., Liu, M. H., Deng, P. Y., Zhang, W. X., Sun, J. Z., Improved compatibility of EVOH/LDPE blends by γ-ray irradiation. *Advances in Polymer Technology* 2009, 28(3), 192–198.

54. Ali, Z. I., Youssef, H. A., Said, H. M., Saleh, H. H., Influence of electron beam irradiation and polyfunctional monomer loading on the physico-chemical properties of polyethylene/polypropylene blends. *Advances in Polymer Tehcnology* 2006, 25, 208–217.

55. Chattopadhyay, S., Chaki, T. K., Bhowmick, A. K., Structural characterization of electron-beam crosslinked thermoplastic elastomeric films from blends of polyethylene and ethylene-vinyl acetate copolymers. *Journal of Applied Polymer Science* 2001, 81, 1936–1950.

56. Chattopadhyay, S., Chaki, T. K., Bhowmick, A. K., Development of new thermoplastic elastomers from blends of polyethylene and ethylene-vinyl acetate copolymer by electron-beam technology. *Journal of Applied Polymer Science* 2001, 79, 1877–1889.

57. Zurina, M., Ismail, H., Ratnam, C., The effect of HVA-2 on properties of irradiated epoxidized natural rubber (ENR-50), ethylene vinyl acetate (EVA), and ENR-50/EVA blend. *Polymer Testing* 2008, 27, 480–490.

58. Zurina, M., Ismail, H., Ratnam, C. T., Effect of trimethylolpropane triacrylate (TMPTA) on the properties of irradiated epoxidized natural rubber (ENR-50), ethylene- (vinyl acetate) copolymer (EVA), and an ENR-50/EVA blend. *Journal of Vinyl and Additive Technology* 2009, 15(1), 47–53.

59. Maziad, N. A., Hassan, M. M., Study of some properties of waste LDPE/waste butyl rubber blends using different compatibilizing agents and gamma irradiation. *Journal of Applied Polymer Science* 2007, 106(6), 4157–4163.

60. Zenkiewicz, M., Dzwonkowski, J., Effects of electron radiation and compatibilizers on impact strength of composites of recycled polymers. *Polymer Testing* 2007, 26, 903–907.

61. Khalf, A. I., Nashar, D. E. E., Maziad, N. A., Effect of grafting cellulose acetate and methylmethacrylate as compatibilizer onto NBR/SBR blends. *Materials and Design* 2010, 31(5), 2592–2598.

62. Zainuddin; Sudradjat, A., Razzak, M. T., Yoshii, F., Makuuchi, K., Polyblend CPP and Bionolle with PP-g-MAH as compatibilizer: I. Compatibility. *Journal of Applied Polymer Science* 1999, 72, 1277–1282.

63. Zainuddin; Razzak, M. T., Yoshii, F., Makuuchi, K., Radiation effect on the mechanical stability and biodegradability of CPP/Bionolle blend. *Polymer Degradation and Stability* 1999, 63, 311–320.

64. Radhakrishnan, C., Alex, R., Unnikrishnan, G., Thermal, ozone and gamma ageing of styrene butadiene rubber and poly(ethylene-co-vinyl acetate) blends. *Polymer Degradation and Stability* 2006, 91, 902–910.

65. Hassan, M. M., Badway, N. A., Gamal, A. M., Elnaggar, M. Y., Hegazy, E.-S. A., Studies on mechanical, thermal and morphological properties of irradiated recycled polyamide and waste rubber powder blends. *Nuclear Instruments and Methods in Physics Research Section B: Beam Interactions with Materials and Atoms* 2010, 268, 1427–1434.

66. Dong, W., Chen, G., Zhang, W., Radiation effects on the immiscible polymer blend of nylon1010 and high-impact strength polystyrene (II): Mechanical properties and morphology. *Radiation Physics and Chemistry* 2001, 60, 629–635.

67. Dong, W., Zhang, W., Chen, G., Liu, J., Radiation effects on the immiscible polymer blend of nylon1010 and high-impact polystyrene (HIPS) I: Gel/dose curves, mathematical expectation theorem and thermal behaviour. *Radiation Physics and Chemistry* 2000, 57(1), 27–35.

68. Van Gisbergen, J. G. M., Hoeben, W. F. L. M., Meijer, H. E. H., Melt rheology of electron-beam-irradiated blends, of polypropylene and ethylene-propylene-diene monomer (EPDM) rubber. *Polymer Engineering and Science* 1991, 31, 1539–1544.

69. Steller, R., Zuchowska, D., Meissner, W., Paukszta, D., Garbarczyk, J., Crystalline structure of polypropylene in blends with thermoplastic elastomers after electron beam irradiation. *Radiation Physics and Chemistry* 2006, 75, 259–267.

70. Przybytniak, G., Mirkowski, K., Rafalski, A., Nowicki, A., Kornacka, E., Radiation degradation of blends polypropylene/poly(ethylene-*co*-vinyl acetate). *Radiation Physics and Chemistry* 2007, 76, 1312–1317.

71. Thomas, S., Gupta, B. R., De, S. K., Mechanical properties, surface morphology and failure mode of gamma-ray irradiated blends of polypropylene and ethylene-vinyl acetate rubber. *Polymer Degradation and Stability* 1987, 18(3), 189–212.

72. Minkova, L., Nikolova, M., Thermomechanical and some mechanical properties of irradiated films prepared from polymer blends. *Polymer Degradation and Stability* 1989, 23(3), 217–226.

73. Minkova, L., Nikolova, M., Melting of irradiated films prepared from polymer blends. *Polymer Degradation and Stability* 1989, 25(1), 49–60.

74. Spenadel, L., Radiation crosslinking of polymer blends. *Radiation Physics and Chemistry* 1979, 14(3–6), 683–697.

75. Zhang, W. X., Liu, Y. T., Sun, J. Z., The relationship between sol fraction and radiation-dose in radiation crosslinking of low-density polyethylene (LDPE) ethylenevinylacetate copolymer (EVA) blend. *Radiation Physics and Chemistry* 1990, 35(1–3), 163–166.

76. Gonzalez, J., Albano, C., Candal, M., Ichazo, M., Hernandez, M., Characterization of blends of PP and SBS vulcanized with gamma irradiation. *Nuclear Instruments and Methods in Physics Research Section B: Beam Interactions with Materials and Atoms* 2005, 236, 354–358.

77. Van Gisbergen, J. G. M., Meijer, H. E. H., Lemstra, P. J., Structured polymer blends: 2. Processing of polypropylene EDPM blends: Controlled rheology and morphology fixation via electron-beam irradiation. *Polymer* 1989, 30(12), 2153–2157.

78. Perera, R., Albano, C., Gonzalez, J., Silva, P., Ichazo, M., The effect of gamma radiation on the properties of polypropylene blends with styrene–butadiene–styrene copolymers. *Polymer Degradation and Stability* 2004, 85, 741–750.

79. Giri, R., Sureshkumar, M. S., Naskar, K., Bharadwaj, Y. K., Sarma, K. S. S., Sabharwal, S., Nando, G. B., Electron beam irradiation of LLDPE and PDMS rubber blends: Studies on the physicomechanical properties. *Advances in Polymer Technology* 2008, 27(2), 98–107.

80. Manshaie, R., Khorasani, S. N., Veshare, S. J., Abadchi, M. R., Effect of electron beam irradiation on the properties of natural rubber (NR)/styrene–butadiene rubber (SBR) blend. *Radiation Physics and Chemistry* 2011, 80(1), 100–106.

81. Zurina, M., Ismail, H., Ratnam, C., Characterization of irradiation-induced crosslink of epoxidised natural rubber/ethylene vinyl acetate (ENR-50/EVA) blend. *Polymer Degradation and Stability* 2006, 91, 2723–2730.

82. Clough, R., High-energy radiation and polymers: A review of commercial processes and emerging applications. *Nuclear Instruments and Methods in Physics Research Section B: Beam Interactions with Materials and Atoms* 2001, 185, 8–33.

83. Senna, M. M., Yossef, A. M., Hossam, F. M., El-naggar, A. W. M., Biodegradation of low-density polyethylene/thermoplastic starch foams before and after electron beam irradiation. *Journal of Applied Polymer Science* 2007, 106, 3273–3281.

84. Tokuda, S., Kemmotsu, T., Electron beam irradiation conditions and foam seat properties in polypropylene-polyethylene blends. *Radiation Physics and Chemistry* 1995, 46, 905–908.

85. Wongsuban, B., Muhammad, K., Ghazali, Z., Hashim, K., Hassan, M. A., The effect of electron beam irradiation on preparation of sago starch/polyvinyl alcohol foams. *Nuclear Instruments and Methods in Physics Research Section B: Beam Interactions with Materials and Atoms* 2003, 211(2), 244–250.

86. Ali, Z., Badr, Y., Eisa, W., Spectroscopic studies of structural changes in irradiated LDPE and its blends. *Journal of Polymer Science: Part B; Polymer Physics* 2007, 45, 850–859.

87. Satapathy, S., Chattopadhyay, S., Chakrabarty, K. K., Nag, A., Tiwari, K. N., Tikku, V. K., Nando, G. B., Studies on the effect of electron beam irradiation on waste polyethylene and its blends with virgin polyethylene. *Journal of Applied Polymer Science* 2006, 101, 715–726.

88. Suarez, J. C. M., Mano, E. B., Characterization of degradation on gamma-irradiated recycled polyethylene blends by scanning electron microscopy. *Polymer Degradation and Stability* 2001, 72, 217–221.

89. Suarez, J. C. M., Mano, E. B., Bonelli, C. M. C., Effects of gamma-irradiation on mechanical characteristics of recycled polyethylene blends. *Polymer Engineering and Science* 1999, 39(8), 1398–1403.

90. Suarez, J. C. M., Mano, E. B., Pereira, R. A., Thermal behavior of gamma-irradiated recycled polyethylene blends. *Polymer Degradation and Stability* 2000, 69(2), 217–222.

91. Razek, T. M. A., Said, H. M., Khafaga, M. R., El-Naggar, A. W. M., Effect of gamma irradiation on the thermal and dyeing properties of blends based on waste poly(ethylene terephthalate) blends. *Journal of Applied Polymer Science* 2010, 117, 3482–3490.

92. Elmaghor, F., Zhang, L., Li, H., Recycling of high density polyethylene/poly(vinyl chloride)/polystyrene ternary mixture with the aid of high energy radiation and compatibilizers. *Journal of Applied Polymer Science* 2003, 88, 2756–2762.

93. Zenkiewicz, M., Czuprynska, J., Polanski, J., Karasiewicz, T., Engelgard, W., Effects of electron-beam irradation on some structural properties of granulated polymer blends. *Radiation Physics and Chemistry* 2008, 77, 146–153.

94. Zenkiewicz, M., Kurcok, M., Effects of compatibilizers and electron radiation on thermomechanical properties of composites consisting of five recycled polymers. *Polymer Testing* 2008, 27, 420–427.

95. Hassan, M. M., Mahmoud, G. A., El-nahas, H. H., Hegazy, E.-S. A., Reinforced material from reclaimed rubber/natural rubber, using electron beam and thermal treatment. *Journal of Applied Polymer Science* 2007, 104, 2569–2578.

96. Meng, Q., Hu, J., A review of shape memory polymer composites and blends. *Composites Part A: Applied Science and Manufacturing* 2009, 40, 1661–1672.

97. Zhu, G., Xu, S., Wang, J., Zhang, L., Shape memory behaviour of radiation-crosslinked PCL/PMVS blends. *Radiation Physics and Chemistry* 2006, 75, 443–448.

98. Hassanpour, S., Khoylou, F., Jabbarzadeh, E., Thermal degradation of electron beam crosslinked polyethylene and (ethylene-vinylacetate) blends in hot water. *Journal of Applied Polymer Science* 2003, 89, 2346–2352.

99. Hui, S., Chaki, T. K., Chattopadhyay, S., Exploring the simultaneous effect of nano-silica reinforcement and electron-beam irradiation on a model LDPE/EVA-based TPE system. *Polymer International* 2009, 58, 680–690.

100. Hui, S., Chattopadhyay, S., Chaki, T. K., Thermal and thermo-oxidative degradation study of a model LDPE/EVA based TPE system: Effect of nano silica and electron beam irradiation. *Polymer Composites* 2010, 31(8), 1387–1397.

101. Martinez-Pardo, E., Vera-Graziano, R., Gamma-radiation induced cross-linking of polyethylene ethylene vinylacetate blends. *Radiation Physics and Chemistry* 1995, 45(1), 93–102.

102. Sharif, J., Aziz, S., Hashim, K., Radiation effects on LDPE/EVA blends. *Radiation Physics and Chemistry* 2000, 58(2), 191–195.

103. Viksne, A., Zicans, J., Kalkis, V., Bledzki, A. K., Heat-shrinkable films based on polyolefin waste. *Die Angewandte Makromolekulare Chemie* 1997, 249, 151–162.

104. Martinez-Pardo, M. E., Zuazua, M. P., Hernandez-Mendoza, V., Cardosa, J., Montiel, R., Vazquez, H., Structure-properties relationship of irradiated LDPE/EVA blend. *Nuclear Instruments and Methods in Physics Research Section B-Beam Interactions with Materials and Atoms* 1995, 105(1–4), 258–261.

105. Yan, S. F., Yin, J. B., Yu, Y., Luo, K., Chen, X. S., Thermo- and pH-sensitive poly(vinylmethyl ether)/carboxymethylchitosan hydrogels crosslinked using electron beam irradiation or using glutaraldehyde as a crosslinker. *Polymer International* 2009, 58, 1246–1251.

106. Abad, L., Relleve, L. S., Aranilla, C. T., Dela Rosa, A. M., Properties of radiation synthesized PVP-kappa carrageenan hydrogel blends. *Radiation Physics and Chemistry* 2003, 68, 901–908.

107. Zhao, L., Mitomo, H., Zhai, M. L., Yoshii, F., Nagasawa, N., Kume, T., Synthesis of antibacterial PVA/CM-chitosan blend hydrogels with electron beam irradiation. *Carbohydrate Polymers* 2003, 53(4), 439–446.

108. Zhao, L., Xu, L., Mitomo, H., Yoshii, F., Synthesis of pH-sensitive PVP/CM-chitosan hydrogels with improved surface property by irradiation. *Carbohydrate Polymers* 2006, 64, 473–480.

109. Tranquilan-Aranilla, C., Yoshii, F., Dela Rosa, A. M., Makuuchi, K., Kappa-carrageenan–polyethylene oxide hydrogel blends prepared by gamma irradiation. *Radiation Physics and Chemistry* 1999, 55(2), 127–131.

110. Zhai, M. L., Yoshii, F., Kume, T., Hashim, K., Syntheses of PVA/starch grafted hydrogels by irradiation. *Carbohydrate Polymers* 2002, 50(3), 295–303.

111. Yang, X., Liu, Q., Chen, X., Yu, F., Zhu, Z., Investigation of PVA/ws-chitosan hydrogels prepared by combined γ-irradiation and freeze-thawing. *Carbohydrate Polymers* 2008, 73, 401–408.

112. Yang, X., Zhu, Z., Liu, Q., Chen, X., Thermal and rheological properties of poly(vinyl alcohol) and water-soluble chitosan hydrogels prepared by a combination of c-ray irradiation and freeze thawing. *Journal of Applied Polymer Science* 2008, 109, 3825–3830.

113. Nurkeeva, Z. S., Mun, G. A., Dubolazov, A. V., Khutoryanskiy, V. V., pH effects on the complexation, miscibility and radiation-induced crosslinking in poly(acrylic acid)-poly(vinyl alcohol) blends. *Macromolecular Bioscience* 2005, 5, 424–432.

114. Bodugöz-Sentürk, H., Güven, O., Enhancement of conductivity in polyaniline-[poly(vinylidene chloride)-co-(vinyl acetate)] blends by irradiation. *Radiation Physics and Chemistry* 2011, 80, 153–158.

115. Güven, O., Radiation-induced conductivity control in polyaniline blends/composites. *Radiation Physics and Chemistry* 2007, 76, 1302–1307.

116. Uzun, C., Ilgın, P., Güven, O., Radiation induced in-situ generation of conductivity in the blends of polyaniline-base with chlorinated-polyisoprene. *Radiation Physics and Chemistry* 2010, 79, 343–346.

117. Faez, R., Schuster, R., Depaoli, M., A conductive elastomer based on EPDM and polyaniline. II. Effect of the crosslinking method. *European Polymer Journal* 2002, 38, 2459–2463.

118. Kang, P. H., Jeun, J. P., Seo, D. K., Nho, Y. C., Fabrication of SiC mat by radiation processing. *Radiation Physics and Chemistry* 2009, 78(7–8), 493–495.

119. Seguchi, T., New trend of radiation application to polymer modification—irradiation in oxygen free atmosphere and at elevated temperature. *Radiation Physics and Chemistry* 2000, 57(3–6), 367–371.

120. Su, Z. M., Tang, M., Wang, Z. C., Zhang, L. T., Chen, L. F., Processing of silicon carbide fibers from polycarbosilane with polypropylene as the additive. *Journal of the American Ceramic Society* 2010, 93(3), 679–685.

121. Idesaki, A., Narisawa, M., Okamura, K., Sugimoto, M., Morita, Y., Seguchi, T., Itoh, M., Application of electron beam curing for silicon carbide fiber synthesis from blend polymer of polycarbosilane and polyvinylsilane. *Radiation Physics and Chemistry* 2001, 60(4–5), 483–487.

122. Narisawa, M., Idesaki, A., Kitano, S., Okamura, K., Sugimoto, M., Seguchi, T., Itoh, M., Use of blended precursors of poly(vinylsilane) in polycarbosilane for silicon carbide fiber synthesis with radiation curing. *Journal of the American Ceramic Society* 1999, 82(4), 1045–1051.

123. Wach, R. A., Sugimoto, M., Yoshikawa, M., Formation of silicone carbide membrane by radiation curing of polycarbosilane and polyvinylsilane and its gas separation up to 250°C. *Journal of the American Ceramic Society* 2007, 90(1), 275–278.

124. Naskar, K., Gohs, U., Wagenknecht, U., Heinrich, G., PP-EPDM thermoplastic vulcanisates (TPVs) by electron induced reactive processing. *Express Polymer Letters* 2009, 3(11), 677–683.

125. Rooj, S., Thakur, V., Gohs, U., Wagenknecht, U., Bhowmick, A. K., Heinrich, G., In situ reactive compatibilization of polypropylene/epoxidized natural rubber blends by electron induced reactive processing: Novel in-line mixing technology. *Polymers for Advanced Technologies* 2011, 22(12), 2257–2263.

126. Wagenknecht, U., Gohs, U., Leuteritz, A., Volke, S., Wiessner, S., Heinrich, G., Modification of particle filled polymers with high energy electrons under in-stationary conditions of melt mixing. *Macromolecular Symposia* 2011, 301, 146–150.

127. Czvikovszky, T., Hargitai, H., Compatibilization of recycled polymers through radiation treatment. *Radiation Physics and Chemistry* 1999, 55(5–6), 727–730.

128. Dahlan, H. M., Zaman, M. D. K., Ibrahim, A., Liquid natural rubber (LNR) as a compatibiliser in NR/LLDPE blends—II: The effects of electron-beam (EB) irradiation. *Radiation Physics and Chemistry* 2002, 64(5–6), 429–436.

10

Directed Assembly of Polymer Blends Using Nanopatterned Chemical Surfaces

Ming Wei and Joey Mead

University of Massachusetts–Lowell
Lowell, Massachusetts

CONTENTS

10.1 Introduction

Patterned polymer structures with different functionalities at the nanoscale have potential applications, such as flash memory devices,[1] quantum dots,[2,3] templates for nanolithography,[4–8] semiconductor transistors[9–11] and LEDs.[12–14] For the majority of these applications, the first challenge with respect to the fabrication is that the polymer structures must be patterned in nonuniform geometries to satisfy the specifications of the application. For example, complex geometries, such as sharp 90° bends, jogs, and T-junctions, are required for integrated circuit layouts.[15] In some cases, arbitrary structures are also required. In addition to needing complex and nonuniform geometries, the other three challenges are (1) fabricating high precision nanostructures (i.e., complete replication of underlying nanopatterns) (2) patterning different geometries with multiple length scales on a single substrate, and (3) fabricating these highly ordered structures in a rapid and one step fashion.

Block copolymers (BCPs) are of significant interest in this area because of their ability to self-assemble into a variety of interesting and useful morphologies for application in nanolithography and nanodevices.[16–19] Obtaining long-range order in block copolymers has received much attention.[20–23] Nealey's group[24,25] has successfully used chemically functionalized templates to pattern BCPs into uniform and nonuniform patterns down to sub-100 nm range. They have demonstrated that the best replication occurred when the phase domain size and pattern periodicity were commensurate and the long-range replication of the pattern required a 10% or less variation between phase domain size and pattern periodicity.[24] Although the assembly of BCPs could achieve the nanostructures with high precision at a small scale, this commensurability requirement and the limited size of the structure dictated by the molecular structure of the polymer (e.g., block length) (1) restricts the range of accessible patterns, (2) does not allow for multiscale patterning on a single substrate, and (3) does not permit the preparation of nonuniform structures using block copolymers alone. To facilitate the assembly of BCPs into patterns with features like angled lines, homopolymers were used.[25] This allowed the assembly of the block copolymer into angles with few defects. Disordered morphologies, however, were still found in the corner area of angles as sharp as 45°.

Polymer blends offer some advantages over block copolymers. Blending two commercially available polymers is more cost efficient and offers a wide range of materials compared to synthesizing new block copolymers. Polymer blends are not limited by a block length as in the BCPs and thus offer more flexibility in terms of pattern size and shape, which facilitates conforming to nonuniform geometries. In addition, polymer blends may enable fabrication of patterns with multiple length scales in a single substrate (or operation), which is potentially more difficult to achieve using block copolymers. Patterning of polymer blends, however, has been less studied in comparison to controlled self-assembly of block copolymers. Cyganik et al.[26–28] used

Microcontact Printing to create a pattern of lines with a periodicity of 4 μm. With this pattern, they illustrated near complete directed assembly of polystyrene (PS)/poly(2-vinylpyridine) (PVP) polymer blends. Their best polymer assembly occurred when the pattern periodicity and phase domain size were commensurate. Ginger and coworkers[29,30] generated uniformly spaced dot patterns with dip pen nanolithography (DPN). They demonstrated assembly for PS/poly-3-hexylthiophene (P3HT) in patterns as small as 150 nm.

In this chapter, two methods for nanomanufacturing of highly ordered polymeric features in nonuniform geometries by directed assembly of polymer blends on chemically patterned surfaces are discussed. In the first method, chemically functionalized templates were used to direct the assembly of a PS/polyacrylic acid (PAA) blend into nonuniform patterns. The patterns were created by combining electron beam lithography (EBL) and self-assembled monolayers of alkanethiols (nanotrench templates). The effect of specific chemical interactions on the ability to control the site-specific deposition of polymer blends is described. The spin coating speed and solution concentration were varied to achieve patterns with different length scales. In the second method, a PS/polymethyl methacrylate (PMMA) blend was assembled into complex nonuniform patterns at multiple length scales on a single substrate with the aid of a solvent annealing step, which was less than 1 hour. In this case the chemically functionalized templates were created using DPN.

Chemically functionalized surfaces prepared using these two approaches have been used to generate a variety of complex geometries including 90° bends, T-junctions, and square and circle arrays at high rates. The chemically patterned surfaces allowed the assembly process to be accomplished in short times (less than 1 hour) without the need for the long annealing times (3 to 7 days) sometimes required in the conventional assembly of block copolymers.[24] Processing conditions, such as spin speed and solution concentration, can be controlled to achieve patterns at smaller scales. Multiscale patterns on a single substrate were also successfully demonstrated using a chemically functionalized surface. These technologies provide a pathway for the preparation of nonuniform and complex patterns using readily available materials and the nanomanufacturing of polymeric nanostructures at high rates and high volumes. These patterned polymeric structures with multiple surface functionalities can be used downstream for the fabrication of microphotonic arrays, biosensors, and other applications.

10.2 Experimental

10.2.1 Materials

11-Amino-1-undecanethiol hydrochloride (MUAM) with 99% purity and 11-Hydroxy-1-undecanethiol (HUT) with 97% purity were purchased from

Asemblon, Inc. (Redmond, Washington). 11-Mercaptoundecanoic acid (MUDA) with 97% purity, 16-mercaptohexadeconic acid (MHA) with 90% purity and 1-octadecanethiol (ODT) with 98% purity were purchased from Aldrich (St. Louis, Missouri). Polystyrene (PS) standard with 18,100 Da M_w, poly (acrylic acid) (PAA) with 2000 Da M_w, polymethyl methacrylate (PMMA) standard with 15,000 Da M_w, PS-PAA block copolymers and the solvents including acetonitrile, acetone, anhydrous ethanol, toluene, and *N,N*-dimethylformamide (DMF) were purchased from Aldrich.

10.2.2 Methods

10.2.2.1 Patterning of Alkanethiols Using the Nanotrench Template

The PMMA trench patterns (300 nm wide with 700 nm spacing) with different geometries were created on a gold substrate as follows. A 120 nm thick Au layer was evaporated onto a Si (100) wafer with 5 nm thick Cr as an adhesion layer. The PMMA resist with 150 nm thickness was spin-coated onto the Au substrate. EBL was used to write previously programmed patterns on the PMMA film. The exposed PMMA regions were developed using methyl isobutyl ketone solution and then treated by oxygen plasma for 5 seconds to remove the residual PMMA resist at the bottom of the PMMA trench.

After the template with PMMA pattern was cleaned by oxygen plasma, it was immersed into mM MUAM, or HUT or MUDA in ethanol for 24 hours to form well-ordered hydrophilic self-assembled monolayers (SAMs) in the PMMA trench area. Then, the template with assembled SAMs was washed by ethanol several times to remove the multiple thiol layers on the top of the SAMs on the surface. Afterwards, the template was extracted by acetone with a Soxhlet extractor for 24 hours to completely remove the remaining PMMA resist in the second area. Finally, the template was immersed into mM ODT in ethanol for 24 hours to grow the well-ordered hydrophobic alkanethiols on the remaining gold area (i.e., the spacing area of the original PMMA pattern).

10.2.2.2 Patterning of Polystyrene/Poly(acrylic acid) Blends on Nanotrench Template

Once the secondary thiol treatment was finished, the chemically patterned template was immediately put on the rotating plate of a spin coater (Specialty Coatings Systems G3P-8) prefilled with high purity nitrogen gas and a droplet of polymer solution, such as 1 wt% PS in DMF, 1 wt% PAA in DMF, or 1 wt% PS/PAA with 70/30 ratio in DMF, was placed on the top of the pattern area of the template. After remaining quiescent for 5 minutes, the polymer solution was spin-coated on the template surface with 3000 rpm rotation speed and 30 seconds endurance time. With the evaporation of DMF, PS, PAA, or PS/PAA blends were assembled on the chemically patterned template.

10.2.2.3 *Patterning of Alkanethiols by Dip Pen Nanolithography (DPN)*

Gold surfaces were prepared by sputtering gold (Denton Vacuum Desk IV Cold Cathode gold sputtering machine) on muscovite mica (Electron Microscopy Sciences). A 52-tip silicon nitride array (Nanoink Inc., A-52 style Nanoink Array) was used to pattern a hydrophilic alkanethiol, MHA. The DPN tips were dipped into the 5 mM MHA in acetonitrile for 5 seconds. The array was then dipped into a solution of deionized water for 5 seconds. The dipping procedure was then repeated a second time and the array was then gently dried with a stream of inert gas (Fisherbrand Super Friendly Air' IT) applied from a direction in-plane to the chip. Following the deposition of the MHA patterns, the patterned substrate was subsequently backfilled with a hydrophobic alkanethiol ODT. The backfilling procedure entailed placing a few droplets of ODT onto the substrate and letting the ODT quiescent on the substrate before the excess ODT was blown off with inert gas.

10.2.2.4 *Patterning of Polystyrene/Polymethylmethacrylate Blends on DPN Template*

To assemble the polymers, the gold-coated substrate containing the MHA pattern and backfilled ODT was placed into a spin coater. The PS/PMMA 1% wt. solution in DMF with 50/50 weight ratio was drop cast and then spun coat at 5000 rpm for 60 seconds. The polymer assembled substrate was then placed in a glass desiccator, which contained toluene, where it was solvent annealed for 45 minutes.

10.2.3 Characterization

Field emission scanning electron microscopy (FESEM) images and energy dispersive spectroscopy (EDS) were performed by JEOL 7401F field emission scanning electron microscopy. Atomic force microscopy (AFM) images were examined by scanned probe microscopy (Veeco NanoScopella, PSIA XE-150 mode, 40N/m tip spring constant) with noncontact mode. To calculate the characteristic lengths of the initial morphologies of PS/PAA blends, fast Fourier transform (FFT) analyses of the corresponding atomic force microscopy (AFM) topography images were performed by AFM image analysis software XEI (PSIA Corp., Version 1.5). The FFT diagrams were used to calculate the characteristic length of the phase separated morphologies of polymer blends.

10.3 Results and Discussions

10.3.1 Directed Assembly of Polystyrene (PS)/ polyacrylic Acid (PAA) Using Nanotrench Templates

EBL is a well-established patterning technology capable of creating the extremely fine patterns down to 10 nm due to the very small spot size (~2 nm) of the electrons.[31] Taking advantage of the EBL capability of patterning PMMA resists on a gold substrate, it was combined with the self-assembly of alkanethiols molecules to create patterned alkanethiols with different chemical functionality. Figure 10.1 shows the schematic diagram of patterning of hydrophilic/hydrophobic alkanethiols.

Figures 10.2a and 10.2b show the FESEM images of the PMMA photoresist patterns on the gold surface and patterned MUAM/ODT stripes on the gold surface after PMMA removal and ODT backfill, respectively. As shown in the figures, the patterned alkanethiols are patterned the same as the original PMMA resist pattern. To study the effect of the chemical functionality of the nanopattern and the selectivity of the specific alkanthiol for the corresponding polymer, a single component (either a PAA or PS solution) was spin-coated on the patterned MUAM/ODT surface. Shown in Figures 10.3a and 10.3b, PS was selectively assembled onto ODT avoiding the MUAM area with high selectivity and PAA was selectively assembled onto the MUAM

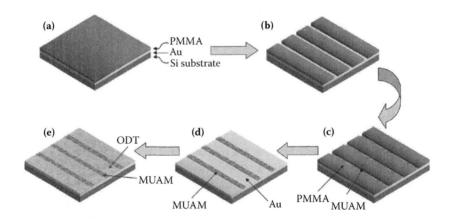

FIGURE 10.1 (See color insert.)
Patterning of hydrophilic/hydrophobic alkanethiols combining electron beam lithography (EBL) and self-assembly of alkanethiol molecules. (a) 150 nm thick polymethyl methacrylate (PMMA) resist spin coated onto the gold deposited on a silicon wafer; (b) Patterned PMMA trenches were defined by electron beam and development; (c) a hydrophilic 11-amino-1-undecanethiol hydrochloride (MUAM) assembled in the PMMA trench area; (d) PMMA resists were removed by acetone to produce patterned MUAM on gold; (e) backfilled by hydrophobic octadecanethiol (ODT) yielding the final chemical pattern. (Reprinted with permission from Wiley.)

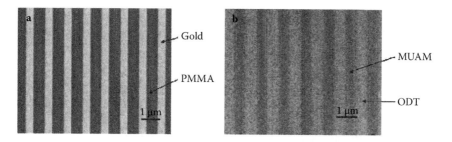

FIGURE 10.2
(a) Field emission scanning electron microscopy (FESEM) images of the patterned polymethyl methacrylate (PMMA) resist; (b) FESEM image of binary patterned monolayers consisting of hydrophilic 11-amino-1-undecanethiol hydrochloride (MUAM) ($-NH_3^+$) (shown as dark) and hydrophobic octadecanethiol (ODT) ($-CH_3$) (shown as light). (Reprinted with permission from Wiley.)

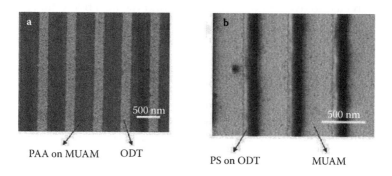

FIGURE 10.3
Field emission scanning electron microscopy (FESEM) images of (a) patterned polyacrylic acid (PAA), (b) patterned polystyrene (PS) on patterned 11-amino-1-undecanethiol hydrochloride (MUAM) ($-NH_3^+$)/octadecanethiol (ODT) ($-CH_3$) monolayers. (Reprinted with permission from Wiley.)

avoiding the ODT area. PAA is a weak polyelectrolyte bringing a negative charge, and MUAM brings a positive charge. The directed assembly of PAA is believed to be driven by the strong electrostatic forces between PAA and MUAM. The hydrophobic interaction is thought to be the main driving force in directing PS to the hydrophobic ODT regions.

Following the successful assembly of each single component of PS or PAA on the patterned alkanethiols, a PS/PAA blend with 70/30 composition ratio was spin coated on the gold template with patterned MUAM/ODT mono-layers in a 300 nm/700 nm scale. Figure 10.4a shows the directed assembly of the blend by the presence of a side-by-side PS/PAA film with the dark areas as PS and the light areas as PAA. The atomic force microscope (AFM) three-dimensional (3-D) image for the patterned PS/PAA film is shown in Figure 10.4b. As with the assembly for the single component, the hydrophobic

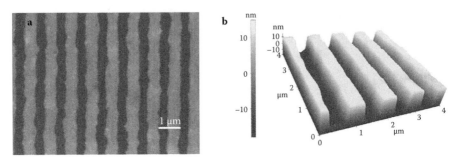

FIGURE 10.4
(a) Field emission scanning electron microscopy (FESEM) and (b) atomic force microscopy
(AFM) three-dimensional image of patterned polystyrene (PS)/patterned polyacrylic acid
(PAA) assembled by patterned 11-amino-1-undecanethiol hydrochloride (MUAM) (–NH₃⁺)/
octadecanethiol (ODT) (–CH₃) monolayers. (Reprinted with permission from Wiley.)

PS was directed to hydrophobic ODT areas and the negatively charged PAA
was directed to the positively charged MUAM area. In conclusion, chemically patterned surfaces were shown to be successful in directing the assembly of a polymer blend solution into nanoscale structures.

The importance to nanomanfucturing is the fact that this assembly process occurred in just 30 seconds and did not require annealing. The short
times make it possible to control the assembly of two polymers into a patterned array on a single patterned surface at high rates, necessary for cost
effective nanomanufacturing. The strong attraction between the macromolecules and chemically patterned surface allow for the rapid assembly of the
polymer blend. The interactions between SAMs and the assembled macromolecules are determined by the terminal functional group of the SAMS
at the surface. These interactions can include hydrogen bonding or van der
Waals forces. To study the utility of this technique and the effect of surface
functionality on the assembly, other types of alkanethiols were patterned by
this technique. In addition to MUAM with positively charged NH_3^+ terminal
groups, 11-hydroxy-1-undecanethiol (HUT) with –OH terminal groups and
11-mercaptoundecanoic acid (MUDA) with –COOH terminal groups were
also patterned for the hydrophilic thiols. All of the hydrophilic thiols had the
same alkane backbone chain length to make the SAMs of similar thickness.
The FESEM images of patterned PS/PAA films directed by MUDA/ODT and
HUT/ODT are shown in Figures 10.5a and 10.5b, respectively. Good patterning can be seen in the films patterned by both MUDA/ODT and HUT/ODT.
Thus, the hydrogen bonding interactions between the –COOH group of PAA
and the –COOH group of MUDA or the –OH group of HUT are thought to be
a critical driving force directing the PAA phase onto MUDA or HUT.

The assembly process is a complex nonequilibrium process and depends
on a number of other thermodynamic and kinetic factors, including the viscosity of polymer systems, thickness of the film, compatibility of polymer

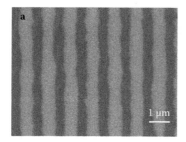
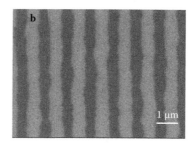

FIGURE 10.5
Field emission scanning electron microscopy (FESEM) images of patterned polystyrene/poly (acrylic acid) blends by patterned (a) 11-mercaptoundecanoic acid (MUDA) (–COOH)/octadecanethiol (ODT) and (b) 11-hydroxy-1-undecanethiol (HUT) (–OH)/ODT. (Reprinted with permission from Wiley.)

phases, molecular weight of the polymer, spin-coating speed and solvent evaporation rate, as well as the interaction of the polymer and the chemical functional groups on the surface. Next, it will be shown that the process parameters such as spin speed and solution concentration can be optimized to allow the assembly at different length scales.[32]

Generally, it is believed that well-ordered directed morphologies are typically obtained only when the characteristic length (L_c) of the polymer is commensurate with the periodicities (λ) of the chemical patterns. The characteristic length represents the average length of vectors, randomly drawn passing through the two phases. It is related to the average domain size of the two polymers.[33–35] The characteristic length (L_c) for a PS/PAA 7:3 weight ratio was found to be 993 ± 66 nm, as shown in the isotropic bicontinuous morphologies in Figure 10.6. When the pattern periodicity was commensurate with the characteristic length (1000 nm ($\sim L_c$)), well-ordered replication was obtained. On the other hand, when the pattern periodicity was smaller than the characteristic length (667 nm ($< L_c$)), the PS (dark) regions were not able to register with the narrower ODT strips (467 nm), crossing into the adjacent MUAM functionalized surface. Similarly, if the pattern periodicity was larger than the characteristic length (1333 nm ($> L_c$)), the PS domains were not well formed on the wider ODT strips.

From these results, for the single-step spin-coating process, it is clear that for good pattern replication it is necessary that the initial (nontemplated) morphology of the polymer blend be on the same length scale as the desired pattern (commensurate). Using this information, increasing spinning speed, solution concentration can be used to reduce the domain size (characterization length) to achieve patterning at a smaller length scale.

Process parameters can play a significant role in controlling the polymer blend morphology. Smaller domains and thus smaller characteristic lengths would be expected at higher spin speeds, because of the shorter times available for the phases to coalesce. Figure 10.7 shows the FESEM images of

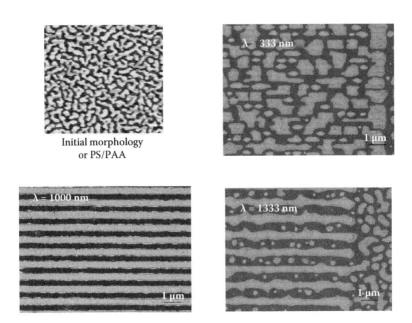

FIGURE 10.6
Nontemplated morphology of polystyrene (PS)/polyacrylic acid (PAA) blends (top left) and template directed morphologies on three pattern periodicities.

directed morphologies of the PS/PAA blend on the MUAM/ODT patterns with different periodicities for different spin speeds. Figures 10.7a through 10.7d show the assembly on patterns with 1333 nm periodicity at different spin speeds. For good pattern replication, it is expected that the characteristic length should match the pattern periodicity of 1333 nm. To fabricate the required characteristic length of 1333 nm the critical spin speed was calculated to be 2210 rpm.[32] As all the spin speeds used (3000 to 9000 rpm) in this work were above the critical spin speed of 2210 rpm, the resultant characteristic lengths were smaller than 1333 nm and resulted in poor replication of the patterns. The effects of spin speed on the directed assembly on patterns with a 1000 nm periodicity are seen in Figures 10.7e through 10.7h. In this case, well-ordered patterned structures of the PS/PAA blend were produced (Figure 10.7e) because the spin speed of 3000 rpm was similar to the calculated critical spin speed of 3314 rpm. Increasing the spin speed above the critical spin speed resulted in characteristic lengths smaller than the pattern periodicity of 1000 nm and increasingly disordered patterned morphologies (Figures 10.7f through 10.7h). The directed assembly on patterns of 667 nm periodicity is seen in Figures 10.7i through 10.7l. For this pattern size the critical spin speed was calculated to be 5862 to obtain the required characteristic length of 667 nm. As expected for spin speeds either higher or lower than the critical spin speed, poor assembly was observed. Good assembly (Figures 10.7j through 10.7k) was found for spin speeds of 5000 and 7000 rpm,

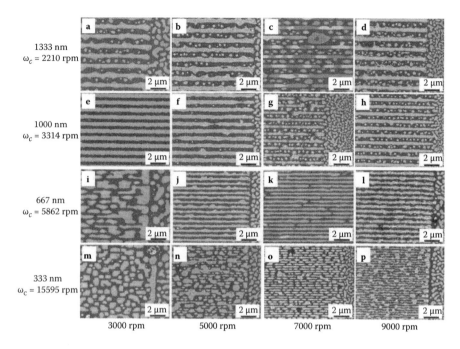

FIGURE 10.7

Field emission scanning electron microscopy (FESEM) images of directed assembly of polystyrene (PS)/polyacrylic acid (PAA) blends using alternative MUAM/octadecanethiol (ODT) patterns with various periodicities: (a–d) 1333, (e–h) 1000, (i–l) 667, and (m–p) 333 nm. The spin speeds were changed: (a,e,i,m) 3000, (b,f,j,n) 5000, (c,g,k,o) 7000, and (d,h,l,p) 9000 rpm. ω_c stands for the critical spin speed for each pattern periodicity. (Reprinted with permission from ACS.)

which produced characteristic lengths of 643 and 565 nm, respectively. The assembly on patterns with a 333 nm periodicity is found in Figures 10.7m through 10.7p. The critical spin speed to obtain the appropriate characteristic length for this pattern size was 15595 rpm. Because this is higher than the spin speeds used in this work, poor assembly occurred.

From these results, it is shown that control of the spin speed is an important processing parameter that can be adjusted to obtain the required characteristic length commensurate with the pattern periodicity. Phase separation and domain coarsening are controlled by solvent evaporation. With solvent evaporation, polymer mobility is reduced; eventually mobility ceases. Increasing spin speed increases the rate of solvent evaporation, reducing the time available for directed assembly. The dewetting/wetting behaviors of the PS/PAA blend on a patterned surface are also time dependent. In addition to spin speed, solution concentration also affects the time available for assembly and polymer mobility. It may also be used to achieve good assembly on patterns when the spin speed is experimentally inaccessible. Lower concentrations allow for the patterning of smaller characteristic lengths.The FESEM images of the

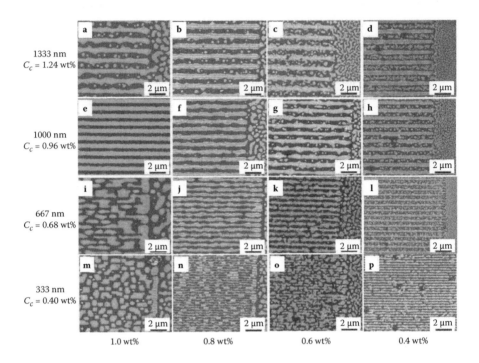

FIGURE 10.8
Field emission scanning electron microscopy (FESEM) images of directed assembly of polystyrene (PS)/polyacrylic acid (PAA) blends using alternative MUAM/octadecanethiol (ODT) patterns with various periodicities: (a–d) 1333, (e–h) 1000, (i–l) 667, and (m–p) 333 nm. The solution concentrations were changed: (a,e,i,m) 1, (b,f,j,n) 0.8, (c,g,k,o) 0.6, and (d,h,l,p) 0.4 wt%. C_c stands for the critical solution concentration for each pattern periodicity. (Reprinted with permission from ACS.)

directed assembly of different solution concentrations of a PS/PAA blend on patterns with different periodicities are shown in Figure 10.8. The required solution concentration was calculated to be 1.24 wt% for a pattern periodicity of 1333 nm (Figures 10.8a through 10.8d). Because the solution concentrations used were smaller than 1.24 wt%, the resulting characteristic lengths were smaller than the pattern periodicity and resulted in disordered morphologies. Figures 10.8e through 10.8h show the assembly for a pattern periodicity of 1000 nm. The critical solution concentration was 0.96 wt% to obtain the commensurate characteristic length. Because this was close to the experimental concentration of 1% (Figure 10.8e), well-ordered morphologies were obtained. For concentrations away from this value (0.8% and below), poor assembly was obtained. Similarly for a pattern periodicity of 667 nm, the critical solution concentration was calculated to be 0.68 wt%, close to the experimental concentrations of 0.6 (Figure 10.8j) and 0.8 wt% (Figure 10.8k), and good patterning was observed for these concentrations. For concentrations smaller (0.4 wt%, Figure 10.8i) or larger (1 wt%, Figure 10.8l) poor assembly was produced. Assemblies on patterns with 333 nm periodicity are presented in Figures 10.8m through 10.8p. Well-ordered assembly was produced when the solution concentration matched the critical solution concentration of 0.4 wt% (Figure 10.8p). Solution concentrations larger than 0.4 wt% (Figures 10.8m through 10.8o) resulted in disordered patterned morphologies. In spite of the fact that for the solution concentration of 0.4 wt%, the characteristic length was close to the pattern periodicity of 333 nm, some defects were still observed. Kinetic factors may still play a role. Solution concentration, along with spin speed, can be effectively used to produce a controlled characteristic length commensurate with the pattern periodicity for directed assembly of a PS/PAA blend.

The ability to reproduce smaller pattern periodicities is improved as the concentration is reduced as shown in Figure 10.8. Consistency of the pattern replication is more challenging for patterning at very small length scales and warrants further investigation. The importance of process parameters on the ability to direct the assembly of polymer blends at different length scales is highlighted by the effects of spinning speed and solution concentration.

The directed assembly of PS/PAA blends to achieve highly ordered periodic lines over a large area in very short times has been demonstrated. The formation of additional features with complex and nonregular geometries is important, however, for many practical applications, such as nanolithography. Typical examples might include circle and square arrays, T-junctions, and 90° bends.[15,36] A set of complex structures generated by the directed assembly of PS/PAA blends is presented in Figure 10.9. Good assembly of periodic lines with 90° bends are shown in Figure 10.9a. The light gray areas in Figure 10.9b demonstrate the assembly into T-type junctions. Additionally, more complicated square and circle arrays are shown in Figures 10.9c and 10.9d, respectively. Through control of the design of different geometries in the PMMA trench template and appropriate chemical functionalization, it is possible to direct the assembly of PS/PAA blends, creating a variety of complex nonuniform patterns using a rapid and simple process. These patterned polymer nanostructures can be created in times less than 1 minute. This approach allows the patterning of many different nonuniform geometries on a single template. Thus, polymer blends provide for the fabrication of structures that are not limited by the polymer chain length and provide for a flexibility in design considerations.

10.3.2 Directed Assembly of Polystyrene (PS)/Polymethyl Methacrylate (PMMA) Using DPN Templates

DPN is a direct writing technique using an ink-coated scanning probe tip to deposit with nanometer precision, organic or inorganic materials onto a substrate.[37] Instead of using a nanotrench template, in this section, DPN was used to produce the patterned alkanethiols with different functionality.

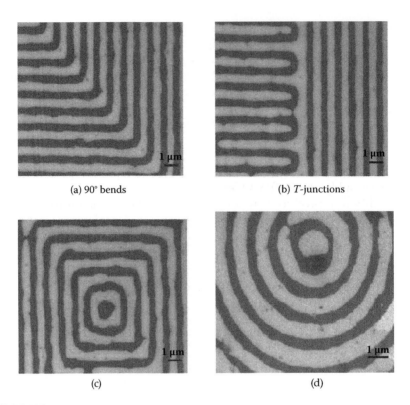

(a) 90° bends

(b) *T*-junctions

(c)

(d)

FIGURE 10.9
A set of generated nonuniform geometries by direct assembly of polystyrene (PS)/poly-acrylic acid (PAA) blends: (a) 90° bends, (b) T-junctions, (c) square arrays, and (d) circle arrays. (Reprinted with permission from Wiley.)

The template design is shown in Figure 10.10a. The template size is 23 µm by 30 µm and consists of a range of uniform and nonuniform geometries including dot, lines, angled lines at angles of 30, 45, 60, 90, and 135°, diamond shapes, and lettering. The largest line width and dot diameter was 300 nm, decreasing in increments of 50 nm to a minimum size of 50 nm. The pattern width and spacing were kept at a 1:1 ratio over the entire template. The entire procedure of chemically functionalizing the gold substrate, assembling the polymer blends, and solvent annealing is depicted in Figure 10.10b.

The FESEM image of a PS/PMMA 50/50 polymer blend assembled onto a chemically patterned MHA/ODT surface is shown in Figures 10.11a, 10.11c, and 10.11e. Two polymer phases are seen: the dark phase is PS and the light color is the PMMA phase. The PMMA (light phase) selectively assembles onto the hydrophilic MHA region while the PS (dark phase) assembles on the hydrophobic ODT region. The PMMA forms continuous polymer domain structures while the PS domains are either discontinuous or droplets. The evaporation of the common solvent and interfacial instabilities results in phase separation. The polymers continue to separate, producing larger

(a)

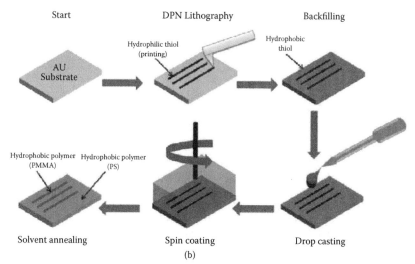

(b)

FIGURE 10.10 (See color insert.)
(a) Pattern geometries from the dip pen nanolithography (DPN) consisting of uniform and nonuniform features (template size 23 μm by 30 μm). (b) Fabrication method for the DPN templates and polymer assembly. (Reprinted with permission from Wiley.)

domains until the polymer mobility is inhibited and the process is terminated. The morphology resulting from this process is a long-lived metastable state, retained until the polymer chains are given mobility. The quick and rapid process of spin coating results in an energetically unfavorable thermodynamic state for the polymer blend molecules. For the conditions represented above there is insufficient time for the PS phases to coalesce into larger domains representative of the underlying pattern, and droplets are formed.

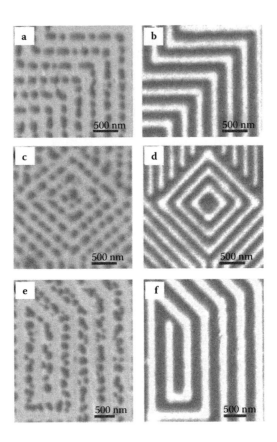

FIGURE 10.11

Field emission scanning electron microscopy (FESEM) images of the directed assembly of polystyrene (PS)/polymethyl methacrylate (PMMA) 50/50 blends on the 16-mercaptohexadeconic acid (MHA) and 1-octadecanethiol (ODT) patterned substrate with a fixed symmetric 1:1 period. The light color is the PMMA on the MHA thiol, and the PS is the dark color on the octadecanethiol (ODT) thiol. (a,c,e) Unannealed; (b,d,f) solvent annealed. (a) and (b) represent nonuniform geometry lines of 90° angle bends at a feature size of 150 nm. (c) and (d) represent nonuniform geometry lines of a diamond shape at a feature size of 100 nm. (e) and (f) represent nonuniform geometry lines of 135° angle bends at a feature size of 200 nm.

One method to produce a more thermodynamically stable state is annealing of the polymer film. This promotes mobility of the polymer chains and allows them to coalesce into larger domains. For this system, solvent annealing was chosen over thermal annealing, in an effort to minimize the thermal exposure of the chemical bond between the SAMs and the gold substrate, which may lack stability at elevated temperatures. Toluene is a well-known solvent for PS and was chosen as the annealing solvent.

Solvent-annealed PS/PMMA polymer blend assembled on a MHA/ODT template is presented in Figures 10.11b, 10.11d, and 10.11f. The micrograph shows that solvent annealing is effective in forming continuous PS polymer

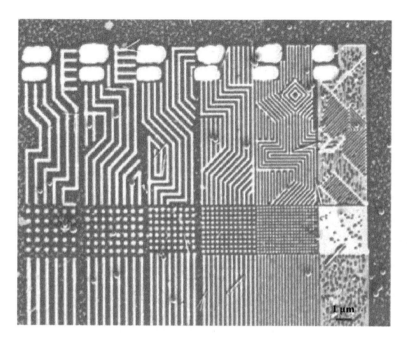

FIGURE 10.12
Field emission scanning electron microscopy (FESEM) image of the full view of the patterned polystyrene (PS)/polymethyl methacrylate (PMMA) polymer structures from 50 nm to 300 nm. (Reprinted with permission from Wiley.)

domains on the ODT surface compared to the isolated domains in the unannealed blend.

From these results it is recognized that the patterned morphology of the polymer blend system shown in Figures 10.11a, 10.11c, and 10.11e represents a nonequilibrium state. Solvent annealing improves the assembly leading to a more desirable equilibrium polymer morphology driven by the chemical interactions between the functionalized surface and the polymer.

The entire FESEM image of the PS/PMMA patterned polymer structures is shown in Figure 10.12. Nonuniform geometries with length scales from 50 to 300 nm are shown replicated on a single template. The solvent annealing step is necessary to facilitate the assembly of PS/PMMA blends over multiple length scales. The solvent annealing step was only 45 minutes, faster than typical for thermal annealing.[24] The use of the solvent annealing step permits the assembly of structures where the phase domain sizes[24,26] (characteristic length) formed initially are not commensurate with the underlying pattern. The use of solvent vapor promotes mobility of the chains and allows the polymer chains sufficient time to be directed by the chemically functionalized surface. Polymer blends offer the unique opportunity to be directly assembled into nonuniform geometries at varying length scales, important for fabrication of structures for a variety of applications. As seen in Figure 10.12 the directed assembly into nonuniform geometries is quite consistent over the entire surface of the template and was repeatable for multiple pens (unpublished results).

In the PS/PMMA system it was possible to consistently replicate lines (uniform and nonuniform) and dots with sizes from 100 to 300 nm. Excellent replication of 30, 45, 60, 90, and 135° angles in lines with widths from 100 to 300 nm was also demonstrated. These results represent smaller nanoscale features and more complex nonuniform geometries than have previously been reported utilizing polymer blends. Block copolymers can also replicate these angles but require the use of a homopolymer. Patterning of a range of sizes is also challenging.[25]

Assembly of sizes down to 50 nm were demonstrated for this blend system, but the assembly was not as consistent as for the larger sizes (see Figure 10.12). As seen in Figure 10.12 the assembly of PS/PMMA blends into 45° angled lines and 90° square array patterns was possible on this length scale; however, the assembly was not as consistent in other regions. One potential reason may be defects in the SAMs patterned area from surface roughness in the sputtered gold substrate. This surface roughness may influence the water meniscus ink diffusion making it more difficult to achieve smaller sizes in the DPN patterned geometry.[38] The cleanliness of the gold surface may also affect the SAMs quality. Last, there is competition between the strength of the attraction between the polymer and the underlying chemical surface pattern and the interfacial tension (leading to larger domains).

10.4 Summary

In summary, two novel and versatile approaches for creating highly ordered polymer nanostructures by template directed assembly of polymer blends using patterned chemical surfaces have been demonstrated. One approach is the directed assembly of PS/PAA using a nanotrench template to give a chemically patterned surface and the second is the use of DPN to directly fabricate a chemically patterned surface for directing the assembly of a PS/PMMA blend. Both approaches can be used to selectively assemble polymer blends into desired patterns in a one-step fashion with high specificity and selectivity. The selective deposition of functional polymer systems on a patterned surface can be based on electrostatic force, hydrogen bonding, as well as hydrophobic forces. For assembly of the PS/PAA on a chemically functionalized surface created from nanotrench templates, the assembly of polymer blends occurred in 30 seconds. For PS/PAA blends, it is possible to pattern a set of complex, nonuniform geometries including 90° bends, T type junctions, square and circle arrays on a single substrate. Process parameters, including spin speed and solution concentration, could be used to control the characteristic length and resulting quality of assembly. Using a different polymer system, PS/PMMA, and a DPN chemically functionalized surface, nonuniform geometries at multiple length scales were achieved on a single template using a short (less than 1 hour) solvent annealing step. Length scales from 300 nm down to 100 nm in uniform and nonuniform geometries were consistently patterned on the same template. Assembly of polymer blends into nonuniform geometries down to 50 nm were demonstrated. In principle, this technology can be extended to the assembly of any other functional polymer system, polyelectrolytes, biomolecules, conducting polymers, colloids, and other nanoparticles. These two approaches for preparation of highly ordered polymer nanostructures demonstrate the breadth of this approach for the fabrication of nanodevices including biosensors, biochips, photonics, nanolithography, and electronics in a high volume, high rate, and one-step fashion.

Acknowledgments

The authors would like to thank the support of the National Science Foundation (Award #NSF-0425826). We also appreciate the Kostas Center for the preparation of templates and Mr. Peter Ryan for computer-aided design (CAD) of patterns.

References

1. K. W. Guarini, C. T. Black, Y. Zhang, I. V. Babich, E. M. Sikorski, L. M. Gignac, Low voltage, scalable nanocrystal FLASH memory fabricated by templated self assembly. *IEEE Tech.Dig.—Int. Electron Devices Meeting 2003*, 541.
2. M. Park, C. Harrison, P. M. Chaikin, R. A. Register, D. H. Adamson, Block copolymer lithography: Periodic arrays of ~1011 Holes in 1 square centimeter. *Science* 1997, *276*, 1401.
3. R. R. Li, P. D. Dapkus, M. E. Thompson, W. G. Jeong, C. Harrison, P. M. Chaikin, R. A. Register, D. H. Adamson, Dense arrays of ordered GaAs nanostructures by selective area growth on substrates patterned by block copolymer lithography. *Appl. Phys. Lett.* 2000, *76*, 1689.
4. G. M. McClelland, M. W. Hart, C. T. Rettner, M. E. Best, K. R. Carter, B. D. Terris, Nanoscale patterning of magnetic islands by imprint lithography using a flexible mold. *Appl. Phys. Lett.* 2002, *81*, 1483.

5. J. Y. Cheng, C. A. Ross, E. L. Thomas, H. I. Smith, G. J. Vancso, Fabrication of nanostructures with long-range order using block copolymer lithography. *Appl. Phys. Lett.* 2002, *81*, 3657.

6. P. Maury, M. Escalante, D. N. Reinhoudt, J. Huskens, Directed assembly of nanoparticles onto polymer-Imprinted or chemically patterned templates fabricated by nanoimprint lithograph. *Adv. Mater.* 2005, *17*, 2718.

7. Y. S. Kim, N. Y. Lee, J. R. Lim, M. J. Lee, S. Park, Nanofeature-patterned polymer mold fabrication toward precisely defined nanostructure replication. *Chem. Mater.* 2005, *17*, 5867.

8. Y. Yin, Y. Xia, Self-assembly of spherical colloids into helical chains with well-controlled handedness. *J. Am. Chem. Soc.* 2003, *125*, 2048.

9. C. H. Yang, T. J. Shin, L. Yang, K. Cho, C. Y. Ryu, Z. N. Bao, Effect of mesoscale crystalline structure on the field-effect mobility of regioregular poly(3-hexyl thiophene) in thin-film transistors. *Adv. Funct. Mater.* 2005, *15*, 671.

10. N. Stingelin-Stuntzmann, E. Smits, H. Wondergem, C. Tanase, P. Blom, P. Smith, D. D. Leeuw, Organic thin-film electronics from vitreous solution-processed rubrene hypereutetics. *Nat. Mater.* 2005, *4*, 601.

11. R. J. Kline, M. D. McGehee, M. F. Toney, Liquid-crystalline semiconducting polymers with high charge-carrier mobility. *Nat. Mater.* 2006, *5*, 328.

12. A. C. Morteani, P. Sreearunothai, L. M. Herz, R. H. Friend, C. Silva, Exciton regeneration at polymeric semiconductor heterojunctions. *Phys. Rev. Lett.* 2004, *92*, 247402.

13. A. C. Morteani, A. S. Dhoot, J. S. Kim, C. Silva, N. C. Greenham, C. Murphy, E. Moons, S. Cina, J. H. Burroughes, R. H. Friend, Barrier-free electron–hole capture in polymer blend heterojunction light-emitting diodes. *Adv. Mater.* 2003, *15*, 1708.

14. F. P. Wenzl, P. Pachler, C. Suess, A. Haase, E. J. W. List, P. Poelt, D. Somitsch, P. Knoll, U. Scherf, G. Leising, The influence of the phase morphology on the optoelectronic properties of light-emitting electrochemical cells. *Adv. Funct. Mater.* 2004, *14*, 441.

15. M. P. Stoykovich, H. M. Kang, K. C. Daoulas, G. L. Liu, C. C. Liu, J. J. D. Pablo, M. Muller, P. F. Nealey, Directed self-assembly of block copolymers for nano-lithography: Fabrication of isolated features and essential integrated circuit geometries. *ACS Nano*, 2007, *1*, 168.

16. N. Hadjichristidis, S. Pispas, S., G. Floudas, Block copolymer morphology, in *Block Copolymers: Synthetic Strategies, Physical Properties, and Applications*, John Wiley and Sons, Hoboken, NJ, 2003.

17. G. M. McClelland, M. W. Hart, C. T. Rettner, M. E. Best, K. R. Carter, B. D. Terris, Nanoscale patterning of magnetic islands by imprint lithography using a flexible mold. *Appl. Phys. Lett.*, 2002, *81*, 1483.

18. D. H. Kim, Z. Lin, H. C. Kim, U. Jeong, T. P. Russell, On the replication of block copolymer templates by poly(dimethylsiloxane) elastomers. *Adv. Mater.*, 2003, *15*, 811.

19. Y. S. Kim, H. H. Lee, P. T. Hammond, High density nanostructure transfer in soft molding using polyurethane acrylate molds and polyelectrolyte multilayers. *Nanotechnology*, 2003, *14*, 1140.

20. L. Rockford, S. G. J. Mochrie, T. P. Russell, Propagation of nanopatterned substrate templated ordering of block copolymers in thick films. *Macromolecules*, 2001, *34*, 1487.

21. X. M. Yang, R. D. Peters, T. K. Kim, P. F. Nealey, S. L. Brandow, M. S. Chen, L. M. Shirey, W. J. Dressick, Proximity x-ray lithography using self-assembled alkylsiloxane films: Resolution and pattern transfer. *Langmuir*, 2001, *17*, 228.
22. E. Schäffer, T. Thurn-Albrecht, T. P. Russell, U. Steiner, Electrically induced structure formation and pattern transfer. *Nature*, 2000, *403*, 874.
23. T. Thurn-Albrecht, J. DeRouchy, T. P. Russell, H. M. Jaeger, Overcoming interfacial Interactions with electric fields. *Macromolecules*, 2000, *33*, 3250.
24. S. O. Kim, H. H. Solak, M. P. Stoykovich, N. J. Ferrier, J. J. DePablo, P. F. Nealey, Epitaxial self-assembly of block copolymers on lithographically defined nanopatterned substrates. *Nature*, 2003, *424*, 411.
25. M. P. Stoykovich, M. Mueller, S. O. Kim, H. H. Solak, E. W. Edwards, J. J. de Pablo, P. F. Nealey, Directed assembly of block copolymer blends into non-regular device-oriented structures. *Science*, 2005, *308*, 1442.
26. J. Raczkowska, P. Cyganik, A. Budkowski, A. Bernasik, J. Rysz, I. Raptis, P. Czuba, K. Kowalski, Composition effects in polymer blends spin-cast on patterned substrates. *Macromolecules*, 2005, *38*, 8486.
27. J. Raczkowska, A. Bernasik, A. Budkowski, P. Cyganick, J. Rysz, I. Raptis, P. Czuba, Pattern guided structure formation in polymer films of asymmetric blends. *Surface Sci.*, 2006, *600*, 1004.
28. P. Cyganick, A. Bernasik, A. Budkowski, B. Bergues, K. Kowalski, J. Rysz, J. Lekki, M. Lekka, Z. Postawa, Phase decomposition in polymer blend "lms cast on substrates patterned with self-assembled monolayers. *Vacuum*, 2001, *63*, 307.
29. D. C. Coffey, D. S. Ginger, Patterning phase separation in polymer films with dip-pen nanolithography. *J. Am. Chem. Soc.*, 2005, *127*, 4564.
30. J. H. Wei, D. C. Coffey, D. S. Ginger, Nucleating pattern formation in spin-coated polymer blend films with nanoscale surface templates. *J. Phys. Chem. B*, 2006, *110*, 24324.
31. K. Kasama, *Optronics* 2003, *256*, 106.
32. L. Fang, M. Wei, C. Barry, J. Mead, Effect of spin speed and solution concentration on the directed assembly of polymer lends. *Macromolecules*, 2010, *43*(23), 9747.
33. H. Sh. Xia, M. Song, Characteristic length of dynamic glass transition based on polymer/clay intercalated nanocomposites. *Thermo. Acta* 2005, *429*, 1.
34. L. T. Yan, J. Sheng, Analysis of phase morphology and dynamics of immiscible PP/PA1010 blends and its partial-miscible during melt mixing from SEM patterns. *Polymer* 2006, *47*, 2894.
35. G. Q. Ma, Y. H. Zhao, L. T. Yan, Y. Y. Li, J. Sheng, Blends of polypropylene with poly(cis-butadiene) rubber. III. Study on the phase structure and morphology of incompatible blends of polypropylene with poly(cis-butadiene) rubber. *J. Appl. Poly. Sci.* 2006, *100*, 4900.
36. G. S. W. Craig, P. F. Nealey, Self-assembly of block copolymers on lithographically defined nanopatterned substrates. *J. Photopolymer Sci. Technol.*, 2007, *20*, 511.
37. R. D. Piner, J. Zhu, F. Xu, S. Hong, C. A. Mirkin, "Dip-pen" nanolithography. *Science*, 1999, *283*, 661.
38. J. Haaheim, R. Eby, M. Nelson, J. Fragala, B. Rosner, H. Zhang, G. Athas, *Ultramicroscopy* 2005, *103*, 117–132.

Index